SCAFFOLDS IN TISSUE ENGINEERING - MATERIALS, TECHNOLOGIES AND CLINICAL APPLICATIONS

Edited by **Francesco Baino**

Scaffolds in Tissue Engineering - Materials, Technologies and Clinical Applications
http://dx.doi.org/10.5772/66016
Edited by Francesco Baino

Contributors

Andrea Peloso, Antonio Citro, Szandra Brambilla, Lorenzo Cobianchi, Graziano Oldani, Lorenzo Piemonti, Ana Colette Maurício, Sílvia Santos Pedrosa, Rita Caseiro, José Domingos, Yaser Greish, Sherif Karam, Abdel-Hamid I. Mourad, Sunitha Puilkkot, Achouak Elghazel, Stefania Raimondo, Shimon Rochkind, Mira M. Mandelbaum-Livnat, Michela Morano, Giulia Ronchi, Nicoletta Viano, Moshe Nissan, Akiva Koren, Tali Biron, Yifat Bitan, Evgeniy Reider, Ofra Ziv-Polat, Abraham Shahar, Stefano Geuna, Mara Almog, Marco Vinícius Chaud, Lindemberg Filho, Marcia Rebelo, Cecilia Torquetti, Patricia Severino, Katiusca Pontes, Thais Alves, Carolina Santos, Venâncio Alves, Luis Atayde, Pedro Olivério Pinto, Carla Mendonça, José Miguel Campos, Mythili Prakasam, Alain Largeteau, Laura Madalina Popescu, Roxana Mioara Piticescu, Miriam V Flores-Merino, Carolina Oliver-Urrutia, Ma. Victoria Dominguez García, Jaime Flores-Estrada, Antonio Laguna-Camacho, Julieta Castillo Cadena, Sakkadech Limmahakhun, Cheng Yan, Jonathan Massera, Amy Nommeots-Nomm, Emad Ewais, Amira Amin, Marcel Ricklefs, Sotirios Korossis, Axel Haverich, Tobias Schilling

Published by InTech
Janeza Trdine 9, 51000 Rijeka, Croatia

Notice

Publishing Process Manager Maja Bozicevic
Technical Editor SPi Global
Cover InTech Design team

First published December, 2017
Printed in Croatia
Legal deposit, Croatia: National and University Library in Zagreb

Additional hard copies can be obtained from orders@intechopen.com

Scaffolds in Tissue Engineering - Materials, Technologies and Clinical Applications, Edited by Francesco Baino
p. cm.
Print ISBN 978-953-51-3641-5
Online ISBN 978-953-51-3642-2

For EU product safety concerns:
IN TECH d.o.o., Prolaz Marije Krucifikse Kozulić 3, 51000 Rijeka, Croatia,
info@intechopen.com or visit our website at intechopen.com.

Meet the editor

Francesco Baino is an Assistant Professor in the Department of Applied Science and Technology, Politecnico di Torino (Italy). He received his M.S. degree in Biomedical Engineering (summa cum laude) and his PhD degree in Materials Science and Technology from Politecnico di Torino. Dr. Baino has published more than seventy-five peer-reviewed journal articles and ten book chapters and is the author of four patents. His current research interests include bioceramics, bioactive glasses and composite biomaterials for tissue engineering and medical implants, as well as processing and testing of advanced ceramics. He is an associate or guest editor of prestigious international journals including the International Journal of Applied Ceramic Technology (Wiley), Materials (MDPI), the Journal of Functional Biomaterials (MDPI), and Frontiers in Bioengineering and Biotechnology (Frontiers).

Contents

Preface

Over the last few decades, we have assisted to an impressive increase of elderly population worldwide associated with age-related pathologies. Therefore, there is more than ever the need for new biomaterials and clinical strategies that can substitute damaged tissues, stimulate the body's own regenerative mechanisms, and promote tissue healing.

Biomaterials used in regenerative medicine are often designed to act as scaffolds, i.e., porous templates that support and stimulate the growth of new healthy tissue in three dimensions (3D) and then safely dissolve once they have performed their functions, thus leaving the body to remodel the tissue to its natural form. The rate and quality of tissue integration are related both to the physical, chemical and biological properties of the scaffold material and to the scaffold 3D architecture, including pore/strut size, pore volume fraction, pore distribution and pore interconnectivity. Strut microstructure and pore geometry also deserve to be carefully taken into account during scaffold design since they play a pivotal role in influencing the entrapment and recruitment of growth factors and cells and strongly determine the biomechanical properties of the scaffold alone as well as of the scaffold-tissue construct in vivo. Scaffolds can also act as multifunctional systems for the controlled release of therapeutic agents in situ, such as ionic dissolution products, drugs, or other biomolecules that can elicit an appropriate response in the organism, when required to treat a specific disease.

This book intends to provide the reader with a comprehensive overview of the current state of the art in the field of tissue engineering scaffolds, highlighting the potential associated to the latest scientific and technological advancements, as well as giving a picture of present and future clinical applications. Given the high multi- and transdisciplinary nature of these topics, the book will be of interest to both academic and industrial scientists working in the broad field of biomaterials for tissue engineering.

The book is divided into two sections: the former focuses on the repair of "hard" tissues (primarily bone) by means of bioactive glass and ceramic scaffolds, and the latter deals with the applications of polymeric scaffolds for regenerating "soft" tissues and structures such as the peripheral nerve, heart and blood vessels, gastric mucosa and pancreas. Special emphasis is given to the challenges associated to scaffold manufacturing, biomimetic properties, cell-scaffold interactions and the role of nanotechnology.

Finally, I would like to express my sincere appreciation and gratitude to all the authors who have contributed to this book.

Francesco Baino
Department of Applied Science and Technology
Politecnico di Torino
Italy

Section 1

Scaffolds for Bone Tissue Engineering

Chapter 1

Fabrication Methodologies of Biomimetic and Bioactive Scaffolds for Tissue Engineering Applications

Mythili Prakasam, Madalina Popescu,
Roxana Piticescu and Alain Largeteau

Additional information is available at the end of the chapter

http://dx.doi.org/10.5772/intechopen.70707

Abstract

Tissue engineering has offered wide technologies for developing functional biomaterials substitutes for repair and regeneration of damaged tissue and organs. Biomimetic materials with their inherent nature to mimic natural materials are possible through the recent advances in the fabrication technology. With the help of porous or dense implants made with biodegradable materials, it is possible to incorporate different vital growth factors, genes, drugs, stem cells and proteins. In this review, we presented various fabrication methodologies of biomimetic and bioactive scaffolds for tissue engineering applications. An overview of the nanocomposites of biomaterials is presented. Further an example of one of the hybrid nanocomposite material is given for additive manufacturing.

Keywords: biomimetic scaffolds, bioactive ceramics, fabrication technology, tissue engineering, 3D printing

1. Introduction

Inspired by the natural evolution mechanisms and their structural, physical and chemical characteristics of biological organisms, researchers worldwide are looking for synthetic materials that mimic the function of the natural materials. Nature offers solutions to many complex technological and materials solutions due to the billon of years of technological evolution. Biomimetics [1] is considered as a future of materials design and productions. The history of biomimetics exploration by humans dates back to early fifteenth century by Leonardo da Vinci speculating the clues of possibility of human air travel following the mechanics of flight of birds. Biomimetic materials have potential applications in various domains such as biology, chemistry, materials science and electronics. In the present scenario, the need for synthetic bone substitutes [2] in the last few decades have increased owing to the number of accidents/trauma and inherent bone defects. The current state-of-art enables the successful fabrication of the materials [3] for biomimicking applications especially for biological applications. Mimicking the 3D structure and extracellular matrix (ECM) [4] of native tissue is critical for successful cell transplantation and growth of artificial tissue. Biomimetic scaffolds are therefore necessary to recapitulate this natural environment and provide various cues to direct cell processes and differentiation. For examples, when considering a biomaterial to be used in implants or bone grafts, various aspects such as biocompatibility, osteogenic properties, bioactivity and its mechanical functions, based on its functionalities have to be studied. To replicate an ideal scaffold as biomimetic for tissue applications, fabrication methodologies play a very major role. Especially, tissue engineering has a great promise in using the biomimetic technologies for organ transplantation [5], reconstructive surgery [6] and artificial extracellular matrices [7] for three dimensional tissue formation. In the field of tissue engineering various approaches are followed such as cell substitution for specified functions, delivery of induced substances such as cell growth and differentiation and growth of cells in three dimensional scaffolds. Bone reconstruction is clinically important due to the large number of patients who have bone defects of critical size, leading to decrease in overall health and quality of life. The main treatment for bone defects remains bone grafting [8]. Approximately, 600,000 bone graft procedures are performed each year in the United States, and about 2.2 million of such procedures are performed worldwide annually, generating a sale of about 2.5 billion dollars per year and representing the second most common tissue transplantation procedure after blood [9–13].

There are a variety of scaffold fabrication methods in tissue engineering, including gas foaming, solvent casting and particulate leaching, emulsion freeze-drying, phase separation, melt molding, membrane lamination, electrospinning, fiber bonding, polymer/ceramic composite foam fabrication or a combination of these techniques [9, 14–16]. In order to replicate the natural bones, scaffolds are engineered to be bioactive or bioresorbable to enhance tissue growth. Scaffolds are often porous that supports mechanically. There are various criteria such as biocompatibility, mechanical properties, pore size and bioresorbability that have to be considered as requirements of an ideal scaffold. Different classes of materials are used for fabrication of scaffolds such as polymers, bioglass, composite and metals. Significant progress was achieved in terms of scaffolds for mechanical support with better osteogenesis and angiogenesis. Various material fabrication methodologies enable the possibility to fabricate scaffolds with complex design [17] to ensure mechanical integrity and scaffold interconnectivity. The objective of this chapter is to review the different fabrication methodologies and the recent advances in the fabrication of biomimetic and bioactive scaffolds are presented with a case study and the possible improvements envisaged are discussed.

2. Biomimetic nanocomposites types and fabrication methodologies

2.1. Various biomimetic nanocomposites based on their design

Human body being a complex and sensitive biological system requires the scaffold materials with diverse and challenging characteristics. Scaffold must combine structural, material, bioactivation, signaling molecules, cells and biological requirements satisfied for different applications. Human bone [18] is classified as the long bones (femur and tibia), short bones (vertebrae and metacarpal bone), flat bones and the spongious bone [19]. The structural property of the corresponding bones is strongly interdependent on the mechanical property [20]. The cortical bone contains 10% porosity and is mostly dense, whereas the spongy bones are highly porous and has less mechanical property. The spongy bones act as a host to soft tissues, cartilage, and meniscus and prevent the stress concentration. The fibrous tissue surrounding the bones is ligaments, which are highly organized fiber tissue composed of collagens, elastin, proteoglycan, water and cells. The mineral inorganic part will be assisting in compression and shear and the collagen matrix provides tensile strength. Anisotropic compact bone has compression strength in the range of 130–225 MPa along the longitudinal direction and has the compression strength in the range of 105–135 MPa in transverse direction [21]. The cancellous/spongy bone has compression strength of 5–10 MPa and elastic modulus of 50–100 MPa. In the current state of art, it can be observed that the mechanical integrity of the synthetic scaffolds is inadequate. Polymers have yielded results close to the cancellous bone properties. Bimodal architectures [22] matching the natural bone is yet in progress. Scaffolds should provide sufficient strength [23] for cell ingrowth in *in vitro* conditions and integrate completely under *in vivo* conditions. In order to fabricate the biomimetic materials various systems [24] such as polysaccharides, proteins, nanocomposites based on calcium phosphates (CaPs) (e.g., β-tricalcium phosphate, hydroxyapatite) with good osteoconductivity, resorbability and biocompatibility are employed. The polymers [25] employed usually assist in structural integrity and promotes cell adhesion. Whereas, CaPs are biocompatible, osteoconductive and biodegradable with limitations on mechanical strength that forbids the load-bearing applications. β-TCP is the high temperature phase of CaPs, but has the chemical similarities with HAp, with different resorbing capability [26]. Resorption of HAp is slow in the biological medium and β-TCP has good reabsorption. Ionic co-substitutions in calcium phosphates are also used that can influence the structure, microstructure, crystallinity and dissolution rate. Various ions [27] such as Sodium, Strontium, Magnesium, Manganese, Silver, Barium Potassium, Zinc, Fluoride, Chlorides and Carbonates are substituted in hydroxyapatite. The substitution of these aforesaid ions have significant role in causing alterations to bone resorption, bone formation, solubility, structure and morphological changes and enhanced surface microstructure. There are also Calcium phosphate based cements [28] that can be used as injectable pastes in the defect bone site for cell delivery and is an upsurging topic of research. Nanocomposites [29] consisting the component of polymer and nanosized CaPs improves the tissue bonding, cell adhesion and cell differentiation. Designing of the 3D scaffold [30] should stimulate the adhesion, proliferation and differentiation mechanisms in addition to the structural complexity of natural bones. The synthetic bone should also enable vascularization to enable biointegration with transport and support of signaling molecules, passage of nutrients and blood vessels. Hence the critical aspects such as highly porous interconnected microstructure and degree of porosity for uniform cell distribution, proliferation and migration *in vitro*. Critical pore sizes [31] are necessary to adjudge their function, because when the pore size is small the cell adherence can block the pores and matrix formation of the scaffold. An optimal pore size of ~150–400 μm can help in bone tissue regeneration. On the other hand, the high porous scaffold will have low Young's modulus and compressive strength. It was reported that the tangent elastic modulus of natural bones decreases with the increase in porosity such as in porous tantalum [32]. The Young's modulus value of wet compact bone is in the range of 6–11 GPa, a porous scaffold of Ni-Ti with porosity in the range of 20–50 Vol % was fabricated by powder metallurgy.

Equally the pore shape, size and orientation are important in adjudging the mechanical properties of the scaffolds. Variation of the mechanical properties [33] of the scaffold with two different pore structures was reported earlier. It was observed that the mechanical properties of the scaffold with spherical pore structure have higher elastic modulus in comparison to the scaffold with cylindrical pore [34]. Similar alteration in the mechanical properties was also reported in the case of polycaprolactone [35]. From the aforesaid it can be concluded that the mechanical properties of the fabricated scaffold can be altered by the variation of the pore shape. Similar influence was reported on the pore size, where an increase in the pore size reduces the mechanical properties. The orientation of the pores also significantly altered the mechanical properties of the scaffold irrespective of the matrix of the material chosen ranging from metals to polymers. When the pore geometry was parallel or aligned in the same direction, the mechanical strength of these scaffolds were high. The mechanical stability of the multiscale porosity can be interesting in terms of crack propagation. This multiscale porosity can be achieved by combining two or three scaffold fabrication methods.

2.2. Materials for fabrication of scaffolds

Materials for fabrication of scaffolds [36] are selected based on their degradability, chemical and physical properties. Materials can be classified based on their source and mode of origin/preparation. Animal and plant based materials such as starch, alginate, chitosan, hyaluronic acid, gelatin, collagen, fibrin, silk, etc., are used as natural biomaterials. Due to the disease transmission and purification, synthetic biomaterials are attracting the interest of the researchers. Synthetic organic and inorganic materials such as HAp, CaPs, glass, polyesters are actively studied for usage as scaffold material. Autograft bone scaffold [37] was considered to the gold standard for bone repair defect. Lack of donors and possible disease infection are few of the major disadvantages of autograft scaffolds. Bone allograft [38] was considered as an alternative to the autograft, but still the transmission of diseases, inflammatory reactions and rejection by the recipient body is prevalent. Hence artificial scaffolds have been proposed as an alternative to the autograft and allograft scaffolds. Usage of metallic implants [39] dates back to 1920. Currently synthetic bone implants made of metals, ceramics, composites and glasses are employed for bone regeneration and bone reconstruction. There are several issues such as poor strength of the polymer scaffolds, poor ion release of ions from the metallic scaffolds, brittleness of the ceramics and difficulty in controlling degradation rate of the composite scaffold to be addressed. Langer and Vacanti [40] were the pioneers of the tissue engineering concept in early nineties. They introduced the concept of introducing the bone marrow cells, different growth factors, gene and drug delivery in to the artificial bone scaffolds. Further investigations by various researchers enabled the possibility of improving the biocompatibility by changes in the topology, nanostructure, chemistry and structural aspects of the scaffolds. Metallic biomaterials are especially preferred for their usage in load-bearing applications due to their good mechanical properties. The porous metallic scaffolds [41] based on the stainless steel, titanium and magnesium are used to eliminate stress shielding and reduce the stiffness to match the natural bone. Interconnected porous metallic scaffolds [42] were fabricated by combining the various rapid prototype techniques such as solvent leaching and 3D printing was reported in the fabrication of Ti scaffolds. Other fabrication techniques such as powder metallurgy, sintering and rapid prototyping to modulate the scaffold design. The mechanical properties of the Ti scaffolds can be used to mimic the human cancellous bone due to its structural flexibility. The elasticity and other vital properties such as deformation force, stress and strain can be modified by alloying with a suitable metal. NiTi alloys were reported to exhibit good *in vivo* compatibility than pure Ti porous scaffolds [43]. The porous metallic scaffold assisted in rapid formation of new bone tissues for load-bearing conditions.

Bioactive glasses [44] were reported to have good osteoconductivity, controlled biodegradability, cell delivery capabilities and inducing osteogenic gene expression for formation of bone minerals and capability for drug delivery. But bioactive glasses yield poor results for load-bearing applications due to their lack of superelastic performance, mismatch of compressive strength and Young's modulus of the natural bone. In order to improve the mechanical properties of the bioactive glasses, modification of structure and composition is achieved during bioglass scaffold fabrication. Biomorphic and mesoporous bioactive scaffolds [45] were shown to have better mechanical properties than in comparison to the classical bioactive glasses. The inorganic component phase in the bioactive glass scaffold is important in addition to the structural design of these bioactive glass scaffolds. Biopolymers and its derived composites are also used for fabrication of scaffold for tissue engineering applications. However their poor mechanical strength makes it difficult to use them in load bearing applications. Biopolymer based composites and hybrids for bone scaffold applications [46] with required strength are prepared by varying the volume fraction of polymer in the composites. ECM structure [47] provides the structural and biochemical support, hence ECM like biopolymers, ECM-HAp composites and ECM bioglass scaffolds are studied for tissue engineering applications. In order to replicate the biomimetic conditions, surface modifications are also carried out. Controlled release of biological molecules is also a key function that the scaffolds play.

2.3. Biomimetic porous scaffolds

Porous biomimetic scaffolds [48] with their 3D structure is advantageous for (1) better cell biomaterial interactions, cell adhesion and growth (2) interconnected porosity for angiogenesis and transport of nutrients, regulatory factors for cell survival, proliferation and differentiation (3) Sufficient structural integration with good tensile strength and elasticity (4) Control degradation and minimal toxicity *in vivo*. Nanocomposites of CaPs and natural polymers such as collagen and gelatin are well known for tissue engineering. Techniques such as foam replica method [49], freeze casting [50], freeze drying [51], phase separation [52], gas foaming [53], rapid prototyping [54] and electrospinning [55] are employed for fabrication. The challenge of nanocomposite scaffolds lies in the ensuring the retention of chemical phases and retaining the porous structure without disturbing their porous structure. A combination of different method of fabrications such as freeze casting and electrospinning are used for fabrication of scaffolds. Other polymers such as poly lactic co-glycolic acid (PLGA), Sodium dodecyl sulfate (SDS), cellulose, poly glycolic acid (PGA), poly (ethylene glycol) and poly l-lactic acid (PLA) are also used for fabricating the nanocomposites due to their excellent biodegradability and biocompatibility. Collagen of type I, the predominantly available protein in mammals can be readily obtained from animal tissues and from human tissues. This extracellular matrix protein collagen can be reconstituted in a different morphology into fibrillary matrix by changing the pH and temperature of precursors. Due to the lack of mechanical strength in collagen for *in vivo* applications, several strategies are employed such as crosslinking with hydrogels [56] or compression so that it can sustain or resist cell-induced contraction. Inherent characteristics of collagen such as dipole moment and alignment under strong magnetic field [57] were demonstrated to induce cell migration and allow preferential growth of neurites along the alignment of the fibril direction. Collagens can hold several cellular receptors that can variate the cell behavior and their biological function can be induced by combining with growth factors for example, vascular endothelial growth factor (VEGF) to improve cardiac function was observed.

Fibrin [58] is a specialized protein network clinically available from autologous sources such as human blood plasma. In the presence of thrombin protease, fibrin matrix is formed spontaneously by polymerization of fibrinogen. Cell migration in fibrin is dependent on the cell-associated proteolytic activity from plasmin and MMPs due to their small diameter. This indeed assists in strong fibril-fibril interaction and the natural network formation and stabilization through covalent bond stabilization. As fibrin matrices are poorly active for most of the cell types, functionalization with ECM peptides or growth factors is necessary. The controlled release of the growth factors and hormones can be efficiently done by covalent bonding of biomolecules. This possibility of delayed/controlled release of the growth factors/hormones [59] is currently employed for clinical evaluation for local bone repair.

Hyaluronan or Hyaluronic acid [60] is a structural protein that is noncovalently attached to the protein core and entwine ECM. Due to their strong anionic nature, these polymers absorb water and hence providing the compressive strength to the ECM. Various chemical hyaluronic acid derivatives have been prepared by controlling the functional group and the type of covalent bond. It is possible to create a wide range of materials with diverse properties. Hyaluronic acid [61] is used for various applications for dermal wound healing, chondrocyte transplantation for tissue repair and for incorporation of other functional biomolecules for improved fibroblast proliferation and wound healing. Other self-assembling polypeptides are also used to form nanofibrillar matrices *in situ*. Self-assembled peptide hydrogels [62] are used as a tool for developing 3D cell culture plates.

To have properties similar to natural ECM, it is necessary to have facilities for cell seeding, adhesion, proliferation, differentiation and new tissue generation. Essential characteristics such as biodegradability and mechanical properties are important to be studied. The biodegradation of the scaffold should be in coherence with the rate of the formation of the new tissue formation that it supports initially to act as a scaffold material to serve as a template. Elastomeric properties of scaffolds [63] are studied to improve their applications in tissue engineering applications. Other than the natural ECM materials, synthetic polymers such as poly (ε-caprolactone) (PCL) and polyurethanes (PU) [64] are investigated for vascular and other tissue engineering applications. Tensile modulus and strength are critical parameters necessary for tendons and ligaments. Natural fibrous protein from silk worm cocoon is a material with excellent tensile and mechanical strength [65]. Though Sericin present in the natural silk causes adverse immune responses and is disadvantageous for tissue engineering applications, natural silk's biocompatibility is improved upon Sericin removal. Silk fibroins have hydrophobic and hydrophilic blocks which forms crystals through hydrophobic interactions and hydrogen bonding resulting in the improved tensile strength. Spider silk fibroin polymers [66] are used for genetic engineering due to their excellent mechanical and cell adhering capacity.

As extracellular proteins have a fibrous structure with diameters in the nanometer or submicrometer scales, various advanced material shaping techniques were developed. Techniques such as electrospinning, self-assembly and phase separation are a few worthy-to-mention. Electro-spinning technique were used to produce nanofibers, but the disadvantage of this technique is to fabricate the complex 3D scaffold structure or to produce intricate pore structures. Various cells were reported to proliferate, differentiate and attach on these electro-spun nanofibers [67]. Phase separation arises when homogeneous multicomponent system tends to be unstable resulting in multiphase system due to the system free energy. This technique is employed in generating porous structures as tissue engineering scaffolds. The physical form of the porosity (closed/open) depends on the system of phase separation. The huge disadvantage of this technique is the lack of interconnected porosity and the lack of control over 3D shapes. Biological effects of nanofibrous scaffold [68] were found to adsorb more human serum proteins in comparison to the scaffolds of smooth pore wall morphology. The cell adhesion proteins were comparatively well detected than the conventional scaffolds.

Hydroxyapatite being the major inorganic compound used in most of the composite scaffolds, it provides good osteoconductivity. In the presence of the composite structures with polymers hydroxyapatite provides structure and design flexibility. Further it assists in improving the protein adsorption capacity, microstructure favoring the bone tissue regeneration and diminishing the cell death. These nanocomposites mimic the features similar to the natural bone with improved mechanical properties. However regarding the biodegradation rate of stoichiometric HAp are less efficient than the partially carbonated apatite. The surface of scaffolds affect cellular response and in turn also the tissue regeneration [69].

2.4. Biomimetic scaffold fabrication

Composites of scaffold materials with organic and inorganic components were designed for mechanical and physiological requirements in tissue engineering. However the brittle nature of ceramics and low mechanical strength inhibits their usage in clinical applications. Composite structure of HAp with porous polymers alongside their pore size, shape and morphology has improved the mechanical strength for tissue engineering applications. With the recent advances in the biodegradable polymers [70], glass and ceramics, it is possible to cater degradation kinetics and resorption *in vivo* after stimulating cellular responses at the molecular level. It is expected that the new generation of scaffolds can perfectly mimic the natural bone in terms of mechanical and structural aspects. With the advances in the manufacturing techniques such as selective laser sintering and rapid prototyping should be helpful in bone tissue engineering applications. These techniques are widely used in the fabrication of temporomandibular joints [71], craniofacial [72] or periodontal structures [73]. The fabrication methodology allows the flexibility of the combination of different materials for increasing mechanical strength. The current state-of-art, however, does not allow to exactly replicate the natural architecture of the extracellular matrix or the natural bone. There is a continuous demand for the improvement and functioning of the scaffold fabrication. Here we will discuss few of the fabrication techniques such as solvent casting, particulate leaching electrospinning, gas foaming, phase separation, fiber meshing and bonding, self-assembly, rapid prototyping, melt molding, lamination and freeze drying.

2.4.1. Solvent casting

A nineteenth century technology, which was initially used for production of thin films, is currently employed for diverse applications ranging from optical applications, photographical film base flexible printed circuits and high temperature resistive films. Solvent casting [74] are generally produced by evaporation of the solvent in order to form the scaffolds. In order to successfully employ the solvent casting, it is necessary that the polymer used should be soluble in a volatile solvent or in water. A stable solution with an adequate solid content and viscosity should be formed. Formation of homogeneous film after removing from cast support must be possible. The main disadvantage of this technique is the denaturation of the protein caused due to the solvent if toxic and influence of the organic material on the solvent. To remove the toxicity of the solvent left over in the scaffold, they are usually dried by vacuum process and dried completely. To make it time efficient, other techniques such as particulate leaching were combined for the fabrication of scaffolds.

2.4.2. Particulate-leaching technique

Being one of the popular techniques in the fabrication of scaffolds for tissue engineering [75] is relatively a technique of ease. Usage of porogens such as salt, sugar and wax is common to produce the pore channels by this technique. Mixture of porogen of the desired size is mixed with the material, then after leaching the porogen, the pores are left behind in the matrix. The control of the pore size is possible by choosing the required size and shape of the porogen. Pore sizes in the range of 300–500 μm is possible with a porosity percentage of around 94–95%, however the control of the interconnected pore is not possible. Thin layers of up to 3 mm thickness is feasible with this technique.

2.4.3. Gas foaming

Gas foaming process uses the utilization of high pressure gases for the fabrication of highly porous scaffolds [53]. While the polymers are used, they get saturated and gets nucleated as a gas bubble with the gas bubble size ranging in the order of 100–500 μm are used. When the gas is injected the gas bubble starts to create the phase separation. A three dimensional porous structure can be obtained by using the gas foaming technique. But the control over the interconnectivity is highly lagging behind in this technique, though is advantageous in terms as a solvent free technique. The porosity will be frequently absent on the external surface.

2.4.4. Phase separation

This technique involves quenching of the polymers which can cause two phases of polymers to separate as polymer rich and polymer poor phases [76]. The polymer poor phase will cause the porous structure network to be formed. The control of the porous microstructure is possible with the help of parameters such as polymer concentration, quenching rate and temperature, solvent concentration, solvent type and dispersion of the solute molecules. The solvent can be removed by extraction, evaporation and sublimation after integrating the bioactive molecules in the scaffold. Nanofibers can be prepared by liquid-liquid phase separation to replicate the architecture similar to type I collagen molecules and other extracellular matrix architectures for 3D cell culture environment. As these experiments are carried out at low temperatures, the incorporation of bioactive molecules is feasible. The phase separation methodology can be used in conjunction with other techniques that can control the porous architecture and also with other rapid prototyping techniques for tissue engineering applications.

2.4.5. Sintering/heat molding

This technique [77] involves filling the mold with the mixture of powder/polymer and porogen and heating/sintering above the glass transition temperature of the polymer or at the porogen evaporation or melting temperature. The external shaping of the sample will take form of the die set up/mold used for sintering or heat molding. When the porogen is leached out and then the scaffold is formed. The inconvenience of this technique is that the possibility of residual porogen and impossibility to incorporate biological molecules due to the involvement of high temperature processing.

2.4.6. Freeze casting

This technique [78] involves the mixing the solvent and the solute followed by decrease of temperature until to reach negative temperature to fabricate ice crystals. The solute molecules are segregated around the ice crystals forming a porous network. In order to remove the solvent, lyophilization is carried out which involves applying lower pressure than the equilibrium vapor phase of the frozen solvent. After sublimation of the solvent ice crystals, the porous network remains intact. The beneficial factors of this technique is the possibility of changing the morphology of the pores by modulating the temperature of ice crystals and the possibility to incorporate bioactive molecules due to the low processing temperature involved in this fabrication methodology. The pore size distribution and the homogeneity of the porosity formed can be controlled with the aid of the processing temperature of the process. Freeze casting can also be used to protect other dry biological samples to retain their bioactivity.

2.4.7. Fiber Mesh

This technique [79] involves interweaving or weaving individual fibers into a three dimensional scaffold with varying pore size. This technique enable the availability of the cell attachment and allows transport of the nutrient for the cell survival and cell growth. Though the practical difficulties such as lack of structural stability exists. In order to increase the structural stability and increase the crystallinity heat treatment is sometimes employed to overcome the structural instability in Fiber Mesh technique.

2.4.8. Fiber bonding

This technique [80] involves combining the nonfiber polymer mixing with a solvent, when the solvent is removed then the solute molecules will be embedded in the polymer matrix. The scaffolds are fabricated by bonding the collagen matrix to polymers with threaded collagen fibers. The fiber bonding encapsulation takes place due to the heat treatment. The resulting scaffold structure has large surface area that allows the cell proliferation by replicating the extracellular matrix. The disadvantage of this technique involves the poor controlling ability of the porosity and the pore size, unavailability of suitable solvents and appropriate melting temperature of the polymers.

2.4.9. Electrospinning

This technique [81] is highly used for fabricating the nanofibrous architecture. Nanofibrous architecture highly favorises cell binding and other cell behavior activities. Electrospinning was successfully employed in fabricating porous scaffolds of nano- and microscale fibers that can replicate the structural and biological functions of the natural extracellular matrix. The method involves electrostatic spraying of polymer coatings. Scaffolds with >90% were obtained by this methodology. An appropriate solvent is required for dissolving the polymer to be loaded in a syringe. The rate of control of polymer flow can be adjusted by constant syringe flow rate. Various parameters are necessary for scaffold fabrication by electrospinning such as viscosity, conductivity, molecular weight and surface tension of the polymers and the processing parameters such as applied voltage, flow rate and temperature. The main advantage of this technique is that it can produce 3D scaffold configuration adaptable for cell and tissue organization, adhesion and spatial cell regeneration with suitable physiological conditions. The main disadvantage of this technique is the impossibility of cell seeding, but cryospinning and templated sacrificial scaffold can create the pore of desired size in the electrospun matrices. Electrospinning is also used in conjunction with other scaffold techniques such as solvent casting and rapid prototyping techniques. Favorable results were observed in cell culture for cartilage and bone tissue engineering applications with electrospun scaffolds.

2.4.10. Rapid prototyping

The group of techniques [82] involving the assistance of computer assisted design (CAD) and computer assisted manufacturing (CAM) such as stereolithography, selective laser sintering, fused deposition modeling, three dimensional printing and three dimensional plotting. The choice of materials ranges from polymers, ceramics to metals. These fabrication techniques are frequently used for fabrication of biomimetic tissue scaffolds. Most of these techniques are capable of yielding scaffolds with good mechanical strength, high accuracy and possibility of incorporation of cells and proteins. There are few limitations such as elevated temperatures, weak bonding between the powder particles, trapped powders in the pores (requiring post treatment measures) and slow fabrication processing in few cases. These materials are prevalently employed in bone, heart valves, adipose tissues and cartilages. Hybrid materials based on calcium phosphates with polymers is also widely used with rapid prototyping also known by terminology as additive manufacturing. In the following section, we will focus on one of the widely used rapid prototyping technique called as 3D printing with an elaborate example as a case study.

2.5. Hybrid materials based on calcium phosphates and polyurethanes for bone tissue engineering – a case study

Additive manufacturing (AM), also known as rapid prototyping or solid freeform fabrication, was developed in the mid-1980s as an advanced technology to overcome the limitations of these conventional methods and have received much attention in recent years due to its ability to deliver a high level of control over the architecture of the construct [13]. With AM techniques, scaffolds with precise geometries can be prepared, using computer-aided design combined with medical imaging techniques (either computed tomography – CT, scanning or magnetic resonance imaging – MRI) to make anatomically shaped implants [83].

According to ISO/ASTM52900-15, there are seven categories of ALM type processes: Binder Jetting, Directed Energy Deposition, Material Extrusion, Material Jetting, Powder Bed Fusion, Sheet Lamination and Vat Photopolymerization [84]. AM techniques may be divided into three sub-groups: (1) laser or other directed energy beam, (2) print or "ink," and (3) nozzle systems [85]. C.K. Chua and W. Y. Yeong have classified AM techniques for the fabrication of tissue scaffolds in two types of methods: (1) direct methods and (2) indirect methods [16], summarized in **Table 1**.

Several Rapid Prototyping (RP) techniques, such as selective laser sintering (SLS), fused deposition modeling (FDM), precision extrusion deposition (PED) and 3-D fiber deposition (3DF) have been used to fabricate polymer–calcium phosphate composites based on poly(hydroxybutyrate-co-hydroxyvalerate), PLLA, PCL or poly(ethylene oxide terephthalate)/poly(butylene terephtalate) (PEOT/PBT) co-polymer and different calcium phosphates, such as carbonated hydroxyapatite, HA or β-tricalcium phosphate (β-TCP) [85].

In 2014, Zhou et al. [86] proposed a new method to increase the printability of calcium phopshate powders, by mixing it with $CaSO_4$ (HA: $CaSO_4$ = 25:75 weight ratio) for scaffold fabrication through 3D printing method. 3D printing technique has been used for fabrication of 3D scaffolds based on hydroxyapatite (HA) and organic polymers such as polyvinyl alcohol (PVA) [87] or collagen [88].

A scaffold made of composite material based on calcium phosphate and collagen was obtained by 3D printing method at low temperatures. For this purpose, 5–20% phosphoric acid was used as binder and Tween 80 was added as a non-cytotoxic surfactant in the proportion of 0.25% in the binder solution. The optimal concentration of the binder solution (8.75%) for which the cytocompatibility and mechanical strength are maximized has been established. Collagen has previously been dissolved in the binder solution to further enhance the fabrication of calcium phosphate-collagen composites by 3D printing.

Type of method		
Direct		**Indirect**
Melt-dissolution deposition techniques	**Particle bonding techniques**	
Fused deposition modeling (FDM)	Selective laser sintering (SLS)	Droplet deposition
Precision extruding deposition (PED)	Three dimensional printing (3D printing) or color jet printing (CJP)	Melt deposition
3D fiber deposition technique (3DF)	TheriForm	Photo-polymerization: – stereolithography – two-photon polymerization
Precise extrusion manufacturing (PEM)		
Low-temperature deposition manufacturing (LDM)		
Multi-nozzle deposition manufacturing (MDM)		
Robocasting		
Pressure-assisted microsyringe (PAM)		
3D bioplotter		
Rapid prototyping robotic dispensing system (RPBOD)		

Table 1. Classification of additive manufacturing (AM) techniques for scaffold fabrication.

In 2013, Nandakumar et al. [88] reported the production of two types of polymer-hydroxyapatite composite scaffold by 3D deposition using a Bioplotter. PolyActive ™ (PA), a commercial copolymer of poly (ethylene oxide-terephthalate)/poly (butylene terephthalate) (PEO/PBT) was used.

In the first embodiment, polymer-ceramic composite filaments with the desired mass ratio (maximum 15% HA) were extruded and further used for scaffold fabrication using a Bioplotter. In the second approach, polymer scaffolds were obtained by 3D deposition, while the ceramic particles were made in the form of columns by sintering the ceramic paste through stereolithography. The two components were then manually assembled by pressing HAp into the pores of the polymer scaffold, thus creating the composite material.

Among 3D printing techniques, *bioplotting* is the most used method in printing of 3D tissue constructs. Bioplotting, first developed in 2000 at the Freiburg Research Centre, is an extrusion-based printing method that is spatially controlled by a robotic system (x, y, and z directions) with microscale resolution, which uses STL files to guide the extrusion head. Nozzles of various diameters are used for dispensing highly viscous or pasty materials applying different actuation mechanisms – mainly screw – or piston based mechanical and pneumatic. The dispensing head moves in three dimensions, while the fabrication platform is stationary. The printed construct from the bioplotter occurs in a laminar fashion by the computer-controlled deposition of material on a surface and could be relatively large (several centimeters in length, width, and height) [82, 89].

Since heating is not required, the system can process thermally sensitive biocomponents, and even cells. The strand thickness can be modulated by varying material viscosity, deposition speed (speed in the planar field), tip diameter, or the applied pressure. The main advantages of bioprinting using a Bioplotter system are the room temperature processing, direct incorporation of cells, and homogenous distribution of cells. The main disadvantages are limited mechanical stiffness, critical time of hardening, specific matching of material, and low resolution [90].

2.5.1. *Fabrication of 3D scaffolds based on hydroxyapatite-polyurethane hybrids*

Hydroxyapatite—polyurethane 3D scaffolds have been obtained using system BioScaffolder SysEng, Germany, (**Figure 1**) connected to a computer on which the Bioscaffolder SW version 3.0 software is installed.

HAp-PU nanostructured hybrid powders synthesized in high pressure hydrothermal conditions (1000 bar and 100 bar, respectively) were mechanically mixed with a carefully selected commercial polymer (Mowiflex, Kuraray Poval) 20% solution, and a crosslinking agent (Baymedix® FD103, Covestro), to obtain HAp-PU-based pastes used as a feedstock in the 3D Bioprinting technique. Some examples of 3D hybrid structures obtained with different nozzles and rotation angle between 2 successive layers (90° and 45°, respectively) are given in **Figure 2**.

2.6. Hydrothermal synthesis of hybrid organic-inorganic nanostructures in high pressure conditions

Hydrothermal method is a well-known and attractive technique for producing pure nanocrystalline, highly homogeneous nanoparticles in a single step, in aqueous medium, with low energy consumption [91–94]. The hydrothermal method has proven to be an effective, convenient and environmentally friendly process. Hydrothermal synthesis at high pressure is characterized by three main advantages: (i) low energy consumption, developed by applying

Figure 1. (a) System BioScaffolder, SYSENG and (b) low temperature deposition head.

Figure 2. 3D printed scaffolds using three types of nozzles: (a) Ø 0.8 mm; (b) Ø 0.6 mm; and (c) Ø 0.4 mm [95].

pressure (for a liquid phase, the same energy is involved on five units for the temperature scale than on 4000 units for the pressure scale); (ii) negative ΔV value between the total molar volume of reaction products and the total molar volume of reactants and (iii) improvement of the chemical reactivity.

2.6.1. Hydrothermal synthesis at 100 bar using Ar gas

In a first step, $Ca(NO_3)_2$*$4H_2O$ (Sigma Aldrich) was dissolved in distilled water. Ammonium dihydrogen phosphate (Sigma Aldrich) was added dropwise into the calcium nitrate solution under continuous stirring. In parallel, the viscous polyurethane-diol solution was dissolved in water, and thus obtained solution was transferred into the initial vessel along with the other reagents. A precipitate was obtained whose pH was adjusted to 10 using 25% ammonia solution. In the second step, the hybrid precursor thus resulted was transferred to the Berghof autoclave, Germany and subjected to hydrothermal treatment at a pressure of 100 bar and a temperature of below 120°C for 3 hours. Pressure is created above the solution vessel of the autoclave using an argon gas bottle [93, 94]. After hydrothermal treatment, the nanostructured powder was washed with distilled water until neutral pH was reached, mixed with polyvinyl alcohol (PVA) 5% solution and spray dried in LabPLANT, United Kingdom spray dryer.

2.6.2. Hydrothermal synthesis at 1000 bar isostatic pressure

In a first step, $Ca(NO_3)_2$*$4H_2O$ (Sigma Aldrich) was dissolved in distilled water. Ammonium dihydrogen phosphate (Sigma Aldrich) was added dropwise into the calcium nitrate solution under continuous stirring. In parallel, the viscous polyurethane-diol solution was dissolved in water, and thus obtained solution was transferred into the initial vessel along with the other reagents. A precipitate was obtained whose pH was adjusted to 10 using 25% ammonia solution. In the second step, the hybrid precursor thus resulted was transferred to the high pressure autoclave HP Systems, France and subjected to hydrothermal treatment in isostatic pressure conditions (1000 bar) and a temperature below 120°C, for 3 hours. Isostatic pressure is created uniformly inside the autoclave by a hydraulic system.

The reaction mixture is perfectly sealed in a special synthesis bag, leading to formation of nanohybrids in microbiologically safe conditions.

After hydrothermal treatment, the nanostructured hybrid powder was washed with distilled water to a pH of 7, mixed with polyvinyl alcohol (PVA) 5% solution and spray dried in LabPLANT, United Kingdom spray dryer.

2.7. Structural characterization

2.7.1. Chemical analysis

Nanostructured hydroxyapatite-polyurethane hybrid powders were characterized by chemical quantitative analysis. Ca content was determined using Flame Atomic Absorption Spectrometry (FAAS) method. P content was measured using *Inductively Coupled Plasma Optical Emission Spectrometry* (ICP-OES) method according to ASTM E 1479/2011.

Ca and P content of some investigated hybrid samples is presented in **Table 2**.

Results presented in **Table 2** confirmed the formation of hydroxyapatite as the major phase of nanostructured powders, namely: Ca:P molar ratio is between 1.54 and 1.75, while in HPU-4 and HPU-8 samples is equal to theoretical value from pure hydroxyapatite.

2.7.2. FT-IR analysis

Powders characterization using Fourier transform infrared spectroscopy (FT-IR) revealed the presence of the following vibration bands: (i) the OH stretching vibration at 3550 cm^{-1} (sharp band) in the HA sample. In the HA-PU hybrid sample the intensity of this band is highly diminished; (ii) stretching vibration of the water molecule (3300 cm^{-1}). It is a broad band, seen in both the HA sample and the hybrid sample; (iii) the deformation vibration of the –OH group (1630 cm^{-1}) from HA, which completely disappears in the case of HA-PU sample. This, correlated with the decrease of the OH band at 3350 cm^{-1}, could be due to the interaction of the two components through the hydroxyl group; (iv) stretching vibrations of $(PO_4)_3$-group at 1094,

Sample name	Sample type	Ca, %	P, %	Ca:P molar ratio (calculated)
HPU-1	80% HA-20% PU	32.9	16.2	1.58
HPU-2	80% HA-20% PU	35.6	17.4	1.59
HPU-4	80% HA-20% PU	35.3	16.5	**1.66**
HPU-5	50% HA-50% PU	35.2	17.8	1.54
HPU-6	80% HA-20% PU	28	12.5	1.75
HPU-7	50% HA-50% PU	38.4	17.1	1.74
HPU-8	80% HA-20% PU	37	17.1	**1.67**

Note: The bold values represents the experimental molar ratio equivalent to theoretical values hydroxyapatite.

Table 2. Chemical analysis results.

1040 and 962 cm^{-1}, being characteristic for hydroxyapatite spectrum (IR fingerprint), can also be observed in the hybrid spectrum. The FT-IR spectra of the analyzed powders (hydroxyapatite, respectively hydroxyapatite-polyurethane based hybrids) are shown in **Figure 3**.

2.7.3. Thermal analysis (DSC-TG)

Powders characterization using thermal analysis (DSC-TG) revealed the presence of several endothermic peaks, as shown in **Table 3**. **Figures 4** and **5** show thermal analyses of HA and HA hybrid samples. Peak 1 could be attributed to the elimination of surface adsorbed water. Peak 2, observed only for HPU hybrid powder, could correspond to the removal of OH groups from water of constitution. Peak 3 in the hybrid sample, although characterized by a very low enthalpy (0.251 J/g), could be associated with breakage of polyurethane chain links. Peak 4 (466°C) of the HPU sample corresponds to the decomposition of soft polyurethane segments. The last two peaks of the hybrid sample are due to the burning of the organic phase. The total mass loss is 14.5%, which means that the polyurethane from the composition of HPU sample has almost completely decomposed. Endothermic peaks at 276 and 336°C observed in the case of hydroxyapatite (HA sample) can be explained by decomposing of the polyvinyl alcohol traces left after drying the sample in the spray dryer. The weight loss for the HA sample is 8.5%.

2.7.4. Particle size distribution by DLS technique

Particle size distribution determination with Malvern Zetasizer ZS90 granulometer is based on a non-invasive dynamic light scattering (DLS) emitted by a laser. The particle assimilated with a sphere is constantly displaced by Brownian motion as a result of statistical collisions with the liquid molecules. In this movement, small particles will move faster than large particles.

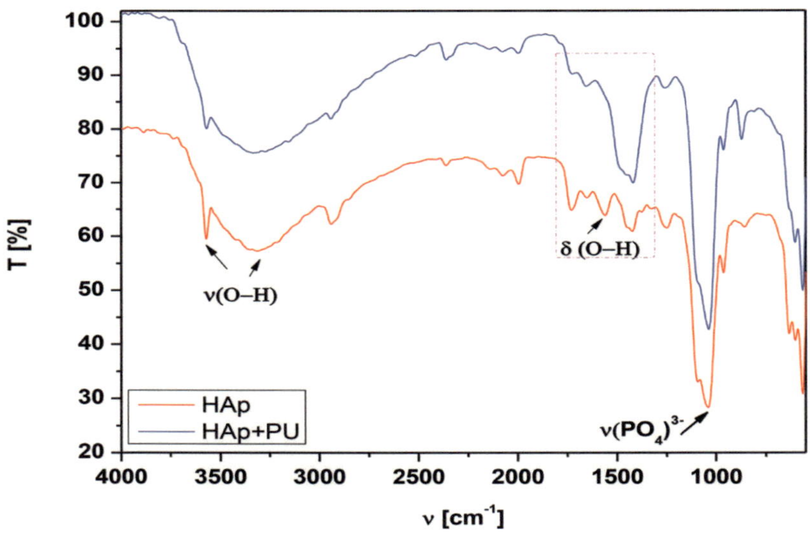

Figure 3. FT-IR spectra of one hybrid sample compared to pure hydroxyapatite.

Sample name	Peak 1 (endothermic)		Peak 2 (endothermic)		Peak 3 (endothermic)		Peak 4 (endothermic)		Peak 5 (endothermic)		Peak 6 (endothermic)	
	T, °C	ΔH, J/g	T, °C	ΔH, J/g	T, °C	ΔH, J/g	T, °C	ΔH, J/g	T, °C	ΔH, J/g	T, °C	ΔH, J/g
HA	58.1	32.019	–	–	276.1	11.963	336.4	21.498	–	–	825.6	13.61
HPU	63.5	49.549	135.8	4.429	246.2	0.251	466.6	4.746	711.4	6.226	784	38.762

Table 3. Results obtained from thermal analysis (DSC-TG).

http://dx.doi.org/10.5772/intechopen.70707

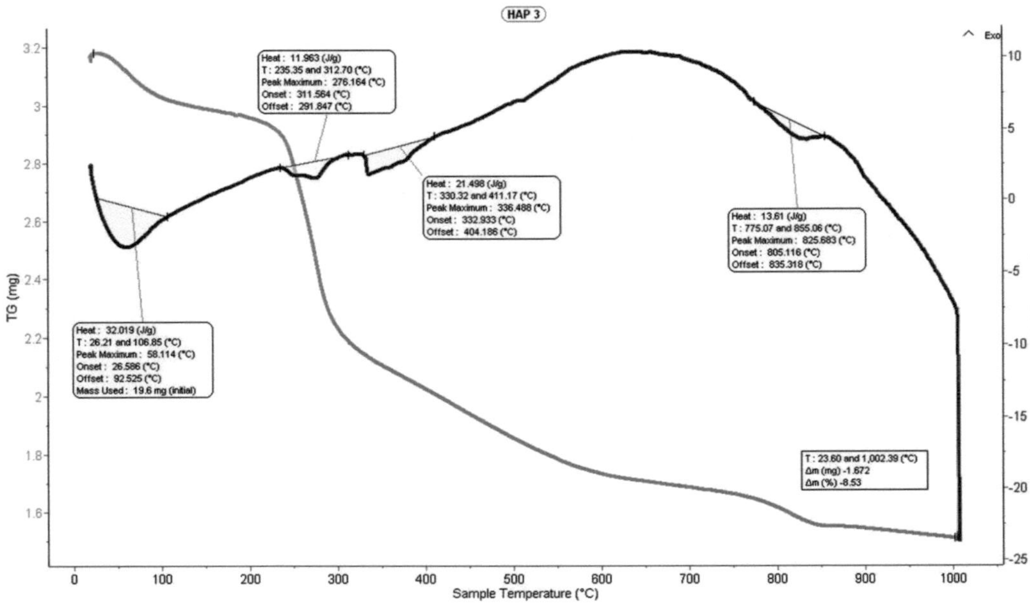

Figure 4. DSC graph and mass loss of pure hydroxyapatite.

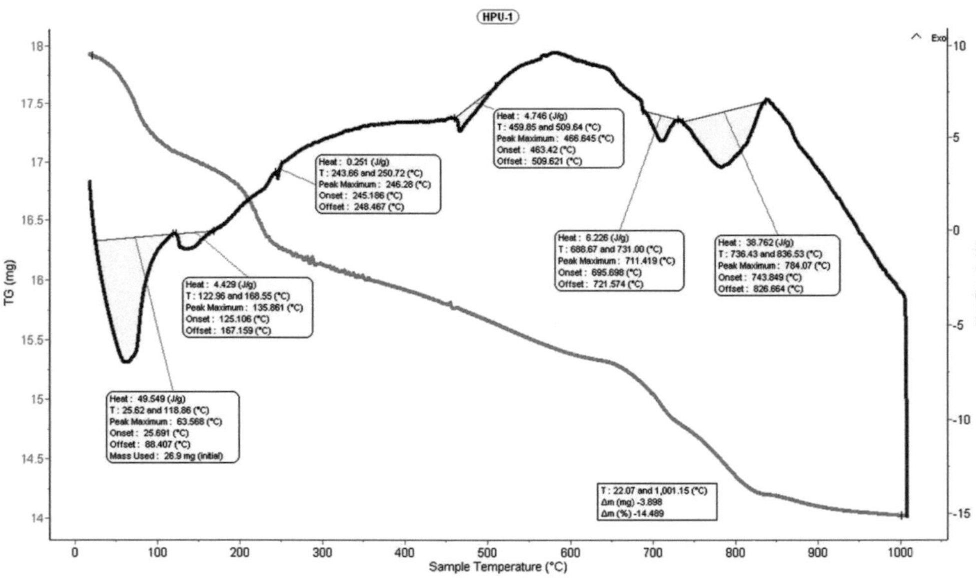

Figure 5. DSC-TG graph and mass loss of one hybrid sample.

The particle size distribution obtained by DLS is a function of the relative intensity of light scattered by particles of different dimensional classes. The time dependence of the intensity fluctuations for the determination of the translational diffusion coefficient (*D*), as well as the hydrodynamic diameter (D_H) are measured by using equation (1). η is the viscosity of the investigated suspension.

$$D_H = \frac{kT}{3\pi\eta D} \quad (1)$$

The result is a distribution of intensity.

Sample preparation: stable, aqueous dispersions with known optical properties (required for size measurement) are prepared. Results are presented in **Table 4** and **Figure 6**.

It can be observed a decrease in the mean particle size of the hybrid material compared to pure hydroxyapatite, both for the filtered and the initial samples.

There is a slight tendency for agglomeration of the hybrid material (**Figure 6**).

Sample name	Mean particle size [nm]	Polydispersity index	Observations
HA	167.5	0.133	
HA filtered	146.3	0.061	Sample filtered through Millipore membrane (d = 0.22 μm)
HPU	119.7	0.238	
HPU filtered	64.65	0.229	Sample filtered through Millipore membrane (d = 0.22 μm)

Table 4. DLS analysis results.

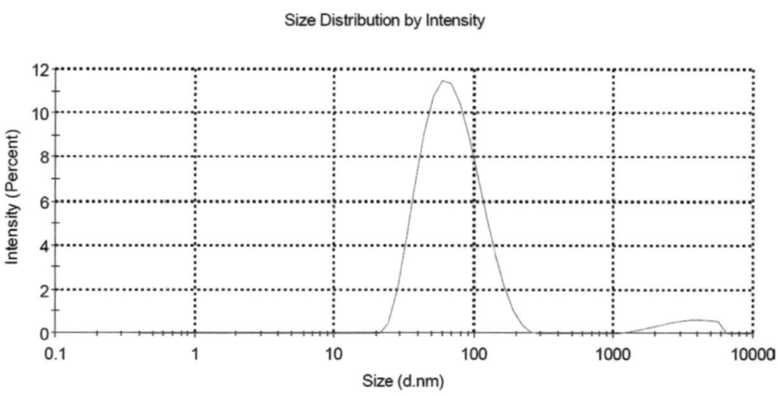

Figure 6. Particle size distribution of one hybrid sample filtered through 0.22 μm Millipore membrane.

2.7.5. HRTEM analysis

HRTEM analysis was performed with a High Resolution Transmission Electron Microscope (HRTEM) Tecnai F30, G2S Twin (1 Å line resolution) – FEI Company.

It can be observed rod like structures of hydroxyapatite inside hybrid structure, typical for hydrothermally prepared HA, as well diffraction lines which confirms sample crystallinity (**Figure 7**).

EDX spectrum (**Figure 8**) reveals Ca, P, O and C content of the investigated sample, in accordance with chemical analysis results. Moreover, it confirms the presence of organic phase (polyurethane), through C and O content.

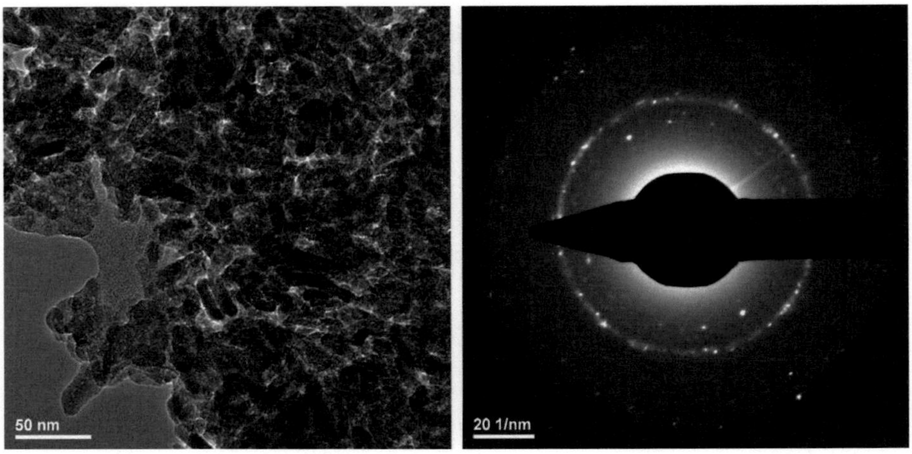

Figure 7. TEM images for one hybrid sample with 80% HA and 20% PU.

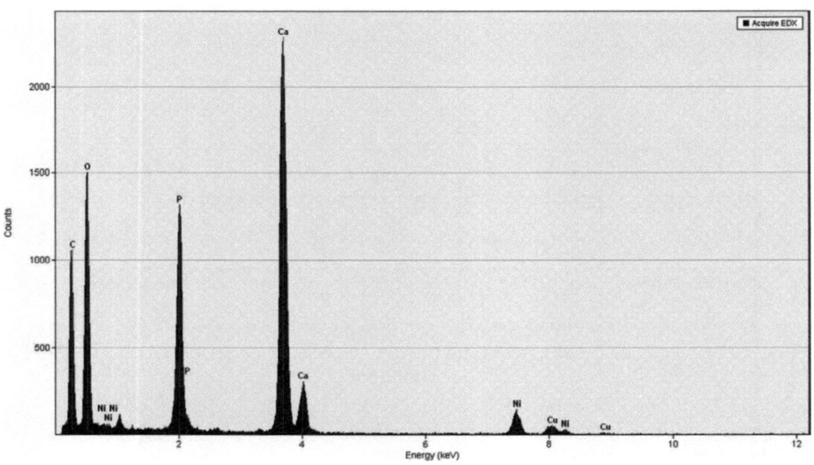

Figure 8. EDX spectra of one hybrid sample with 80% HA and 20% PU.

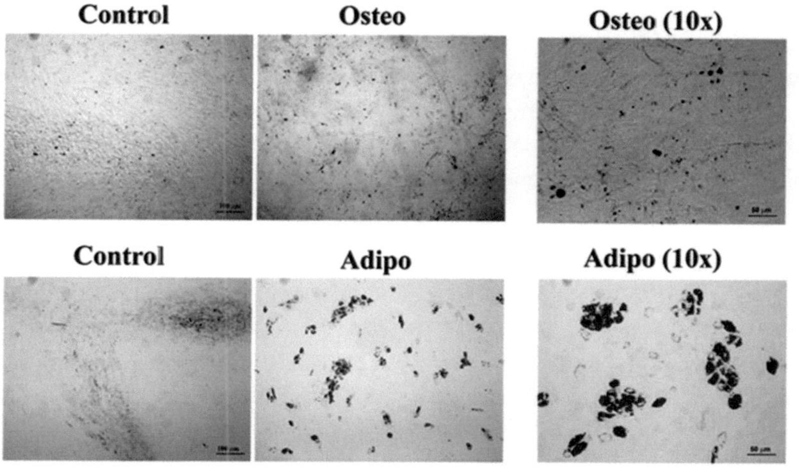

Figure 9. Optical microscopy illustrating the ability of human MSC to generate osteoblasts (up) and adipocytes (down) *in vitro*. UP: (left) Human MSC cultivated in normal growth medium (DMEM + 10% SBF-MSC qualified) and von Kossa stained; (Middle and right) human MSCs increased 2 weeks in osteogenic differentiation environment; The formation of aggregates strongly colored with silver nitrate is observed (von Kossa method). Down (left) Human MSC cultured in normal growth medium (DMEM + 10% SBF-MSC qualifies) and stained with Oil Red O; (Middle and right) human MSCs increased 2 weeks in adipogenic differentiation environment; There is evidence of accumulation of red-colored lipid charges with Oil Red O.

2.8. Biocompatible properties

MSC cells were characterized in terms of their multipotent potential, demonstrating their ability to differentiate *in vitro* into adipocytes and osteoblasts (**Figure 9**). Dilutions tested were: 1/10, 1/20, 1/50, 1/100 and 1/200. Diluted solutions were added to both cultured fresh cells and after they were maintained at 37°C under 5% CO_2 (pH equilibration). Also, for each condition, cell interaction studies were performed in culture, both 24 hours after cell adhesion and 48 hours after cell adhesion (assuming that in the first case we can see the effect of HAp on cellular proliferation, and in the latter case, HAp effect on cell viability).

3. Conclusion

The current chapter deals with the various fabrication methodologies of biomimetic and bioactive scaffolds that can be employed for the tissue engineering applications. Biomimetic nanocomposites scaffolds have the ability to mimic the structural and mechanical properties of native tissues. Various fabrication methodologies are available to replicate and produce the biomimetic biomaterials for tissue engineering applications. Various advantages and disadvantages of these fabrication methodologies are discussed here. Different classes of materials are used for biomimetic applications and designing of these biomimetic materials are cited with examples. Recent advances and research on scaffolds with controlled nano/micro architecture, pore distribution and pore density reveal the potential of the nanocomposites for tissue engineering applications. Technologies such as additive manufacturing, electrospinning and other combined recent advanced manufacturing technologies holds promise for new advances in the field of tissue engineering. Due to space constraints the present chapter has given one of the examples of the possibility to fabricate biomimetic materials by additive manufacturing. Further developments are required in the field of biomaterials fabrication to exactly replicate the biomimetic conditions.

Acknowledgements

The authors acknowledge Grant Agreement 692216 SUPERMAT for receiving funding from the European Union's Horizon 2020 research and innovation program.

Author details

Mythili Prakasam[1]*, Madalina Popescu[2], Roxana Piticescu[2] and Alain Largeteau[1]

*Address all correspondence to: mythili.prakasam@icmcb.cnrs.fr

1 CNRS, Université de Bordeaux, ICMCB, Pessac, France

2 National R&D Institute for Nonferrous and Rare Metals, Pantelimon, Romania

References

[1] Shin H, Jo S, Mikos AG. Biomimetic materials for tissue engineering. Biomaterials. 2003;**24**(24):4353-4364

[2] Peter V. Giannoudis, Haralambos Dinopoulos, Eleftherios Tsiridis, Bone substitutes: An update, Injury, 2005;**36**(3):S20-S27, ISSN: 0020-1383

[3] Boris Michael Holzapfel, Johannes Christian Reichert, Jan-Thorsten Schantz, Uwe Gbureck, Lars Rackwitz, Ulrich Nöth, Franz Jakob, Maximilian Rudert, Jürgen Groll, Dietmar Werner Hutmacher, How smart do biomaterials need to be? A translational science and clinical point of view, Advanced Drug Delivery Reviews, 2013;**65**(4):581-603, ISSN: 0169-409XX

[4] Mouw JK, Ou G, Weaver VM. Extracellular matrix assembly: A multiscale deconstruction. Nature Reviews Molecular Cell Biology. 2014;**15**:771-785

[5] Orlando G, Baptista P, Birchall M, De Coppi P, Farney A, Guimaraes-Souza NK, Opara E, Rogers J, Seliktar D, Shapira-Schweitzer K, Stratta RJ, Atala A, Wood KJ, Soker S. Regenerative medicine as applied to solid organ transplantation: Current status and future challenges. Transplant International. 2011;**24**(3):223-232

[6] Teven CM, Fisher S, Ameer GA, He T-C, Reid RR. Biomimetic approaches to complex craniofacial defects. Annals of Maxillofacial Surgery. 2015;**5**(1):4-13

[7] Kim B-S, Park I-K, Hoshiba T, Jiang H-L, Choi Y-J, Akaike T, Cho C-S. Design of artificial extracellular matrices for tissue engineering. Progress in Polymer Science. 2011;**36**(2):238-268

[8] Dutta SR, Passi D, Singh P, Bhuibhar A. Ceramic and non-ceramic hydroxyapatite as a bone graft material: A brief review. Irish Journal of Medical Science (1971–). 2015;**184**(1):101-106

[9] Park HJ, Lee OJ, Lee MC, Moon BM, Ju HW, Lee J, Kim JH, Kim DW, Park CH. Fabrication of 3D porous silk scaffolds by particulate (salt/sucrose) leaching for bone tissue reconstruction. International Journal of Biological Macromolecules. 2015;**78**:215-223

[10] Wang P, Zhao L, Liu J, Weir MD, Zhou X, Xu HHK. Bone tissue engineering via nanostructured calcium phosphate biomaterials and stem cells. Bone Research. 2014;**2**:14017-14013

[11] Fernandez-Yague MA, Abbah SA, McNamara L, Zeugolis DI, Pandit A, Biggs MJ. Biomimetic approaches in bone tissue engineering: Integrating biological and physicomechanical strategies. Advanced Drug Delivery Reviews. 2015;**84**:1-29

[12] Yunus Basha R, Sampath Kumar TS, Doble M. Design of biocomposite materials for bone tissue regeneration. Materials Science and Engineering: C. 2015;**57**:452-463

[13] Tang D, Tare RS, Yang L-Y, Williams DF, Ou K-L, Oreffo ROC. Biofabrication of bone tissue: approaches, challenges and translation for bone regeneration. Biomaterials. 2016;**83**:363-382

[14] de Azevedo Gonçalves Mota RC, da Silva EO, de Lima FF, de Menezes LR, Thiele ACS. 3D printed scaffolds as a new perspective for bone tissue regeneration: Literature review. Materials Sciences and Applications. 2016;**7**:430-452

[15] Do AV, Khorsand B, Geary SM, Salem AK. 3D printing of scaffolds for tissue regeneration applications. Advanced Healthcare Materials. 2015;**4**(12):1742-1762

[16] Chee Kai Chua, Wai Yee Yeong, Bioprinting. Principles and Applications, World Scientific Publishing 2015, 296, ISBN: 978-981-4612-10-4.

[17] Costa PF, Vaquette C, Zhang Q, Reis RL, Ivanovski S, Hutmacher DW. Advanced tissue engineering scaffold design for regeneration of the complex hierarchical periodontal structure. Journal of Clinical Periodontology. 2014;**41**:283-294

[18] White TD, Folkens PA. The Human Bone Manual. Academic Press; Massachusetts, USA. 2005

[19] Clarke B. Normal bone anatomy and physiology. Clinical Journal of the American Society of Nephrology. 2008;**3**(3 Suppl):S131-S139

[20] Currey JD. The Mechanical Adaptations of Bones. Princeton University Press; New Jersy, USA, 2014

[21] Keller TS, Mao Z, Spengler DM. Young's modulus, bending strength, and tissue physical properties of human compact bone. Journal of Orthopaedic Research. 1990;**8**(4):592-603

[22] Hsu YH, Turner IG, Miles AW. Fabrication of porous bioceramics with porosity gradients similar to the bimodal structure of cortical and cancellous bone. Journal of Materials Science: Materials in Medicine. 2007;**18**(12):2251-2256

[23] Hutmacher DW. Scaffolds in tissue engineering bone and cartilage. Biomaterials. 2000;**21**(24):2529-2543

[24] Ma PX. Biomimetic materials for tissue engineering. Advanced Drug Delivery Reviews. 2008;**60**(2):184-198

[25] Puppi D, Chiellini F, Piras AM, Chiellini E. Polymeric materials for bone and cartilage repair. Progress in Polymer Science. 2010;**35**(4):403-440

[26] Combes C, Rey C. Amorphous calcium phosphates: Synthesis, properties and uses in biomaterials. Acta Biomaterialia. 2010;**6**(9):3362-3378

[27] Boanini E, Gazzano M, Bigi A. Ionic substitutions in calcium phosphates synthesized at low temperature. Acta Biomaterialia. 2010;**6**:1882-1894

[28] Ishikawa K. Calcium phosphate cement. Advances in Calcium Phosphate Biomaterials. Berlin Heidelberg: Springer; 2014. p. 199-227

[29] Ramay HRR, Zhang M. Biphasic calcium phosphate nanocomposite porous scaffolds for load-bearing bone tissue engineering. Biomaterials. 2004;**25**:5171-5180

[30] Hutmacher DW. Scaffold design and fabrication technologies for engineering tissues—State of the art and future perspectives. Journal of Biomaterials Science, Polymer. 2001;**12**:107-124

[31] Karageorgiou V, Kaplan D. Porosity of 3D biomaterial scaffolds and osteogenesis. Biomaterials. 2005;**26**:5474-5491

[32] Shimko DA, Shimko VF, Sander EA, Dickson KF, Nauman EA. Effect of porosity on the fluid flow characteristics and mechanical properties of tantalum scaffolds. Journal of Biomedical Materials Research Part B: Applied Biomaterials. 2005;**73**(2):315-324

[33] Havaldar R, Pilli SC, Putti BB. Insights into the effects of tensile and compressive loadings on human femur bone. Advanced Biomedical Research. 2014:**3**;1-15

[34] Khodaei M, Meratian M, Savabi O, Razavi M. The effect of pore structure on the mechanical properties of titanium scaffolds. Materials Letters. 2016;**171**:308-311

[35] Zhu Y, Wan Y, Zhang J, Yin D, Cheng W. Manufacture of layered collagen/chitosan-polycaprolactone scaffolds with biomimetic microarchitecture. Colloids and Surfaces B: Biointerfaces. 2014;**113**:352-360

[36] Keane TJ, Badylak SF. The host response to allogeneic and xenogeneic biological scaffold materials. Journal of Tissue Engineering and Regenerative Medicine. 2015;**9**(5):504-511

[37] Li X, Liu X, Zhang G, Dong W, Sha Z, Feng Q, Watari F. Repairing 25 mm bone defect using fibres reinforced scaffolds as well as autograft bone. Bone. 2008;**43**:S94

[38] Lord CF, Gebhardt MC, Tomford WW, Mankin HJ. Infection in bone allografts. Incidence, nature, and treatment. Journal of Bone and Joint Surgery. 1988;**70**(3):369-376

[39] Chen Q, Thouas GA. Metallic implant biomaterials. Materials Science and Engineering: R: Reports. 2015;**87**:1-57

[40] Langer RS, Vacanti JP. Tissue engineering: The challenges ahead. Scientific American. 1999;**280**:86-89

[41] Ryan G, Pandit A, Apatsidis DP. Fabrication methods of porous metals for use in orthopaedic applications. Biomaterials. 2006;**27**(13):2651-2670

[42] Liu X, Won Y, Ma PX. Surface modification of interconnected porous scaffolds. Journal of Biomedical Materials Research Part A. 2005;**74**:84-91

[43] Shishkovsky IV, Volova LT, Kuznetsov MV, Morozov YG, Parkin IP. Porous biocompatible implants and tissue scaffolds synthesized by selective laser sintering from Ti and NiTi. Journal of Materials Chemistry. 2008;**18**(12):1309-1317

[44] Hoppe A, Güldal NS, Boccaccini AR. A review of the biological response to ionic dissolution products from bioactive glasses and glass-ceramics. Biomaterials. 2011;**32**(11):2757-2774

[45] Yoo D. New paradigms in hierarchical porous scaffold design for tissue engineering. Materials Science and Engineering: C. 2013;**33**:1759-1772

[46] Van Vlierberghe S, Dubruel P, Schacht E. Biopolymer-based hydrogels as scaffolds for tissue engineering applications: A review. Biomacromolecules. 2011;**12**:1387-1408

[47] Badylak SF, Freytes DO, Gilbert TW. Extracellular matrix as a biological scaffold material: Structure and function. Acta Biomaterialia. 2009;**5**:1-13

[48] Panzavolta S, Torricelli P, Amadori S, Parrilli A, Rubini K, Della Bella E, Fini M, Bigi A. 3D interconnected porous biomimetic scaffolds: In vitro cell response. Journal of Biomedical Materials Research Part A. 2013;**101**:3560-3570

[49] Washbourne C. Method of making a ceramic fiber replica of a body of reticulated organic foam. U.S. Patent 3,939,002, issued February 17, 1976

[50] Deville S. Freeze-casting of porous ceramics: A review of current achievements and issues. Advanced Engineering Materials. 2008;**10**(3):155-169

[51] Haugh MG, Murphy CM, O'Brien FJ. Novel freeze-drying methods to produce a range of collagen–glycosaminoglycan scaffolds with tailored mean pore sizes. Tissue Engineering Part C: Methods. 2009;**16**:887-894

[52] Liu X, Ma PX. Phase separation, pore structure, and properties of nanofibrous gelatin scaffolds. Biomaterials. 2009;**30**:4094-4103

[53] Nam YS, Yoon JJ, Park TG. A novel fabrication method of macroporous biodegradable polymer scaffolds using gas foaming salt as a porogen additive. Journal of Biomedical Materials Research. 2000;**53**:1-7

[54] Landers R, Pfister A, Hübner U, John H, Schmelzeisen R, Mülhaupt R. Fabrication of soft tissue engineering scaffolds by means of rapid prototyping techniques. Journal of Materials Science. 2002;**37**:3107-3116

[55] Li W-J, Laurencin CT, Caterson EJ, Tuan RS, Ko FK. Electrospun nanofibrous structure: A novel scaffold for tissue engineering. Journal of Biomedical Materials Research Part A. 2002;**60**:613-621

[56] Drury JL, Mooney DJ. Hydrogels for tissue engineering: Scaffold design variables and applications. Biomaterials. 2003;**24**:4337-4351

[57] Kakade MV, Givens S, Gardner K, Lee KH, Chase DB, Rabolt JF. Electric field induced orientation of polymer chains in macroscopically aligned electrospun polymer nanofibers. Journal of the American Chemical Society. 2007;**129**(10):2777-2782

[58] Janmey PA, Winer JP, Weisel JW. Fibrin gels and their clinical and bioengineering applications. Journal of the Royal Society Interface. 2009;**6**:1-10

[59] Jain KK. Drug delivery systems-an overview. Drug Delivery Systems. 2008;**437**:1-50

[60] Necas J, Bartosikova L, Brauner P, Kolar J. Hyaluronic acid (hyaluronan): A review. Veterinarni Medicina. 2008;**53**:397-411

[61] Zhao N, Wang X, Qin L, Zhai M, Yuan J, Chen J, Li D. Effect of hyaluronic acid in bone formation and its applications in dentistry. Journal of Biomedical Materials Research. Part A. 2016;**104**(6):1560-1569

[62] Zhou M, Smith AM, Das AK, Hodson NW, Collins RF, Ulijn RV, Gough JE. Self-assembled peptide-based hydrogels as scaffolds for anchorage-dependent cells. Biomaterials. 2009;**30**: 2523-2530

[63] Stankus JJ, Guan J, Wagner WR. Fabrication of biodegradable elastomeric scaffolds with sub-micron morphologies. Journal of Biomedical Materials Research Part A. 2004;**70**:603-614

[64] Ma PX. Biomimetic materials for tissue engineering. Advanced Drug Delivery. 2008;**60**:184-198

[65] Altman GH, Diaz F, Jakuba C, Calabro T, Horan RL, Chen J, Lu H, Richmond J, Kaplan DL. Silk-based biomaterials. Biomaterials. 2003;**24**:401-416

[66] Hofmann S, Wong Po Foo CT, Rossetti F, Textor M, Vunjak-Novakovic G, Kaplan DL, Merkle HP, Meinel L. Silk fibroin as an organic polymer for controlled drug delivery. Journal of Controlled Release. 2006;**111**:219-227

[67] Chronakis IS. Novel nanocomposites and nanoceramics based on polymer nanofibers using electrospinning process—A review. Journal of Materials Processing Technology. 2005;**167**:283-293

[68] Shin M, Yoshimoto H, Vacanti JP. In vivo bone tissue engineering using mesenchymal stem cells on a novel electrospun nanofibrous scaffold. Tissue Engineering. 2004;**10**:33-41

[69] Stevens MM, George JH. Exploring and engineering the cell surface interface. Science. 2005;**310**:1135-1138

[70] Wu W, Wang W, Li J. Star polymers: Advances in biomedical applications. Progress in Polymer Science. 2015;**46**:55-85

[71] Aryaei A, Vapniarsky N, Hu JC, Athanasiou KA. Recent tissue engineering advances for the treatment of temporomandibular joint disorders. Current Osteoporosis Reports. 2016;**14**

[72] Zhang W, Yelick PC. Craniofacial tissue engineering. Cold Spring Harbor Perspectives in Medicine. 2017;a025775

[73] Ivanovski S, Vaquette C, Gronthos S, Hutmacher DW, Bartold PM. Multiphasic scaffolds for periodontal tissue engineering. Journal of Dental Research. 2014;**93**:1212-1221

[74] Thadavirul N, Pavasant P, Supaphol P. Development of polycaprolactone porous scaffolds by combining solvent casting, particulate leaching, and polymer leaching techniques for bone tissue engineering. Journal of Biomedical Materials Research Part A. 2014;**102**:3379-3392

[75] Liao C-J, Chen C-F, Chen J-H, Chiang S-F, Lin Y-J, Chang K-Y. Fabrication of porous biodegradable polymer scaffolds using a solvent merging/particulate leaching method. Journal of Biomedical Materials Research Part A. 2002;**59**:676-681

[76] Nam YS, Park TG. Porous biodegradable polymeric scaffolds prepared by thermally induced phase separation. Journal of Biomedical Materials Research. 1999;**47**:8-17

[77] Oh SH, Kang SG, Kim ES, Cho SH, Lee JH. Fabrication and characterization of hydrophilic poly (lactic-co-glycolic acid)/poly (vinyl alcohol) blend cell scaffolds by melt-molding particulate-leaching method. Biomaterials. 2003;**24**:4011-4021

[78] Deville S, Saiz E, Tomsia AP. Freeze casting of hydroxyapatite scaffolds for bone tissue engineering. Biomaterials. 2006;**27**:5480-5489

[79] Van den Dolder J, Farber E, Spauwen PHM, Jansen JA. Bone tissue reconstruction using titanium fiber mesh combined with rat bone marrow stromal cells. Biomaterials. 2003;**24**: 1745-1750

[80] Mikos AG, Bao Y, Cima LG, Ingber DE, Vacanti JP, Langer R. Preparation of poly (glycolic acid) bonded fiber structures for cell attachment and transplantation. Journal of Biomedical Materials Research Part A. 1993;**27**:183-189

[81] Zong X, Bien H, Chung C-Y, Yin L, Fang D, Hsiao BS, Chu B, Entcheva E. Electrospun fine-textured scaffolds for heart tissue constructs. Biomaterials. 2005;**26**:5330-5338

[82] Billiet T, Vandenhaute M, Schelfhout J, Van Vlierberghe S, Dubruel P. A review of trends and limitations in hydrogel-rapid prototyping for tissue engineering. Biomaterials. 2012;**33**:6020-6041

[83] Melchels FPW, Domingos MAN, Klein TJ, Malda J, Bartolo PJ, Hutmacher DW. Additive manufacturing of tissues and organs. Progress in Polymer Science. 2012;**37**:1079-1104

[84] ISO / ASTM52900-15. Standard Terminology for Additive Manufacturing – General Principles – Terminology. West Conshohocken, PA: ASTM International; 2015 Available from: www.astm.org

[85] Cox SC, Thornby JA, Gibbons GJ, Williams MA, Mallick KK. 3D printing of porous hydroxyapatite scaffolds intended for use in bone tissue engineering applications. Materials Science and Engineering C. 2015;**47**:237-247

[86] Zhou Z, Buchanan F, Mitchell C, Dunne N. Printability of calcium phosphate: Calcium sulfate powders for the application of tissue engineered bone scaffolds using the 3D printing technique. Materials Science and Engineering C. 2014;**38**:1-10

[87] Inzana JA, Olvera D, Fuller SM, Kelly JP, Graeve OA, Schwarz EM, Kates SL, Awad HA. 3D printing of composite calcium phosphate and collagen scaffolds for bone regeneration. Biomaterials. 2014;**35**:4026-4034

[88] Nandakumar A, Cruz C, Mentink A, Birgani ZT, Moroni L, van Blitterswijk C, Habibovic P. Monolithic and assembled polymer–ceramic composites for bone regeneration. Acta Biomaterialia. 2013;**9**:5708-5717

[89] Jana S, Lerman A. Bioprinting a cardiac valve. Biotechnology Advances. 2015;**33**:1503-1521. DOI: http://dx.doi.org/10.1016/j.biotechadv.2015.07.006

[90] Chia HN, Wu BM. Recent advances in 3D printing of biomaterials. Journal of Biological Engineering. 2015;**9**:4-18

[91] Deriu MA, Popescu ML, Ottaviani MF, Danani A, Piticescu RM. Iron oxide/PAMAM nanostructured hybrid systems. Combined computational and experimental studies. Journal of Materials Science. 2016;**51**:1996-2007

[92] Popescu LM, Piticescu RM, Petriceanu M, Ottaviani MF, Cangiotti M, Vasile E, Dîrtu MM, Wolff M, Garcia Y, Schinteie G, Kuncser V. Hydrothermal synthesis of nanostructured hybrids based on iron oxide and branched PEI polymers. Influence of high pressure on structure and morphology. Materials Chemistry and Physics. 2015;**161**:84-95

[93] Popescu LM, Piticescu RM, Antonelli A, Rusti CF, Carboni CE, Sfara C, Magnani M, Badilita V, Vasile E, Trusca R, Buruiana T. Recent advances in synthesis, characterization of hydroxyapatite/polyurethane composites and study of their biocompatible properties. Journal of Materials Science: Materials in Medicine. 2013;**24**:2491-2503

[94] Popescu LM, Rusti CF, Piticescu RM, Buruiana T, Valero T, Kintzios S. Synthesis and characterization of acid polyurethane-hydroxyapatite composites for biomedical applications. Journal of Composite Materials. 2013;**47**:603-612

[95] Popescu LM, Piticescu RM, Motoc AM, Voinea LM, Gradinaru SL(Istrate), Ulieru D, Topor A. Three-dimensional structures based on hydroxyapatite and polyurethane diol obtained through 3d printing technology, National patent request OSIM A/00102/2017

Chapter 2

Glass and Glass-Ceramic Scaffolds: Manufacturing Methods and the Impact of Crystallization on *In-Vitro* Dissolution

Amy Nommeots-Nomm and Jonathan Massera

Additional information is available at the end of the chapter

http://dx.doi.org/10.5772/intechopen.70242

Abstract

Synthetic biomaterials mimicking bone morphology have expanded at a tremendous rate. Among all, one stands out: bioactive glass. Bioactive glasses opened the door to a new genre of research into materials able to promote the regeneration of functioning bone tissue. However, despite their ability to promote cell attachment, proliferation and differentiation, these materials are mainly used as granules. However to promote loaded and sustained bone repair, a 3D structure, with open and highly interconnected pores, is desirable. 3D scaffolds are generally produced into green bodies via various techniques. The particles are then bound together via sintering. However, the highly disrupted silica network of the typical bioactive glasses composition leads to crystallization. Therefore, sintering of the most commonly used bioactive glass compositions (i.e. 45S5 and S53P4) leads to partly to fully crystallize bodies. The impact of crystallization on bioactivity still leads to large debate among the scientific community. Does crystallization reduce or suppress the materials bioactivity? Within this chapter, the processing routes for scaffold manufacture are presented, as well as an introduction to the thermal processing of glasses to form glass and glass-ceramics and the consequent effect on bioactivity is discussed.

Keywords: scaffold, bioactive glass, crystallization, glass-ceramic, *in-vitro* dissolution

1. Introduction

With the ageing and continuous growth of the population, surgeons face an increasing number of operations aiming to replace and/or repair tissue that has been damaged through disease and trauma [1]. According to the International Osteoporosis Foundation, incidence of fracture, for example, is anticipated to increase by 240% in women and 310% in men by 2050 [http://www.mdbuyline.com/research-library/articles/bone-graft-substitutes-bone-matrix-cost/, visited last April 2016]. This factor is expected to drive the rise in the global bone graft market. Bone regeneration market was evaluated at USD 2.35 billion in 2014 and is expected to rise to USD 3.48 billion by 2023, as reported in transparency market research [http://www.transparencymarketresearch.com/pressrelease/bone-grafts-substitutes-market.htm, visited June 2017]. The gold standard in bone grafts is, at present, the use of allograft. An allograft is a tissue received from a donor which is demineralised before use, thus, genetically different than an autograft (where tissue comes from the patient itself). However, a risk of infection is associated with such practices. To overcome the infection risk, the allograft tissue must be pre-treated. Techniques used are tissue freezing, freeze drying and sterilization. The average cost per procedure was estimated to be USD 3154.

Bioceramics, pertaining to their excellent biocompatibility and higher mechanical properties compared to polymers or metals, have always been regarded as the most promising biomaterials for hard tissue repair. The advantage of bioceramics lies in their wide tissue/implant response: from being nearly inert to bioactive depending on their composition. A subset of bioceramics which have expanded at a tremendous rate in the past decade are bioactive glasses. These materials support not only osteoconduction (growth of bony tissue at the surface of the implant) but also osteoinduction (acceleration of new bone formation by chemical means) [2]. Bioactive glasses have, now, been used widely clinically especially in dental and bone repair applications.

For the repair and regeneration of bone defects, tissue engineering strives to produce 3D constructs that can support mechanical load and provide a template in which cells can migrate and colonized. The optimum attribute of a 3D scaffolds are summarized as follow:

- The biomaterials should not only be biocompatible but must support cell adhesion and proliferation. The material should be at minimum osteoconductive but ideally osteoinductive.
- The biomaterials should resorb over time. The resorption rate of the implant should match as closely as possible the regrowth of the new tissue.
- The implant should be porous with an open interconnect structure to allow fluid penetration, angiogenesis and cell proliferation. In general, interconnected pores of at least 100 μm and an open porosity of over 50% is considered the minimum requirement for tissue ingrowth [3-5]. Optimally, it is reported that the construct should have pores within 100–500 μm and an overall porosity of over 90%. However an increase of porosity comes at the expense of mechanical strength [6].
- The mechanical properties, which is mainly function of the porosity and the pores organisation, should be tailored to the site of implantation, i.e. similar to cancellous bone in non-load bearing applications and similar to cortical bone in load-bearing applications. In all case, the mechanical properties should be maintained throughout the bone reconstruction.

2. Scaffold fabrication

2.1. Foamed morphology processing

2.1.1. Porogen

Porogen processing is a method in which glass powders are mixed with a polymer bead used as a 'space holder', and packed into a mould. The polymer is then burnt away in a two stage sintering process to create porous scaffold. This process relies on using a combustable polymer and sufficient oxygen flow to gain complete burn out during sintering, if complete burn out is not achieved it can result in inhomogeneous scaffolds being formed with a porosity gradient. The scaffolds tend to have limited porosity of up to 50% [7].

2.1.2. Freeze casting

Freeze casting is a traditional ceramic foam processing method, which is based upon the formation of an aqueous slurry where ice crystals are formed in the direction of freezing. Under vacuum, the ice crystals are then sublimated to leave a porous green body which is then sintered [8]. This was initially developed in 2011 for use with hydroxyapatite, but has since been used with various glass and glass-ceramic compositions [9-11]. The scaffolds produced tend to have small pores (up to 100 μm) with low levels of interconnection, with consequently high strengths. Due to the small pores, they have not been significantly investigated in either the glass or glass-ceramic field for bone repair.

2.1.3. Foam replica process

The foam replica process employs a polymeric foam, commonly made of polyurethane, to act as a template for a glass slurry to create scaffolds with a foamed morphology. The foam is soaked in a glass suspension, and dried. By completing multiple cycles of infiltration and drying the wall thickness of the scaffolds can be increased, however, the shape obtained is still dependent upon the foam itself. The sintering process is then completed in two stages, firstly to burn out the polymer foam, and secondly to sinter the particles to form a continuous construct. The foam template is key within this process, its pore shape, size, and distribution all control the outcome of the scaffolds produced. The mechanical properties of the scaffolds produced via this technique are predominately dependent upon the particle distribution within the foam prior to burnout, which is controlled by the stability and concentration of the suspension used.

The process was first adopted to produce glass ceramics from the 45S5 composition in 2006. Scaffolds were produced from 45S5 glass powder, then sintered at 1000°C for 1 hour, forming fine crystals of $Na_2Ca_2Si_3O$ [12]. It was then adapted by Fu et al. [13] using the 13-93 glass composition which were kept amorphous post sintering (54.6 SiO_2, 22.4 CaO, 6 Na_2O, 1.7 P_2O_5, 7.9 K_2O, 7.7 MgO in mol%). The scaffolds met many of the requirements for bone regeneration, with a reported porosity of 85 ± 2%, interconnected pores between 100 and 500 μm, and compressive strengths of 11 ± 1 MPa. These values are similar to those of human trabecular bone and outperformed other glass-ceramic foams reported in literature at the time [13]. This could be attributed to their optimisation of glass loading within the slurry. Their particle size distribution (modal size of 2 μm) allowed for good particle packing within the green body enhancing sintering. Combining this with the sintering window provided by the 13-93 composition allowing viscous flow without crystallisation during sintering, leads to the formation of dense struts with limited distortion of the foams morphology.

Table 1 summarises the variety of glass and glass ceramics that have been produced via this method; as shown there are large differences between the properties obtained but it highlights the flexibility of the processing method. A limitation of the foam replica process is that excess slurry can become trapped inside the sacrificial polymer foam after infiltration, which can be difficult to remove prior to burnout and sintering. It has to be squeezed out and if excess slurry remains, the structure produced will be heterogeneous with a porosity gradient being formed. Even with optimum processing, the foam replica method can leave struts with hollow centres which reduce the mechanical properties and performance of the scaffolds [14].

2.1.4. Gel-cast foaming

Sepulveda and Binner [15] first used gel-cast foaming to create porous ceramics. Rather than using a foam template, as used within the foam replica process, the ceramic slurry was foamed under vigorous agitation with the aid of a surfactant. The process used in-situ polymerisation of an organic monomer, such as methacrylamide, to form a gel and stabilise the foam. The polymer was then burnt out during the sintering cycle. Wu et al. [16] adapted this gel-casting process to manufacture scaffolds using the composition ICIE16 (49.46% SiO_2, 36.27% CaO, 6.6% Na_2O, 1.07% P_2O_5 and 6.6% K_2O, in mol%). The process was based upon in situ polymerisation of acrylamide by using a cross-linker, surfactant and initiator. Although this alternative processing method avoided the 'hollow strut' issues associated with the foam replica process it was limited in its ability to be scaled up to higher volume manufacture due to the small gelation window of the polymer used, it was also shown to precipitate sulphur rich crystals upon the glass surface. The scaffolds produced by Wu et al. [16] had porosities of 79%, with modal pore diameters of 379 μm and compressive strength of 1.9 MPa. These values were a factor of 10 lower than those reported by Fu et al. for foam replica scaffolds manufactured from 13-93 [13].

The gel-casting process was later adapted by Nommeots-Nomm et al. by using the thermally controlled gelation of gelatin to increase the processability of the system and stop unwanted sulphur byproducts. Amorphous scaffolds with foam morphologies were produced with porosities of 75% with modal pore interconnects between 100 and 150 μm. Scaffolds were produced from three glass compositions ICIE16, PSrBG (44.5 SiO_2, 17.8 CaO, 4 Na_2O, 4.5 P_2O_5, 4 K_2O, 7.5 MgO and 17.8 SrO mol%) and 13-93 with compressive strengths of 3.4 ± 0.3, 8.4 ± 0.8 and 15.3 ± 1.8 MPa respectively, higher than values presented in literature for equivalent foamed scaffolds. *In-vivo* analysis within a femoral head defect in a rabbit model supported growth across 12 weeks; however, the rate of scaffold degradation raised some questions about its suitability for a human model [26].

This process has also been adapted for the production of glass ceramics and more recently glass ceramics from pre-ceramic polymers. Fiocco et al. used pre-ceramic polymers to produce wollastonite diopside scaffolds with 77% open porosity and compressive strengths of 1.8 ± 0.3 MPa [27].

Composition name	Composition used/ mol %	Pore size/μm	Porosity/%	Compressive strength/MPa	Reported bioactivity
Glass-ceramics					
CEL2 [17]	45% SiO_2, 3% P_2O_5, 26% CaO, 7% MgO, 15% Na_2O, 4% K_2O	100–300	75	1 ± 0.4	HA reported after 1 week
Borosilicate® [18]	P_2O_5–Na_2O–CaO–SiO_2	200–500	95	0.06 ± 0.01	HA reported after with 72 hours
[19]	$57SiO_2$–34CaO–$6Na_2O$–$3Al_2O_3$ crystal formation: $CaSiO_3$	Mean: 240	56 ± 6	18	Not reported
45S5 [12]	Crystallised phase: $Na_2Ca_2Si_3O_9$	510–720	89–92	0.27–0.42	HA reported after 3 days
45S5 [20]	Crystallised phase: $Na_2Ca_2Si_3O_9$	Not reported	90	0.4	Not tested
45S5 [21]	$Na_6Ca_3Si_6O_{18}$ and $Na_2Ca_4(PO_4)_2SiO_4$	25–600	61 ± 3	13.78 ± 2.43	HA reported after 7 days
Glass					
Zn-doped borosilicate [22]	$6Na_2O$, $8K_2O$, 8MgO, 22CaO, $36B_2O_3$, $18SiO_2$, $2P_2O_5$; Zn substitution at 1.5, 5 and 10 wt.% ZnO	200–400	80–92 ± 7	Not reported	HA reported after 90 days,8 weeks in a rat calvarial defects, glass doped with 5% Zn enhanced bone formation
13-93 doped with copper [23]	$6Na_2O$, $8K_2O$, 8MgO, 22CaO, $36B_2O_3$, $18SiO_2$, and $2P_2O_3$ (mol%) Copper doping at 0, 1 and 3 wt.%	100–300	85 ± 3	Not reported	HA reported after 90 days
13-93 doped with cobalt [14]	wt.%: $53SiO_2$, $6Na_2O$, $12K_2O$, 5MgO, 20CaO, and $4P_2O_5$ Cobalt doping at 0 wt.%, 1 wt.% and 5 wt.% CoO	200–400	89–91	2.3–4.2	HA reported after 3 days
13-93 [13]	$53SiO_2$, $6Na_2O$, $12K_2O$, 5MgO, CaO, $4P_2O_5$ wt.%	100–500	85 ± 2	11 ± 1	Within 7 days
13-93B1 [24]	6% Na_2O, 8% K_2O, 8% MgO, 22% CaO, 18% B_2O_3, 36% SiO_2, 2% P_2O_5	Mean 500	78	5.1 ± 1.7	HCA within 30 days, to non critical osseous defect model in a rabbit, one in the femoral head, one in radius bone supported regrowth
[25]	22 CaO, 6 Na_2O, 8 MgO, 8 K_2O, 18 SiO_2, 36 B_2O_3, and 2 P_2O_5	250–500	72 ± 3	6.4 ± 0.1	HA reported after 7 days

Table 1. Summary of the glass and glass-ceramic scaffolds reported in literature produced via the foam replica method.

2.2. Additive manufacturing techniques

2.2.1. *Selective laser sintering*

Selective laser sintering (SLS) is a layer-by-layer additive manufacture technique capable of manufacturing green bodies. Glass particles are most commonly mixed with a thermoplastic binder after which, a laser locally heats the areas of interest creating a layered green body, the powder bed then drops down and another layer of particulate is dispersed and then selectively bound. Unbound particles are removed and the green body is sintered using a two stage process to remove the binder and merge the particles together. As with most powder based 3D printing technique, its advantages are that complex or challenging geometries can be manufactured. However, it is limited by the accuracy of the laser, the size of construct that can be produced, the detail of the features that can be formed, and the ability to completely burn out the binder without leaving residual porosity [28].

One study utilising SLS by Velez et al. reports scaffolds of 13-93 glass, with particles of sizes up to 75 μm, using stearic acid as a binding agent. The scaffold with 50% porosity had strengths of 40 ± 10 MPa post-sintering [29]. Work by Kolan et al. [28] reported mechanical properties within the same range as Velez et al. [29] of 41 MPa for 50% porosity. Their work looked at optimising the process by reducing the binder concentration and particle size, and increasing the laser power to increase processing speed. However, the time to produce scaffolds compared with competing 3D printing techniques, such as direct ink writing was still much longer.

2.2.2. *Stereolithography*

Stereolithography is a slurry-based 3D printing technique; a bed is flooded with a homogenous dispersed slurry of particles containing a photocurable polymer. A laser is then used to cure the polymer locally to build a layer of the desired design. The bed is then dropped down and flood again with the slurry, and a second layer is cured into place. The stereolithography process offers the freedom to design and fabricate very complex shapes, but again as with selective laser sintering the processing times are slow.

45S5 glass-ceramic scaffolds have been successfully printed into a variety of pore design and geometries [30]. They found that, for a diamond-like structure, reducing the pore size from 700 to 400 μm while maintaining the same overall porosity (~60 vol%) increased the mechanical strength from 3.5 to 6.7 MPa respectively.

2.2.3. *Robocasting*

Robocasting is another '3D printing' method being used to manufacture porous scaffolds from bioactive glasses. Originally developed in 1998 by Cesarano et al., it can be described as a solution based extrusion process controlled by computer aided design [31]. The robocasting technique relies on the formulation of glass loaded inks which can be extruded through fine diameter nozzles. An example of the scaffolds produced is shown in **Figure 1**. The first development of this technique relied on manipulating the interparticle forces within particle loaded suspensions to create inks with the correct rheological properties for printing [31]. Now binder 'inks' are formulated with the correct rheological properties to act as a carrier of

http://dx.doi.org/10.5772/intechopen.70242

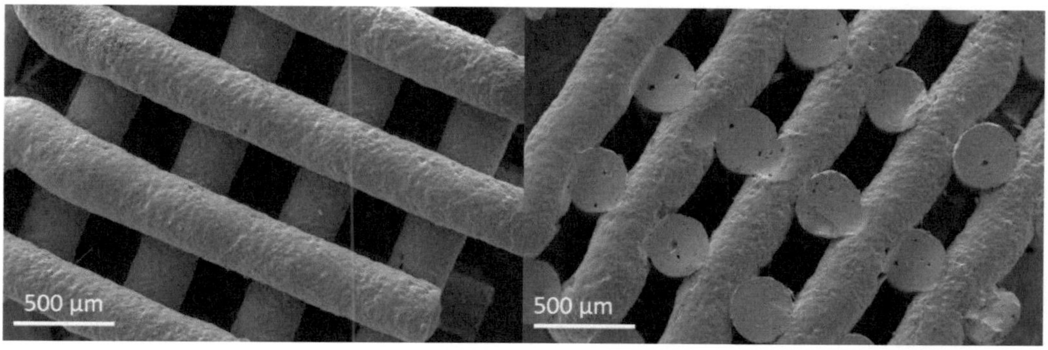

Figure 1. An example of robocast scaffolds produced from ICIE16 bioactive glass (authors own).

the particles which bind them together prior to sintering. An ideal ink would be easy to mix, allow for high glass powder loadings while maintaining a printable viscosity, have pseudo-plastic rheological properties, and yield strength and storage modulus high enough to be able to withstand the weight of multiple layers of scaffolds and the spanning distance.

Robocasting is the most commonly used 3D printing technique for bioactive glasses. This is for a number of reasons: the speed of fabrication; the ease at which inks can be produced; and the capability and accessibility of the machines. This enables the printing of inks with high glass loading, which produce scaffolds with low strut porosity, enhancing their mechanical properties.

The first robocasting paper published from a bioactive glass was by Fu et al. using 13-93 glass and Pluronic F-127 surfactant as a carrier ink [32]. Pluronic F-127 is a block co-polymer surfactant with thermally reversing rheological behaviour, made up of poly(ethylene oxide)-poly(propylene oxide)-poly(ethyleneoxide) tri-blocks (PEO-PPO-PEO) ($HO(C_2H_4O)a(C_3H_6O)b(C_2H_4O)c$). Pluronic F-127 can be dissolved in water, forming a stable suspension via steric repulsion of the –OH groups [33-35]. Due to its thermally reversible properties it enables easy mixing of glass particle at low temperatures while enabling the formation of high modulus, and high stiffness pseudoplastic inks at room temperature. Since Fu's initial work, groups have worked with a variety of different glass and binder chemistries, summarised in **Table 2**.

Glass used	Max glass loading/vol %	Ink chemistry and (dispersant)	Strut size/μm	Pore size/μm	Porosity/%	Compressive strengths/MPa
6P53B [32]	30	F-127 (water)	100	500	60	136 ± 22
13-93B3 13-93 [38]	45	Ethyl cellulose/PEG (ethanol)	300 ± 20	420 ± 30	48 ± 3	65 ± 11 142 ± 20
13-93 [39]	40	F-127 (water)	330	300	47	86.9 ± 9
45S5 [40, 41]*	45	Carboxymethyl cellulose (water)	250–300**	Not available	63 ± 3	13 ± 1

*Not amorphous post sintering.

**Values not reported, therefore, measured using imageJ from the published SEM images available.

Table 2. Summary of current literature on scaffolds produced via direct ink writing.

The homogeneity of the ink, particle packing, sintering process, strut size and spacing all contribute to the scaffolds overall mechanical properties. Due to the periodically arrange structure the strengths achieved are substantially higher than other scaffold processing methods, offering strengths in the range of cortical bone instead of cancellous [36].

Glass-ceramic scaffolds have also been produced via this technique. Roohani-Esfahani et al. produced scaffolds of various designs from a heat treated sol-gel Sr doped $Ca_2ZnSi_2O_7$ (HT) composition. They reported high strengths in excess of 100 MPa with high fatigue and crack resistance created by the multiphase composition [37].

3. Sintering and crystallization

One point, that all scaffold processing technique have in common, is the necessity for a sintering step. Sintering is a thermally induced phenomenon where, initially, the particles will undergo viscous flow. Viscous flow will ultimately lead to reorientation of the glass particles and neck growth. At a later stage, the contact between particles will grow and the pores will shrink resulting in a dense body where, usually, some closed pores will become trapped. Trapped pores are characteristically in the sub-micron range. However, in the case of glass and more particularly bioactive glasses, the heating process may induce crystallization.

The crystallization of the two most used and FDA approved glass, i.e. 45S5 and S53P4, has been studied in details by Massera et al. [42]. In this study, the activation energy for viscous flow as well as the activation energy for crystallization was quantify for fine (<45 μm) and coarse (300–500 μm) glass particles. As expected, the activation energy for viscous flow was independent of the glass particle size and ranged between 750 and 800 kJ/mol. On the other hand, the activation energy (Ec) for crystallization was found to be constant with respect to glass particles size, for the glass S53P4 with a value around 300 kJ/mol. In the case of the glass 45S5, Ec was found to decrease, from 338 to 230 kJ/mol, with increasing particles size. This is typically an indication of a change in the crystallization dimensionality with respect to the crystal size. The Jonhson-Mehl-Avrami exponent, which gives an indication of the dimensionality of the primary crystal phase, was calculated. While a value of ~1 was constantly derived for the glass S53P4, independently of the technique used (Augis and Bennett equation, Ozawa and isochronal method), the value was found to change from 0.8 to 2.8 when using the Augis and Bennett equation. Not only the 'n' exponent was found to change as a function of particles size but it also varied depending on the method employed. Such variation in the JMA exponent calls for questioning the Johnson-Mehl-Avrami crystallization model. The JMA model validity was tested using the methods proposed by Malek and Mitsuhashi [43]. While the JMA model was validated for the glass S53P4, the JMA model was found to be unsuitable to the crystallization of glass 45S5. Overall, both glass S53P4 and 45S5 were found to be fragile glass, i.e. a small change in temperature leads to a significant change in viscosity. The activation energy for viscous flow is high indicating that sintering or fibre drawing at moderate temperature will be virtually impossible. Crystallization is surface initiated (n ~ 1) for the glass S53P4. **Figure 2a** presents a SEM image of the glass surface after a heat treatment at 730°C for 2 hours and **Figure 2b** presents the crystallized layer thickness as a function of heat treatment time.

http://dx.doi.org/10.5772/intechopen.70242

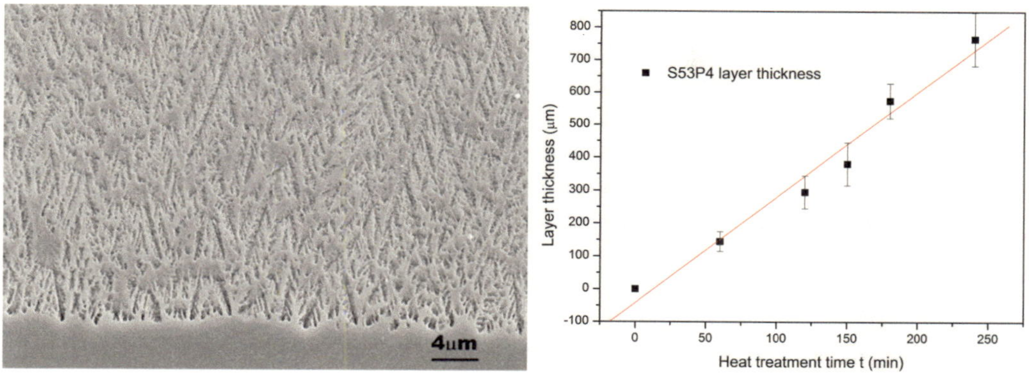

Figure 2. SEM image of a glass S53P4 heat treated for 2 hours at 730°C (a) and layer thickness as a function of heat treatment time at 730°C (b) (authors own).

As shown in the SEM image the crystallization initiates at the glass surface with a growth rate of about 3.5 μm/min. Such crystallization kinetic will prevent proper sintering of the glass particles. Furthermore, one should keep in mind that during scaffolds processing, it is generally accepted, that small particles size should be used. Decreasing the particles size leads in an increase in the surface area and therefore an increase in nucleation sites. Therefore, this glass may not be suited for the sintering of amorphous scaffolds.

The crystallization of the glass 45S5, described in [42] seemed to be initiated within the bulk. However, it appears that the glass becomes phase separated, with one amorphous phase crystallizing at lower temperatures than the secondary phase. Upon increasing the duration of the heat treatment the crystals grew via surface controlled crystallization thereby the elements of the secondary phase, at the interface with the crystals, 'feeds' the crystal. The formation of large pores was also evident during sintering of this glass composition.

The kinetics driving the crystallization of the two most known and clinically used bioactive glasses does not permit complete sintering of scaffolds without formation of crystallites.

4. Impact of crystallization on bioactivity

The impact of crystallization on the bioactivity or, more precisely, on the ability of the glass to precipitate a HA layer, has been first studied by Peitl Filho et al. [44]. Their conclusion was that below 60% crystallinity the ability of the glass 45S5 to precipitate a reactive layer was not impaired whereas above 60% crystallinity the reactive layer precipitated at a slower rate. This was then confirmed with glass S53P4 by Fagerlund et al. [45]. They demonstrated that when the glass particles were heat treated at temperature to maintain the glass fully amorphous a clear formation of Si-rich layer, a mixed layer and a Ca–P with a Ca/P = 1.6–1.7 was formed. When the glass started to crystallize, the precipitated reactive layer was much thinner and disappeared with an increase in heat treatment temperature. Interestingly, at temperature above 750°C a secondary crystal phase was formed. Upon immersion in simulated body fluid the primary crystal phase, belonging to the combeite composition dissolved, leading to the formation of thick mixed-layer. However, no Ca–P reactive layer precipitated at the surface of the particles. Furthermore, the secondary phase was found insoluble in simulated body fluid. Therefore, it can be concluded that while the bioactivity is not fully suppressed in this silicate-bioactive glasses a clear decease in the ability of the glass to form a HA layer was greatly slowed down. One should keep in mind that the precipitation of such layer is only a small contribution of the glass towards the bioactivity and, as such, is not sufficient to make firm conclusion regarding the bioactivity of the materials *in-vivo*. Nonetheless, it remains that one of the most interesting aspects of bioactive glasses is the release of therapeutic ions in a control fashion and this cannot be controlled when the glass is crystallized in an uncontrolled manner, as it is the case during sintering of rapidly crystallizing glasses. Such uncontrolled crystallization also raises the question about the ability to confidently reproduce similar microstructure, which would be key in production for the clinic.

Other well-known and promising bioactive glasses for bone regeneration belongs to the phosphate composition. Ahmed et al. [46] Neel et al. [47], as well as Massera et al. [48–50], have developed phosphate glass composition that not only can promote cell attachment and proliferation but that can also be shaped into fibres. Nevertheless, a thorough understanding of the glass crystallization is necessary in order to understand the possibilities and limitations of these glasses. Massera et al., developed a glass within the $50P_2O_5$–(40–x)CaO–xSrO–$10Na_2O$ that can promote the activity of human gingival fibroblasts [49]. The crystallization kinetics of these glasses along with the *in-vitro* dissolution in simulated body fluid of partially to fully crystallized glasses has been studied in [50]. If heat treated at temperature below their crystallization, the reactive layer precipitation rate remained unchanged. Particles that were partially crystallized were ground to 125–250 μm particles size. The crystallization was found to initiate at the surface of the glass particles. During the grinding process the particles were fractured and revealed surfaces that were crystallized and some surfaces that were amorphous. **Figure 3** shows the SEM image of crushed glass particles with an amorphous surface and a rounded crystallized surface after 72 hours of immersion in simulated body fluid.

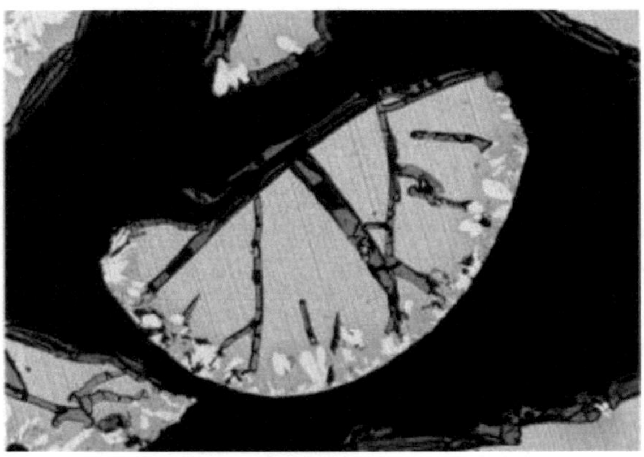

Figure 3. SEM micrograph of glass with composition $50P_2O_5$–20CaO–20SrO–$10Na_2O$ heat treated at 40°C below the crystallization temperature Tp (from DTA) for 30 min and immersed for 72 hours in simulated body fluid (reproduced from Massera et al. [49] with permission).

It appears clear on the SEM micrographs that the precipitation of the reactive layer is not suppressed. However, the reactive layer is only visible at the amorphous surface. This indicates that fully crystallized phosphate glasses will not support the precipitation of reactive layer.

A fully crystallized phosphate glass ($50P_2O_5$–40CaO–$10Na_2O$) was also immersed in simulated body fluids. No reactive layer could be detected at the surface of the glass particles. The phosphate and calcium ions released within the solution were quantified by ICP and are presented in **Figure 4**.

As expected the amorphous glass released phosphate ions and consume the calcium. This is attributed to the congruent dissolution of the phosphate glass and subsequent precipitation of a calcium phosphate reactive layer with a Ca/P ratio of 1. The crystallized glass, however, released a greater amount of P whereas the Ca remained constant within the accuracy of the measurements. From XRD the fully crystallized glass is composed of $NaCa(PO_3)_3$ crystals and $Ca(PO_3)_2$. The $NaCa(PO_3)_3$ crystals which is the primary crystal phase is more soluble than the secondary phase. The large release of phosphorus ions leads to a drastic decrease in the pH, which suppress the precipitation of the reactive layer [50].

Therefore, one can say that the impact of crystallization on the *in-vitro* dissolution can either reduce or suppress the precipitation of the CaP reactive layer. Also, one should keep in mind that the ion release of the partially to fully crystallized glass will be different than the original amorphous composition. The impact of crystallization on the bioactivity is a function of the crystal composition, crystal density and the composition of the remaining amorphous phase.

Alternative composition, which can be processed into amorphous porous scaffolds, are available. One example, among others, is the glass 13-93 [51]. This glass has been found to be bioactive and to have a wide hot forming window and a viscosity/temperature relationship enabling sintering prior to the crystallization. More recently, another family of glass composition which is gaining interest are borosilicate. Fu et al., taken as an example, processed 13-93 glass composition were the SiO_2 was partially to fully replaced by B_2O_3 [52]. They demonstrated that the

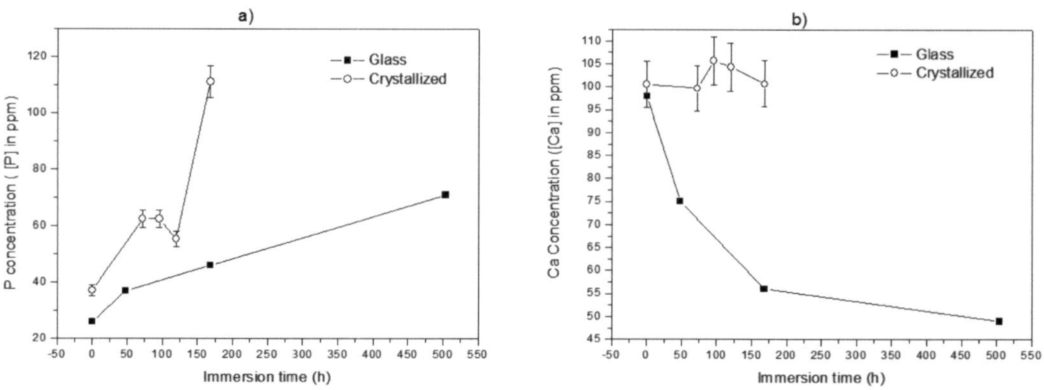

Figure 4. [P] (a) and [Ca] (b) concentration in the simulated body fluid when the glass and the corresponding crystallized are immersed for various times (authors own).

precipitation of the hydroxyapatite was accelerated with the presence of boron. However, the cell proliferation, *in-vitro*, was greatly reduced both in static and dynamic test. Nevertheless, the *in-vivo* outcome of these materials was positive [52]. Other researcher are attempting to develop borosilicate glass based on the fast dissolving bioactive 45S5 and S53P4 [53].

5. Conclusion

The current need for an off the shelf bone repair product is evident, and due to the current ageing demographic that need is growing. Literature presents a wide variety of bioactive glass and glass-ceramic compositions with tailored bioactivity and therapeutic effects suitable for bone repair. A huge wealth of different scaffold processing methods have been employed in the last 10 years, offering different morphologies, porosities and mechanical strengths. Previously these materials have been limited by their mechanical strength - porosity relationships. However, since 2011, with the growth of 3D printing, scaffolds with compressive strengths in the range of cortical bone have now been developed.

The common point of all scaffolds processing technique is the need for a sintering step at medium to elevated temperature. Sintering of bioactive glass is known to often lead to glass crystallization. Understanding the relationship between crystallization and bioactivity of bioactive glasses is of paramount importance. The available results demonstrate that the impact of the crystallization can have various effect on bioactivity depending upon the glass composition in question. While partial to full crystallization of the most known and widely used silicate bioactive glasses (S53P4 and 45S5) leads to a decrease in the bioactivity. Metaphosphate bioactive glasses have been shown only to precipitate HA on remaining amorphous surfaces.

Regardless of the glass composition, partial to full crystallization leads to drastic change in the ions releasing rate and the dissolution mechanism. The dissolution behaviour of the crystallized phases is then different to that of the original glass. The crystallization of bioactive glasses is difficult to control, therefore leading to large variation from samples to sample disabling full prediction of the materials degradation.

The impact of glass crystallization on the bioactivity is dependent upon the crystals composition, size and density. Therefore, a better understanding of the crystallization mechanism of these glasses may allow for predictable ion release and enhanced mechanical properties.

Author details

Amy Nommeots-Nomm and Jonathan Massera*

*Address all correspondence to: jonathan.massera@tut.fi

BioMediTech Institute and Faculty of Biomedical Sciences and Engineering, Tampere University of Technology, Tampere, Finland

References

[1] O'Brien FJ. Biomaterials & scaffolds for tissue engineering. Materials Today. 2011;**14**:88-95. DOI: https://doi.org/10.1016/S1369-7021(11)70058-X

[2] Hench LL, et al. Bonding mechanisms at the interface of ceramic prosthetic materials. Journal of Biomedical Materials Research. 1971;**2**:117-141. DOI: 10.1002/jbm.820050611

[3] Fu Q, et al. Bioactive glass scaffolds for bone tissue engineering: State of the art and future perspectives. Materials Science and Engineering C: Materials for Biological Applications. 2010;**31**:1245-1256. DOI: https://doi.org/10.1016/j.msec.2011.04.022

[4] Hollinger JO, et al. Role of bone substitutes. Clinical Orthopaedics and Related Research. 1996;**324**:55-65

[5] Karageorgiou V, et al. Porosity of 3D biomaterial scaffolds and osteogenesis. Biomaterials. 2005;**26**:5474-5491. DOI: https://doi.org/10.1016/j.biomaterials.2005.02.002

[6] Gerhardt LC, Boccaccini AR. Bioactive glass and glass-ceramic scaffolds for bone tissue engineering. Materials. 2010;**3**:3867-3910. DOI: 10.3390/ma3073867

[7] Jones JR. Review of bioactive glass: From Hench to hybrids. Acta Biomaterialia. 2013;**9**(1): 4457-4486. DOI: http://dx.doi.org/10.1016/j.actbio.2012.08.023

[8] Deville S, Saiz E, Tomsia AP. Freeze casting of hydroxyapatite scaffolds for bone tissue engineering. Biomaterials. 2006;**27**(32):5480-5489. DOI: https://doi.org/10.1016/j.biomaterials.2006.06.028

[9] Huang TS, et al. Freeze extrusion fabrication of 13-93 bioactive glass scaffolds for repair and regeneration of load-bearing bones. Ceramic Transactions. 2011;**228**:45-55. DOI: 10.1007/s10856-011-4236-4

[10] Liu X, Rahaman MN, Fu Q. Oriented bioactive glass (13-93) scaffolds with controllable pore size by unidirectional freezing of camphene-based suspensions: Microstructure and mechanical response. Acta Biomaterialia. Acta Materialia Inc. 2011;**7**(1):406-416. DOI: 10.1016/j.actbio.2010.08.025

[11] Mallick KK. Freeze casting of porous bioactive glass and bioceramics. Journal of the American Ceramic Society. Blackwell Publishing Inc. 2009;**92**:S85-S94. DOI: 10.1111/j.1551-2916.2008.02784.x

[12] Chen QZ, Thompson ID, Boccaccini AR. 45S5 Bioglass®-derived glass-ceramic scaffolds for bone tissue engineering. Biomaterials. 2006;**27**(11):2414-2425. DOI: https://doi.org/10.1016/j.biomaterials.2005.11.025

[13] Fu Q, et al. Mechanical and in vitro performance of 13-93 bioactive glass scaffolds prepared by a polymer foam replication technique. Acta Biomaterialia. 2008;**4**(6):1854-1864. DOI: 10.1016/j.actbio.2008.04.019

[14] Hoppe A, et al. Cobalt-releasing 1393 bioactive glass-derived scaffolds for bone tissue engineering applications. ACS Applied Materials and Interfaces. 2014;**6**(4):2865-2877

[15] Sepulveda P, Binner JG. Processing of cellular ceramics by foaming and in situ polymerisation of organic monomers. Journal of the European Ceramic Society. 1999;**19**(12):2059-2066. DOI: 10.1016/S0955-2219(99)00024-2

[16] Wu ZY, et al. Melt-derived bioactive glass scaffolds produced by a gel-cast foaming technique. Acta Biomaterialia. Acta Materialia Inc. 2011;**7**(4):1807-1816. DOI: 10.1016/j.actbio.2010.11.041

[17] Vitale-Brovarone C, et al. Development of glass-ceramic scaffolds for bone tissue engineering: Characterisation, proliferation of human osteoblasts and nodule formation. Acta Biomaterialia. 2007;**3**(2):199-208. DOI: https://doi.org/http://dx.doi.org/10.1016/j.actbio.2006.07.012

[18] Desimone D, et al. Biosilicate(®)-gelatine bone scaffolds by the foam replica technique: Development and characterization. Science and Technology of Advanced Materials. 2013;**14**(4):45008. DOI: https://doi.org/10.1088/1468-6996/14/4/045008

[19] Baino F, Vitale-Brovarone C. Mechanical properties and reliability of glass-ceramic foam scaffolds for bone repair. Materials Letters. 2014;**118**:27-30. DOI: https://doi.org/http://dx.doi.org/10.1016/j.matlet.2013.12.037

[20] Chen Q, Mohn D, Stark WJ. Optimization of Bioglass® scaffold fabrication process. Journal of the American Ceramic Society. 2011;**94**(12):4184-4190. DOI: https://doi.org/10.1111/j.1551-2916.2011.04766.x

[21] Aguilar-Reyes EA, et al. Processing and in vitro bioactivity of high-strength 45S5 glass-ceramic scaffolds for bone regeneration. Ceramics International. 2017;**43**(9):6868-6875. DOI: https://doi.org/10.1016/j.ceramint.2017.02.107

[22] Wang H, et al. Three-dimensional zinc incorporated borosilicate bioactive glass scaffolds for rodent critical-sized calvarial defects repair and regeneration. Colloids and Surfaces B: Biointerfaces. 2015;**130**:149-156. DOI: https://doi.org/http://dx.doi.org/10.1016/j.colsurfb.2015.03.053

[23] Wang H, et al. Influence of Cu doping in borosilicate bioactive glass and the properties of its derived scaffolds. Materials Science and Engineering: C. 2016;**58**:194-203. DOI: https://doi.org/https://doi.org/10.1016/j.msec.2015.08.027

[24] Gu Y, et al. Bone regeneration in rat calvarial defects implanted with fibrous scaffolds composed of a mixture of silicate and borate bioactive glasses. Acta Biomaterialia. 2013;**9**(11):9126-9136. DOI: https://doi.org/10.1016/j.actbio.2013.06.039

[25] Fu H, et al. In vitro evaluation of borate-based bioactive glass scaffolds prepared by a polymer foam replication method. Materials Science and Engineering: C. 2009;**29**(7):2275-2281. DOI: http://doi.org/10.1016/j.msec.2009.05.013

[26] Nommeots-Nomm A, et al. Highly degradable porous melt-derived bioactive glass foam scaffolds for bone regeneration. Acta Biomaterialia. 2017;**57**:449-461. DOI: https://doi.org/10.1016/j.actbio.2017.04.030

[27] Fiocco L, et al. Bioactive wollastonite-diopside foams from preceramic polymers and reactive oxide fillers. Materials. 2015;**8**(5):2480-2494. DOI: 10.3390/ma8052480

[28] Kolan KCR, et al. Effect of material, process parameters, and simulated body fluids on mechanical properties of 13-93 bioactive glass porous constructs made by selective laser sintering. Journal of the Mechanical Behavior of Biomedical Materials. Elsevier. 2012;**13**:14-24. DOI: 10.1016/j.jmbbm.2012.04.001

[29] Velez M, et al. In vivo evaluation of 13-93 bioactive glass scaffolds made by selective laser sintering (SLS). In: Narayan R, Bose S, Bandyopadhyay A, editors. Biomaterials Science: Processing, Properties and Applications II. Hoboken, NJ, USA: John Wiley & Sons, Inc.; 2012. pp. 91-99. DOI: 10.1002/9781118511466.ch10

[30] Thavornyutikarn B, et al. Porous 45S5 Bioglass®-based scaffolds using stereolithography: Effect of partial pre-sintering on structural and mechanical properties of scaffolds. Materials Science and Engineering: C. 2017;**75**:1281-1288. DOI: https://doi.org/10.1016/j.msec.2017.03.001

[31] Cesarano III J, Segalman R, Calvert P. Robocasting provides moldless fabrication from slurry deposition. Ceramic Industry. 1998;148(4):94. Available from: https://ezp.lib.unimelb.edu.au/login?url=https://search.ebscohost.com/login.aspx?direct=true&db=edsggo&AN=edsgcl.20872588&site=eds-live&scope=site

[32] Fu Q, Saiz E, Tomsia AP. Direct ink writing of highly porous and strong glass scaffolds for load-bearing bone defects repair and regeneration. Acta Biomaterialia. 2011;**7**(10):3547-3554. DOI: 10.1016/j.actbio.2011.06.030.

[33] Cohn D, Sosnik A, Garty S. Smart hydrogels for in situ generated implants. Biomacromolecules. 2005;**6**(3):1168-1175. DOI: 10.1021/bm0495250

[34] Franco J, et al. Direct write assembly of calcium phosphate scaffolds using a water-based hydrogel. Acta Biomaterialia. Acta Materialia Inc. 2010;**6**(1):218-228. DOI: 10.1016/j.actbio.2009.06.031

[35] Lenaerts V, et al. Temperature-dependent rheological behavior of Pluronic F-127 aqueous solutions. International Journal of Pharmaceutics. 1987;**39**(1):121-127. DOI: http://dx.doi.org/10.1016/0378-5173(87)90206-7

[36] Grimal Q, et al. Assessment of cortical bone elasticity and strength: Mechanical testing and ultrasound provide complementary data. Medical Engineering & Physics. 2009;**31**(9):1140-1147. DOI: 10.1016/j.medengphy.2009.07.011

[37] Roohani-Esfahani S-I, Newman P, Zreiqat H. Design and fabrication of 3D printed scaffolds with a mechanical strength comparable to cortical bone to repair large bone defects. Scientific Reports. Nature Publishing Group. 2016;**6**(February 2015):19468. DOI: 10.1038/srep19468

[38] Deliormanli AM, Rahaman MN. Direct-write assembly of silicate and borate bioactive glass scaffolds for bone repair. Journal of the European Ceramic Society. Elsevier Ltd. 2012;**32**(14):3637-3646. DOI: 10.1016/j.jeurceramsoc.2012.05.005

[39] Liu X, et al. Mechanical properties of bioactive glass (13-93) scaffolds fabricated by robotic deposition for structural bone repair. Acta Biomaterialia. 2013;**9**(6):7025-7034. DOI: 10.1016/j.actbio.2013.02.026

[40] Eqtesadi S, et al. A simple recipe for direct writing complex 45S5 bioglass 3D scaffolds. Materials Letters. 2013;**93**:68-71. DOI: https://doi.org/10.1016/j.matlet.2012.11.043

[41] Eqtesadi S, et al. Robocasting of 45S5 bioactive glass scaffolds for bone tissue engineering. Journal of the European Ceramic Society. 2014;**34**(1):107-118. DOI: https://doi.org/10.1016/j.jeurceramsoc.2013.08.003

[42] Massera J, et al. Crystallization behavior of the commercial bioactive glasses 45S5 and S53P4. Journal of the American Ceramic Society. 2012;**95**:607-613. DOI: 10.1111/j.1551-2916.2011.05012.x

[43] Malek J, Mitsuhashi T. Testing method for the Johnson-Mehl-Avrami equation in kinetic analysis of crystallization processes. Journal of the American Ceramic Societies. 2000;**83**:2103

[44] Peitl Filho O, LaTorre GP, Hench LL. Effect of crystallization on apatite-layer formation of bioactive glass 45S5. Journal of Biomedical Research Part A. 1996;**30**:509-514. DOI: 10.1002/(SICI)1097-4636(199604)30:4<509

[45] Fagerlund S, et al. Phase composition and in vitro bioactivity of porous implants made of bioactive glass S53P4. Acta Biomaterialia. 2012;**8**:2331-2339. DOI: https://doi.org/10.1016/j.actbio.2012.03.011

[46] Ahmed I, et al. Processing, characterisation and biocompatibility of iron-phosphate glass fibres for tissue engineering. Biomaterials. 2004;**25**:3223-3232. DOI: https://doi.org/10.1016/j.biomaterials.2003.10.013

[47] Neel EAA, et al. Bioactive functional materials: A perspective on phosphate-based glasses. Journal of Materials Chemistry. 2009;**19**:690-701. DOI: 10.1039/B810675D

[48] Massera J, et al. Processing and characterization of novel borophosphate glasses and fibers for medical applications. Journal of Non-Crystalline Solids. 2015;**425**:52-60. DOI: 10.1016/j.jnoncrysol.2015.05.028

[49] Massera J, et al. The influence of SrO and CaO in silicate and phosphate bioactive glasses on human gingival fibroblasts. Journal of Materials Science: Materials in Medicine. 2015;**26**:196. DOI: 10.1007/s10856-015-5528-x

[50] Massera J, et al. Crystallization behavior of phosphate glasses and its impact on the glasses' bioactivity. Journal of Materials Science. 2015;**50**:3091-3102. DOI: 10.1007/s10853-015-8869-4

[51] Brink M. The influence of alkali and alkaline earths on the working range for bioactive glasses. Journal of Biomedical Research Part A. 1997;36:109-117. DOI: 10.1002/(SICI)1097-4636(199707)**36**:1<109

[52] Fu Q, et al. Silicate, borosilicate, and borate bioactive glass scaffolds with controllable degradation rate for bone tissue engineering applications. I. Preparation and in vitro degradation. Journal of Biomedical Research Part A. 2010;**95A**:164-171. DOI: 10.1002/jbm.a.32824

[53] Fabert M, et al. Crystallization and sintering of borosilicate bioactive glasses for application in tissue engineering. Journal of Materials Chemistry B. 2017;**5**:4514-4525. Available online. DOI: 10.1039/C7TB00106A

Chapter 3

Bioceramic Scaffolds

Amira M. M. Amin and Emad M. M. Ewais

Additional information is available at the end of the chapter

http://dx.doi.org/10.5772/intechopen.70194

Abstract

Millions of peoples in the world suffer from their bone damage tissues by disease or trauma. Every day, thousands of surgical procedures are performed to replace or repair these tissues. The availability of these tissues is a big problem, and their costs are expensive. The repair of these defects has become a major clinical and socioeconomic need with the increase of aging population and social development. The emerge of tissue engineering (TE) is considered as a glimmer of hope to contribute in solving this problem. It aims at the regeneration of damaged tissues with restoring and maintaining the function of human bone tissues using the combination of cell biology, materials science, and engineering principles. In this chapter, the current state of the tissue engineering in particular bioceramic scaffolds was discussed. Concept of tissue engineering was explored. Bioceramic scaffold materials, their processing techniques, challenges taken into consideration the design of the scaffolds, and their in-vitro and in-vivo studies were highlighted. The scaffolds with extra-functionalities such as drug release ability and clinical applications were mentioned.

Keywords: bioceramics, scaffolds, classifications, processing, in-vivo, in-vitro, applications, challenges

1. Introduction

Bone defects and its functional disturbance have become a huge health care problem in the worldwide [1, 2]. The repair of these defects has become a major clinical and socioeconomic need with the increase of aging population and social development [3, 4]. Every day, thousands of surgical procedures are performed to replace or repair tissue that has been damaged through disease or trauma. The availability of these tissues is a big problem and their cost is expensive. The emerge of tissue engineering (TE) is considered as a glimmer of hope to contribute in solving this problem. It aims to regenerate damaged tissues. Through this approach, the regeneration of damaged tissue is achieved by combining cells from the human body with highly porous biomaterial scaffolds. These biomaterials scaffolds are metal, polymer, or ceramics which act as templates for growing the new viable tissues [5]. The suggested materials for tissue engineering (scaffolds) must interact with cells and culture media from the in-vitro stage to their implantation. They have to (i) host a sufficient amount of cells and (ii) support their viability for several weeks. Like any implanted biomaterials, there are some characteristics that must be found in the ideal scaffolds such as: (1) they are porous structure to allow cell penetration, tissue in-growth, (2) biocompatible, i.e., compatible without causing any toxic reactions or an inflammatory response, (3) bioactive, i.e., form strong bonding with the host bone, (4) has sufficient mechanical properties. A major difficulty in the design of scaffolds is to simultaneously tailor these requirements due to their competing nature in fulfilling host tissue demands. Namely, if a specific requisite is achieved, another one is negatively affected. It can be said, although some scaffolds have a highly porous structure with interconnected pores and have features like degradability, biocompatibility, and bioactivity, they cannot be used under a heavy load because of their poor mechanical properties. Thus, the biomaterials scaffolds with suitable mechanical properties, bioactivity, biocompatibility, and biodegradability in order to become the bottleneck. Therefore, recently many researchers have tried to find new solutions that tackle this bottleneck for achieving abovementioned requirements. They have tried to achieve that through many ways for example, (i) proposing new biomaterials like bioceramics, or composite based bioceramics (bioceramic-polymers composites, or bioceramic metal composites), (ii) using new processing techniques or developing the current processing techniques [6–9].

To verify the bone job of TE tissue engineering in restoring and maintaining the function of human bone tissues using the combination of cell biology, materials science, and engineering principles, the scaffolds must have the following criteria:

Biocompatibility: The biocompatibility of the scaffolds means that their ability to perform as the 3D substrates that will have surface chemistry (with the facilitation of molecular and mechanical signaling system) to promote cell adhesion, proliferation, and migration in vitro [10]. And after implantation, the scaffold must not induce any undesirable immune reaction that may reduce healing or cause rejection by the body [11].

Biodegradability: The gradual degradation of scaffolds helps to make space for new growing tissues to deposit their own matrix and hence avoids the necessity of second surgery to remove the implant [12]. Thus, it is one of the crucial factors for scaffolds. The degradation of an ideal scaffold must occur with time in-vivo, and its rate must proportional to the rate of the tissue formation. The biodegradation products should be nontoxic to other tissues in-vivo.

Bioactivity: Stimulation of rapid tissue attachment to the implant surface (without formation of fibrous tissue) and creation of a stable long-term bonding that prevents micromotion at the interface and the onset of an inflammatory response [13].

Structural requirements: An ideal scaffold should have void volume for vascularization, neo tissue formation and remodeling, necessary to facilitate host tissue integration on implantation [14]. Biomaterials should be processed to provide a highly porous structure with interconnected porosity for transporting oxygen, nutrients, and waste metabolites in and out of the scaffold without significantly compromising the mechanical stability of the scaffold [14]. If a scaffold has too small pore size, it may enact the cells to penetrate the scaffold initially and subsequently to migrate through these pores to the other regions of the scaffolds. But, if it has too large pore size, it may inhibit the effective neo-tissue regeneration by disabling the cells to bridge pores during cell proliferation [13].

Manufacturing technology/commercialization potential: The fabrication technique of the scaffolds is a crucial factor in the production of TE scaffolds. It is a challenge to produce a large quantity of scaffolds at a relatively low or reasonable cost, i.e., to be easily offered to the market [13, 14].

This chapter defines the current state of tissue engineering regarding bioceramic scaffolds. In addition, the complexity of this field. In other words, the following items will be highlighted in details:

It discusses

- Concept of tissue engineering
- Bioceramic scaffold materials
- Processing techniques
- Challenges
- In vitro and in vivo studies of bioceramic scaffold materials
- Scaffolds with extra-functionalities such as drug release ability
- Clinical applications

2. Concept of tissue engineering

Over 8 million surgical operations for treating the organ failure or tissue loss are performed annually in the US [15]. Despite of the success made by the organ transplantation and reconstruction surgery in the life quality, and in some cases the life save, there are still problems associated with the patients. These operations, in most cases, need either organ donation from donor individual or tissue transplantation from a second surgical site in the individual being treated. The generated problems from the use of organ transplantation are the drastic shortage of donor organs. For example, in 1996, 20,000 donor organs were only available, and the number of patients waiting the organs were 50,000. This means that patients are more likely to die while they are waiting for a human donor than in the first 2 years after transplantation [15]. The problems associated with the second surgical sites are pain and morbidity. Accordingly, organs development, tissues, and synthetic materials outside of the body ready for future transplant use have emerged [5, 15–21]. The estimated market of these products is approximately $5 billion worldwide [16].

Tissue engineering may be defined as the application of biological, chemical, and engineering principles toward the repair, restoration, or regeneration of living tissue by using biomaterials, cells, and growth factors alone or in combination. In the early 1990s, the emerged tissue engineering started to address the limitations of tissue grafting and alloplastic tissue repair. The concept was to transplant a biofactor like cells, genes and/or proteins within a porous degradable material known as a scaffold. The biofactors are used to stimulate tissue repair. They include stem cells and gene therapy approaches. The tissue/organ repair has been considered the ultimate goal of surgery from ancient times until now. Generally, repair is achieved through two approaches: (i) tissue grafting and organ transplantation and (ii) alloplastic or synthetic material replacement. Since 2000 BC gold has been used in cranial defects as repair; however, the grafting of the tissue has been used at the earlier of 1660s. Both approaches mentioned above have limitations. The grafting needs second surgical sites with associated morbidity and is confined by finite amounts of material, especially for organ replacement. One of the efforts made to solve the problems associated with the use of the autologous allograft, and bone cements is the finding appropriate materials to replace lost or missing tissues from the human body [22, 23].

The definition of 'Tissue Engineering' term according to the NSF workshop held in 1988 is "the application of principles and methods of engineering and life sciences toward the fundamental understanding of structure-function relationships in normal and pathological mammalian tissues and the development of biological substitutes to restore, maintain, or improve tissue function" [24]. This definition hold some promises such as (1) driving out the re-operations by using biodegradable biological substitutes, (2) encourage the use of biological substitutes as a natural regeneration process to repair or replace lost or damaged tissues i.e. to provide long term solutions, (3) solving the generated problems from the immune rejection of implants, infections or diseases transmission pertinent allografts and xenografts, and organ donation shortage, (4) offering practical solutions for currently untreatable cases [25, 26].

Since emergence of tissue engineering in the mid-1980s, it has continued to evolve as an exciting and multidisciplinary field that aiming to develop biological substitutes to restore, replace, or regenerate defective tissues. Key component of tissue engineering are cells, scaffolds, and growth-stimulating signals which are generally referred to as the tissue engineering triad (**Figure 1**) [27]. Porous 3D scaffolds are generally seeded with cells and occasionally with signaling molecules or subjected to biophysical stimuli in the form of a bioreactor [28]. These cell-seeded scaffolds are either subjected to a pre-implantation differentiation culture in-vitro to synthesize tissues and then transplanted or are directly implanted into the injured site, using the body's own systems, where tissue regeneration is induced in-vivo [11]. These approaches with porous scaffolds are shown in below by **Figure 2** [29].

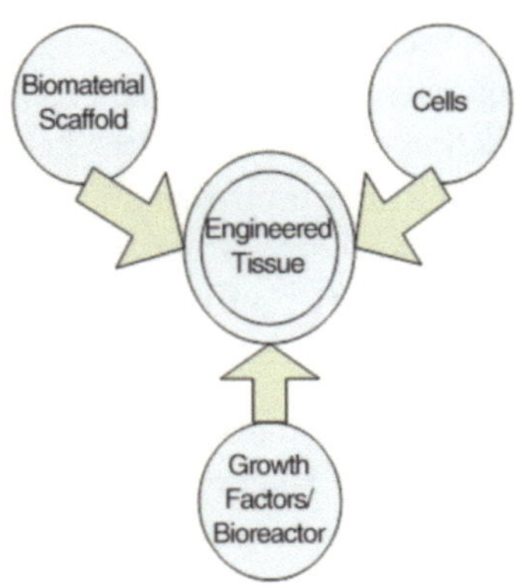

Figure 1. Tissue engineering triad [11].

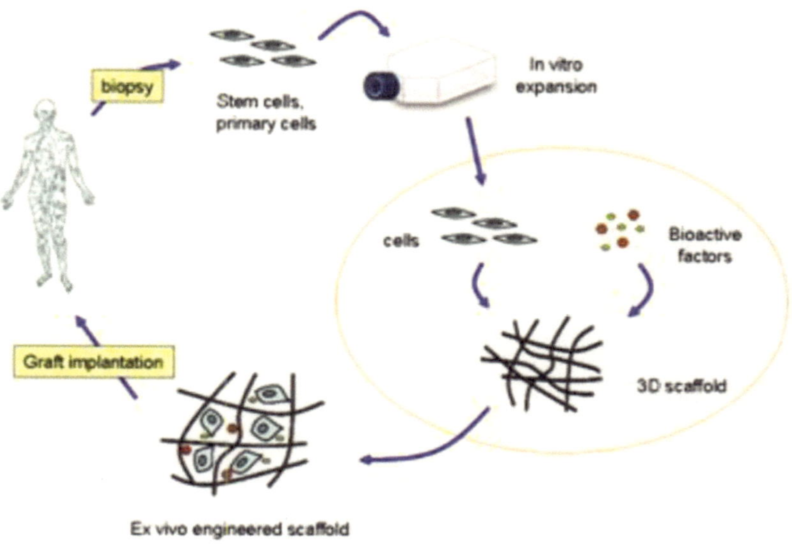

Figure 2. Tissue engineering approaches with porous scaffolds [30].

3. Bioceramic scaffold materials

Today, tissue engineering has emerged as a rapidly expanding approach to overcome the drawbacks of the classical treatments by regenerating damaged tissues, instead of replacing them [11, 30]. This approach leads to the development of biomaterials to prepare porous 3D scaffolds as biological substitutes to restore, maintain, or improve defective tissues [14]. Various materials have been proposed for tissue engineering including different types of biomaterials (metal, polymers, and ceramics) to overcome the problems associated with natural bone grafts in reconstructive surgery. Although porous metallic scaffolds are considered as the most suitable implants for hard tissue engineering in load bearing areas, they have some limitations such as (1) lacking of biological recognition or bioactivity [31, 32], (2) lacking of the integration of biomolecules [33], (3) nonbiodegradable [33], (4) releasing of toxic ions [33], (5) corrosion or wear, and (6) The architecture control [34]. One of the essential requirements for using a biocompatible metal in tissue engineering scaffold is the surface modification by coating with bioactive ceramic materials, where it reduces some of the limitations of metallic scaffolds.

Another primary materials studied mostly to fabricate scaffolds are polymers such as polylactic acid [35, 36], polyglycolic acid [37], polyurethane [38], and a number of copolymers [39–41]. Natural polymer-based scaffolds have excellent bioactivity, biodegradability but poor mechanical properties. These characteristics reveal their successful use in soft tissue engineering and limit their use in the load bearing applications. Moreover, there is still immunological concern associated with naturally derived polymers. An additional problem limits the application of the natural polymeric materials scaffolds, there are still a question mark on their structure homogeneity and reproducibility [11, 35–37, 42–48]. Metal and polymer drawbacks mentioned above lead to emerging of a new type of materials prevent the production of the wear debris and can be designed to more closely match the material properties of natural bone. Such materials must be mechanically stronger than polymers and play a critical role in providing mechanical stability to construct prior to synthesis of new bone matrix by cells. The materials nominated to fit this purpose are called bioceramics.

Ceramic scaffold possesses many aspects like being bioactive, biocompatible, biodegradable, mechanically stiff (Young's modulus) [49], less elastic and brittle. They also exhibit shaping difficulties. Bioceramics can be classified into three groups as given in the following **Table 1** [50]:

Groups	Phases
1. Bioinert	e.g., Aluminum oxide (Al_2O_3) and zirconium oxide (ZrO_2).
2. Surface bioactive	e.g., Sintered hydroxyapatite(s-HA) at high temperature, bioglass
3. Bioresorbable	e.g., Sintered hydroxyapatite(u-HA) at low temperature, tricalcium phosphate (α-TCP and β-TCP), tetracalcium phosphate (TTCP), octacalcium phosphate (OCP).

Table 1. Bioceramics classification.

Later group (3) is used in bone tissue engineering, various calcium phosphates (CaPs) specially HA, β-TCP, and biphasic calcium phosphate, BCP (mixture of HA and β-TCP) has long been studied as porous scaffold materials. As natural bone composed of large amounts of HA ($Ca_{10}(PO_4)_6(OH)_2$), so it might be useful to use HA, β-TCP as they closely simulate the chemical and crystalline nature of the mineral phase of the native bone [51, 52], and hence, they will be biocompatible. Hydroxyapatite (HA) is known by its bioactivity, biocompatibility, nontoxicity, noninflammatory, osteoconductivity, and biodegradability. By its comparison with β-TCP, it degrades slowly in following order: OCP > α-TCP > β-TCP > u-HA>> s-HA [53] and after implantation, it undergoes little conversion to a bone like material [54]. For the same porosity, β-TCP scaffolds often exhibit lower mechanical strength than HA scaffolds, limiting their use in the load bearing applications [55]. The degradation rate and other properties can be influenced by varying HA to β-TCP ratios in BCP. Interestingly, researchers have shown that dopant addition in the scaffolds of CaPs can control the biocompatibility, densification behavior, dissolution rates, and mechanical strength [33, 56].

Recently, calcium phosphates containing materials used for tissue engineering are nominated as bioactive glasses, silicate bioactive glasses, borate bioactive glasses, phosphate bioactive glasses, and akermanite. Each one will be explained as the following:

Bioactive glasses have already shown their excellence as promising biomaterials for tissue engineering due to their ability to enhance bone cell growth, bonding to both hard and soft tissues [27], ability to restore defect sites and controllable degradation rate in vivo. Glass compositions and of the scaffolds and their microstructure play an important role in the determination of the degradation rate and conversion to an HA-like material, mechanical properties, and response to cells.

Recently, doped bioglasses with various elements, such as Cu, Zn, and Sr, promote the healthy bone growth that have been developed [58]. These types of the bioglass showed an enhancement of angiogenesis (formation of blood vessels) [55, 59] and soft tissue wound healing [55]. And this capacity of bioglasses has provided an alternative approach to the use of expensive growth factors for stimulating neovascularization of engineered tissues [55].

45S5 glass has long been established as highly bioactive, biocompatible [60], and biodegradable. The composition of 45S5 glass is 45% SiO_2, 6% P_2O_5, 24.5% CaO, 24.5% Na_2O and the low SiO_2 content (<55% SiO_2), high content of network modifiers like Na_2O and CaO, high CaO/P_2O_5 ratio contributes to the bioactivity of 45S5 glass. The immersion of this form of glass is body fluid, and it forms HCA layer (carbonate substituted HA, typical bone composition) on its surface that significantly promotes osteoblast activity. However, there is a difficulty in the processing of this glass into a porous 3D scaffold due to its low mechanical strength, slow degradation rate, and conversion to an HA mineral [52]. Recently, it was found that by heating this type of glass to high temperatures (>950°C), its phases crystallize with strong mechanical strength and become bioactive glasses. In addition, it converts to a biodegradable, amorphous calcium phosphate at the body temperature, and in a biological environment [61]. This process enables the mechanical competence and biodegradability to be incorporated in a single scaffold, making it promising as tissue engineering scaffold [51].

Borate bioactive glass: Researchers have indicated that borate or borosilicate bioactive glasses promote cell proliferation and differentiation in vitro, as well as tissue infiltration in vivo [55]. Borate bioactive glasses degrade as faster as than 45S5 glasses. They completely convert to an HA-like material because of their lower chemical stability [62]. The degradation rate can be controlled by manipulating the glass composition [62, 63]. Besides, there is a concern about the toxicity of boron released into the solution as borate ions $(BO_3)_3$ [55].

Phosphate Bioactive glass: [55] It forms networks, where CaO and Na_2O act as network modifiers. It shows a chemical affinity toward bone due to the existed ions in the organic mineral phase of the bone. The degradability of this glass can be controlled by modifying their composition. Its flexibility displayed has made it potential resorbable biomaterials for tissue engineering.

Akermanite ($Ca_2MgSi_2O_7$): Recently, it has received more attention due to its controllable mechanical properties and degradation rate [64, 65]. In previous studies, marrow-derived or adipose-derived stem cells and osteoblasts have displayed good activities of proliferation and osteogenesis on akermanite compared by β-TCP [66–70]. The recent studies suggest that this Mg-containing silicate ceramic as a bone graft material may meet the requirement of bone regeneration than b-TCP. However, the mechanism of akermanite's bioactivity is still unknown. The materials chemistry of biomaterials is one of the main factors in the proliferation and differentiation of various cells.

In tissue engineering, biomaterials play a critical role. They act as a 3D template, supply mechanical support and allow artificial extracellular matrix environment (ECM) for neo-tissue formation. This means that one type of biomaterials is not sufficient to compromise hard and soft tissue engineering. Therefore, each type from the biomaterial types such as metals, ceramics, and polymers has its own importance in making tissue engineering scaffolds. Therefore, the composite materials have been emerged. Sometimes biocompatibility and biodegradability of ceramics are not sufficient. Moreover, ceramics are very brittle, and too stiff, while the polymers are found to be biocompatible and biodegradable with low mechanical strength. So, this biological and mechanical mismatch can be overcome by blending the ceramics with natural or synthetic polymers or metal. Recently, various composites were explored such as synthetic polymers/natural polymers, synthetic polymers/bioceramics, polymers/metals, metals/ceramics, ...etc. However, novel metal/polymer/ceramic composites have also been suggested for load bearing applications [34]. Composite materials are necessary approach to obtain optimal biological, structural, mechanical, and chemical properties of scaffolds. Thus, bioceramics/polymers are commonly used composites.

Finally, needless to say, yet, there exists no polymers or metals that can effectively bond to bone. On the other hand, it does not exist ceramic materials that can sufficiently in mechanical properties. Therefore, composites of biodegradable polymers and hard metals with bioactive ceramic composites are still a promising approach.

4. Processing techniques

It is a key point to obtain porous structure with proper mechanical properties to create a micro-environment for cell adhesion and proliferation. Nature bone has multi-level 3D pore structure size ranging from several nano to hundreds of micrometers [71]. This level of pore sizes meets the requirements of a tissue growth. Pore sizes in the range of 150–800 μm prevent the growth of the bone tissue and the vessels of the blood. However, pore sizes in the range of 10–100 μm are useful for the growth of the blood capillaries, nutrients exchange, and waste products excretion. Nano pores are larger specific surface area and more active targets. They are good for the formation of apatite and the attachment of protein or osteoblast [72]. Meanwhile, they are also important for the adjustment of cell adhesion and proliferates.

Many fabrication techniques are available to produce ceramic scaffolds with varying architectural features. There are two main types of fabrication techniques: conventional techniques and advanced techniques. Conventional techniques for the fabrication of porous structure mainly include replica; sacrificial template; and direct foaming as seen in **Figure 3** [3].

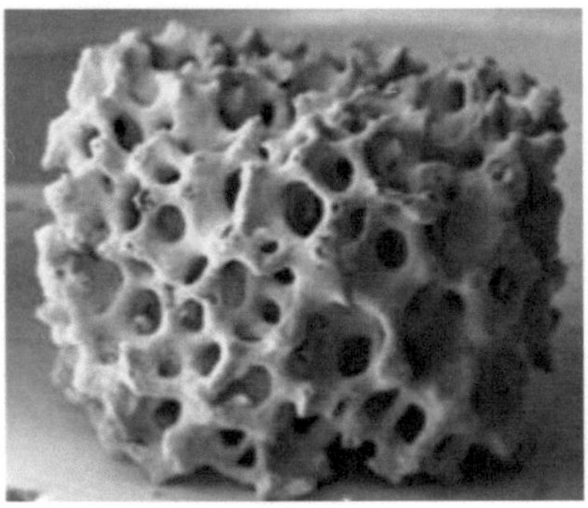

Figure 3. Calcium phosphate–based scaffold (Willis [57]).

The replica technique employs a synthetic or natural template that is impregnated with a ceramic suspension. After drying, the template is removed thus creating a replica of the original template structure [73]. Many synthetic and natural cellular structures can be used as templates to fabricate macroporous ceramics through the replica technique [74].

The sacrificial template method incorporates some sort of pore former or sacrificial material to act as a place holder within the ceramic powder or slurry. Once the green body is formed, the pore former is removed to leave behind pores which are empty (**Figure 1(b)**) [73]. This method leads to porous materials displaying a negative replica of the original sacrificial template, as opposed to the positive morphology obtained from the replica technique described above [74].

Direct foaming is a process where gas bubbles are incorporated into a ceramic suspension, and once the slurry is set and dried, the ceramic retains the resulting spherical pores (**Figure 4(c)**) [73]. To obtain high-strength ceramic foams, the dried objects are then sintered at high temperatures. The total porosity of the obtained foam is proportional to the amount of gas incorporated into the suspension or liquid medium during the foaming process. The sizes of the pores depend on the stability of the wet foam before setting [74].

Freeze casting is considered as one of the promising techniques for manufacturing of porous structure. It utilizes growing ice crystals in a ceramic slurry to form the pores in a ceramic

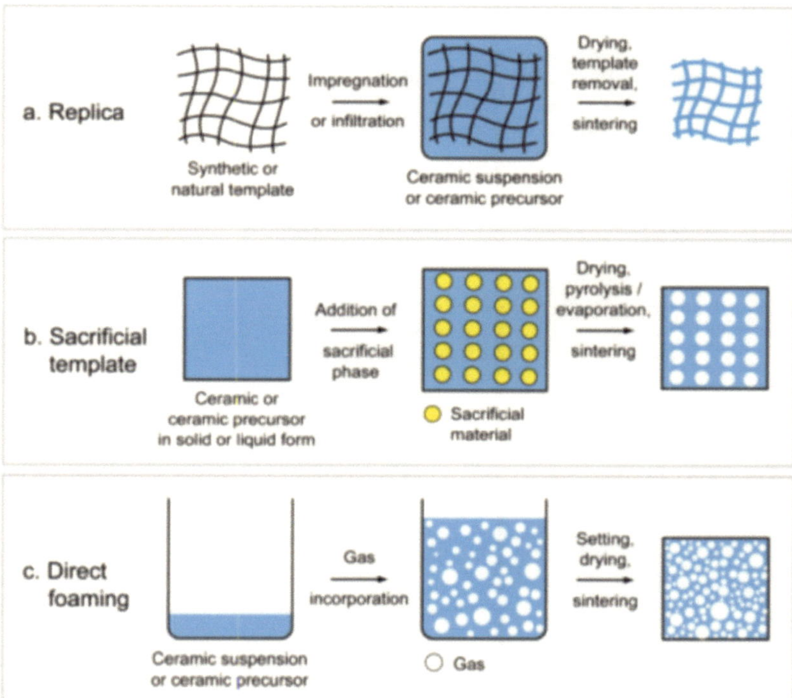

Figure 4. Scheme of possible processing routes used for the production of macroporous ceramics [77].

body. It is a simple technique to produce porous complex-shaped ceramic or polymeric parts. It has first been developed as a near net shape forming technique, yielding dense ceramics parts with fine replicate of the mold details. In this technique, a ceramic suspension is poured into a mold and then frozen. The frozen solvent acts as a temporary binder to hold the part together, see **Figure 5**. The de-molded part is subjected to freeze drying to sublimate the frozen solvent under vacuum, avoiding the stresses and shrinkage that might lead to cracks and warping during normal drying. After drying, the parts are sintered to obtain a scaffold with (1) a complex and often anisotropic porous microstructure and (2) proper strength and stiffness. By controlling the direction of ice crystals growth direction, it is possible to tailor a preferential orientation for the porosity in the ultimate products [75].

Human cortical bone has a compressive strength of 100–150 MPa and toughness of 2–12 MPa $m^{1/2}$, while human trabecular bone has a compressive strength of 2–12 MPa and toughness of 0.1–0.8 MPa $m^{1/2}$. The question is: How can design ceramic scaffolds and mimic the structure and properties of natural bone as closely as possible? This means the scaffold fabrication must be carried out with high accuracy taking into consideration the effective and functional properties such as the microstructure, mechanical properties as well as the biocompatibility [73].

In some cases, it is difficult to achieve the targeted pore for tissue growth such as interconnectivity, pore size, and pore geometry when traditional processing techniques are used. The limited control over the pore characteristics generates closed pores and leads to lack of interconnecting pores. In addition, it gives low strength and variable properties. Recently, additive

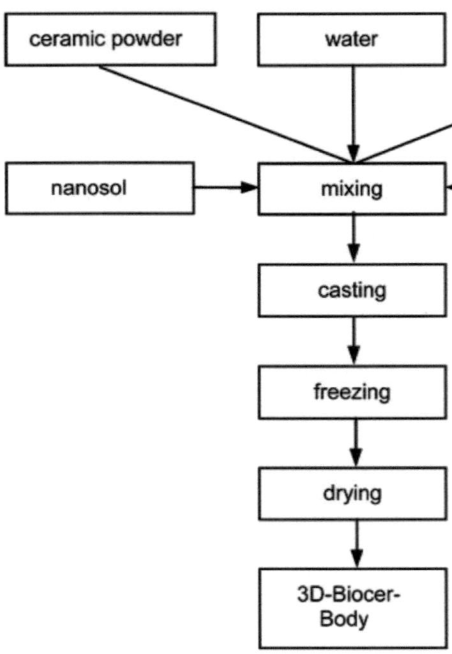

Figure 5. Freeze cast process for bioceramic preparation [78].

http://dx.doi.org/10.5772/intechopen.70194

manufacturing (AM) techniques, the rapid prototyping (RP) and electrospinning, have been proposed for the fabrication of porous ceramic scaffolds. They have several potential benefits over traditional techniques [73].

The rapid prototyping (RP) techniques: RP techniques are referred to as a solid free-form (SFF) manufacturing. They are precise and reproducible for controlling the internal pore size, porosity, pore interconnectivity, mechanical performance, and overall dimensions of tissue engineering scaffolds [76, 77]. Based on programmed 3D images, they are defined as automated deposition of each tomographic layer sequence into the desired architecture through an additive layer-by-layer method [78]. One of the main requirements for translational applications is a high productivity using automated method and possibility to produce patient-specific constructs; therefore, an RP-based method can potentially be used to fabricate such customized tissues [79]. The benefits of RP technology are numerous, e.g. the versatility of modeling software allows for the fabrication of the desired parts without the need for expensive molds, and the process is usually achieved in only a few steps. It could easily be used in the manufacturing bio-scaffolds with fit needs of a specific individual and match the surrounding bone which may vary from person to another depending on the age, health, condition of the surrounding bone, or location within the intended recipient, see **Figure 6** [73].

The most relevant RP techniques in the design of 3D scaffolds for tissue engineering are 3D printing (3DP), selective laser sintering (SLS), stereolithography (SLA), robocasting (RC), and

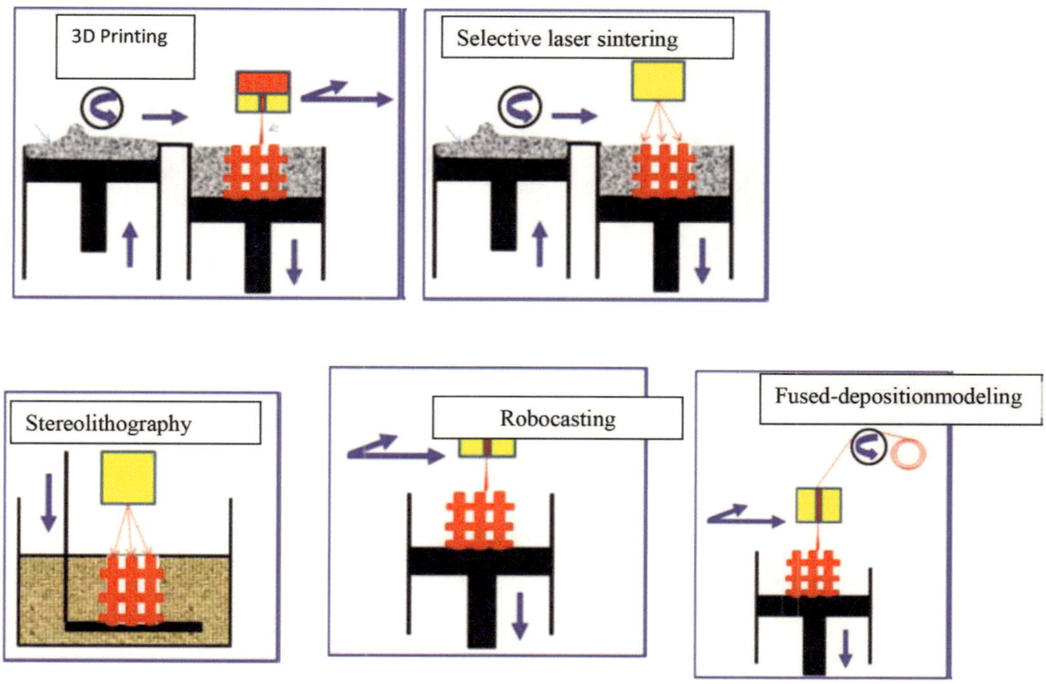

Figure 6. Main RP-based techniques relevant for tissue engineering applications [93].

fused deposition modeling (FDM). **Figure 6** shows a schematic diagram of the processing of each RP based-technique [80].

3D printing (3DP): The inkjet head of this device prints droplets of a binder fluid onto a powder bed. This process is repeated for every layer until the 3D scaffold structure is printed, and the remaining powder is removed. It has been used to create scaffolds for use in bone tissue engineering. One of the important benefits of this method is the powder bed support by itself for each successive layer. The fragility of the obtained parts is considered a drawback [81–88].

Selective laser sintering (SLS): It is a heat dependent. This method uses a CO_2 laser beam to selectively sinter polymer or composite powders to form material layers. The laser beam is directed onto the powder bed by a high precision laser scanning system [89].

Stereolithography (SLA): This technique is light dependent. The laser beam selectively initiates solidification in a thin layer of liquid photopolymer [90].

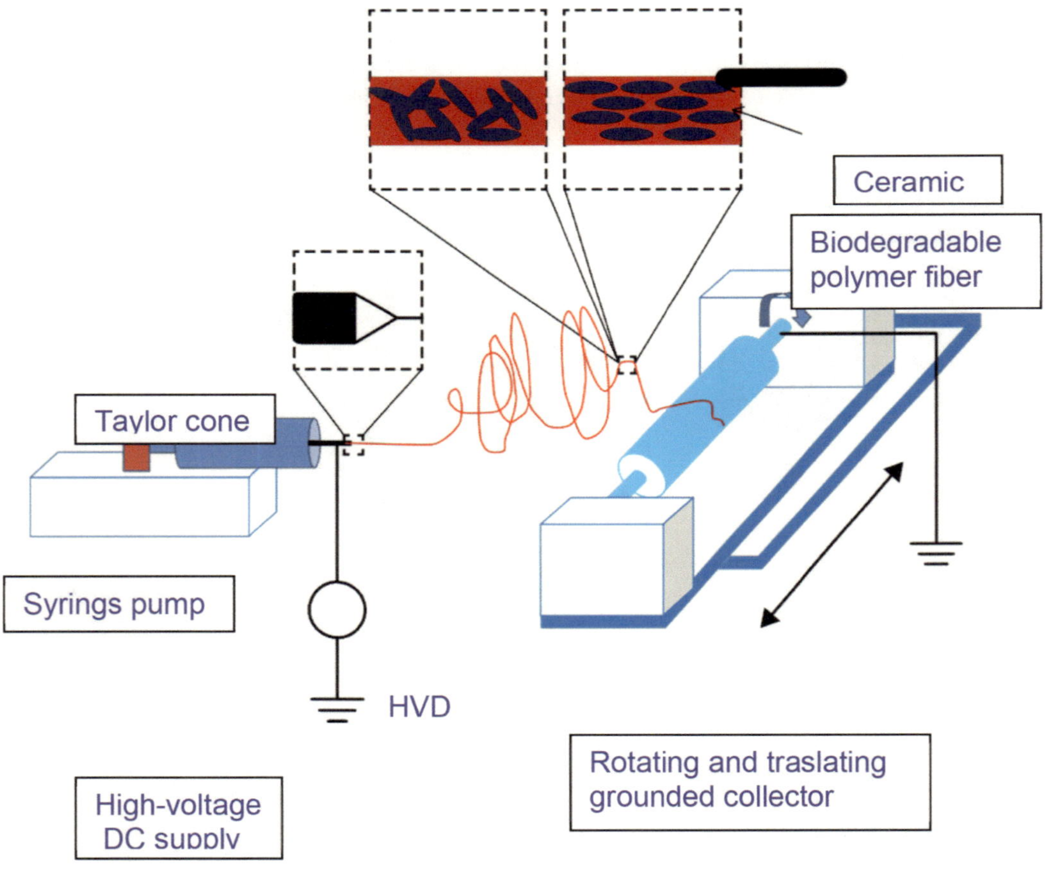

Figure 7. Schematic of a typical electrospinning system.

http://dx.doi.org/10.5772/intechopen.70194

Robocasting (RC): This technique is slurry dependent. It consists of the robotic deposition of a highly concentrated colloidal suspension (inks) [91, 92].

Electrospinning: This technique is easy to use. It is used to produce nano to microfibers by subjection of a solution of polymeric materials or ceramic/polymeric composites to an electric field. Solid fibers are produced from electrified jets using high voltage. These fibers are continuously elongated because of the electrostatic repulsion between the surface charges and the solvent evaporation, see **Figure 7** [94].

This technique allows obtaining high surface area scaffolds, which simulate the size scale of fibrous proteins found in the natural ECM [95–97]. One of the great interests in this methodology is the capacity to easily produce materials at the biological length scale for tissue engineering and drug delivery applications [98]. Also, this technique is able to form nonwoven fibrous mats, which ensure fiber production from a broad range of precursor materials including synthetic polymers, natural polymers, semiconductors, ceramics, or their combinations [99, 100]. As mentioned above, the strength and the great merit of electrospinning technology are the ability to conduct fiber size, porosity, and shape using processing variables, such as applied voltage, polymer melt flow rate, capillary/collector distance, polymer/ceramic concentration, and solvent conductivity and volatility [94]. **Figure 7** displays schematic of a typical electrospinning system.

Recently, inorganic nanoparticles, such as HA, bioactive glass, and carbon nanotubes, have been widely co-electrospun into polymer nano fibers to enhance their mechanical properties and their biocompatibility response [101–106]. Although the bioceramic needle oriented in polymer fiber is challenging, but it is very important for enhancing the mechanical properties of the scaffolds.

5. Challenges

The design and fabrication of the synthetic tissue scaffold and the engineering of tissue constructs in vitro and in vivo are big challenges. Various materials like metals, ceramics, natural and synthetic polymers, and even their composites have been explored as TE scaffolds. Bioceramics and polymers are suitable for bone TE, where the native bone composed mainly of a naturally occurring polymer and biological apatite. Since ceramics are brittle and the mechanical properties of the polymers are not sufficient, the applications of these materials are limited, in particular, in load-bearing areas. Although metals have high mechanical strength are suitable for load-bearing applications, but they are bio-inert and their biodegradability are none. Therefore, they are not suitable for soft tissue engineering. Generally, the most important challenge of tissue engineering is to mimic what happens in nature. Attempts are being made to engineer in vitro practically every tissue and organ in the body [11]. In addition, the following items are playing an important role in the properties of the bioceramic scaffolds:

(1) **Macrostructure:** It is important to determine the geometry of the regenerating tissue and in turns to be capable to re-shape it.

(2) **Mechanical properties:** the scaffolds should have sufficient mechanical strength to provide temporary function in a defect until the tissue regenerates. If the mechanical properties

of the native bone are used as a guideline in the scaffolds designing, they must exhibit linear elastic properties with a moduli of hundreds of megapascals and microstructures of preferred orientations due to bone anisotropy. In addition, it is important to know that the scaffold mechanical properties will decrease with the scaffold degradation. Thus, if the scaffolds have sufficient mechanical properties at the time of implantation, the change of their mechanical properties during the degradation could be expected and affected on the function within the tissue defect [107, 108].

(3) **Pore size, porosity and interconnectivity:** The pore size is an important variable to stimulate cell ingrowth and new bone formation [109, 110], while the interconnected porous network and porosity are critical in maintaining spatially uniform cell distribution, cell survival, proliferation, and migration in vitro. Moreover, the scaffold's porosity (exceeding 60%) and degree of pore interconnectivity directly affects the diffusion of physiological nutrients and gases [111, 112]. Interestingly, the pores with smaller sizes than 1 μm are appropriate to interact with proteins and are mainly responsible for inducing the formation of an apatite-like layer in contact with simulated blood fluids. Pores of sizes from 1 to 20 μm are important in cellular development, where the cells are attached and the orientation and directionality of cellular in-growth. Pore of sizes between 100 and 1000 μm are essential to assure nutrient supply, waste removal of cells and promoting the in-growth of bone cells. Finally, the presence of pores of sizes >1000 μm will play an important role in the implant functionality [113, 114].

6. In-vitro and in-vivo studies of bioceramic scaffolds

The in-vitro and in-vivo responses of bioceramic scaffolds are dependent on their composition and their pore architecture (microstructure). Various studies were handled biological response of the bioceramic scaffolds, some of them discussed below.

6.1. In-vitro and in-vivo studies of bioactive glass scaffolds

The ability of bioactive glass scaffolds to support cell proliferation and function in-vitro and tissue ingrowth in-vivo has been shown in numerous studies [115–122]. Fu et al. showed that 13–93 bioactive glass scaffolds prepared using a polymer foam by replica method supported the attachment and proliferation of MC3T3-E1 preosteoblastic cells both on the surface and within the interior pores of the scaffold [115].

Poh et al. prepared and estimated the in-vitro response of two different types of bioactive glass composite scaffolds. They are polycaprolactone with 45S5 glass (PCL/45S5) and strontium-substituted glass with polycaprolactone (PCL/SrBG). These two types of bioactive glasses were incorporated into polycaprolactone (PCL) and fabricated by additive manufacturing technology. The in vitro results showed that the rates of degradation of these scaffolds were PCL/SrBG > PCL/45S5 > PCL scaffolds. It was found that the degradation rate of PCL/SrBG scaffolds was faster than PCL/45S5 scaffolds. This is due to the substitution of Sr^{2+} of larger ionic radius (1.12A°) by Ca^{2+} of lower ionic radius (0.99 A°) leading to the expansion of the silicate glass network [7, 35, 36]. Such expansion weakens the glass network and increases the

http://dx.doi.org/10.5772/intechopen.70194

dissolution rates of the SrBG. The cytotoxicity test indicated that all scaffolds (PCL, PCL/45S5, and PCL/SrBG) were noncytotoxic and are able to support cell attachment, growth, and proliferation; at day 7 and 14, PCL/SrBG (control and osteo group) show a significantly higher degree of mineralization compared to all other groups (PCL/45S5 and PCL); indicating that PCL/SrBG can stimulate earlier matrix mineralization [123]. Melchers et al. investigated the effect of alumina from 0.5 to 15 mol% in mesoporous bioactive glasses based on composition 80% SiO_2–15%CaO -5% P_2O_5. Sol–gel method in combination with a structure directing agent for the formation of mesopores was used. It was found that the incorporation of Al_2O_3 in a range of 1 to 10 mol% reduces the order of the mesostructure, while the further increase of doped amount of Al_2O_3 to 15 mol% creates well-ordered mesopores again. In addition, pore diameter, pore volume, and specific surface area decrease only slightly on the incorporation of Al_2O_3. In-vitro bioactivity tests of these glasses, a decrease in their bioactivities upon the incorporation of small amounts of alumina was observed, while a sudden drop was noticed beyond the addition of 3 mol% of Al_2O_3. These results back to the strong interaction of Al^{3+} and PO_4^{3-}, which could be proven by multinuclear single and double resonance solid state nuclear magnetic resonance (NMR) spectroscopy [124].

6.2. In-vitro and in-vivo studies of hydroxyapatite scaffolds

Generally, hydroxyapatite (HA) is a material which most often induces osteogenesis both in-vivo and in-vitro. Adding HA to other materials (either natural or synthetic) could, therefore, modulate the osteogenic potential and mechanical properties of the subsequent mixture. Therefore, many papers have been devoted for a combinatorial approach of HA with another supporting material [45]. The choice of supporting material is often paramount. The main reasons of combination of HA with another material are improved strength, increased porosity, altered cell binding abilities, and so forth. These materials are divided to natural materials and synthetic materials. Natural materials (collagen, gelatin, fibrinogen) tend to have good cellular adhesion remodeling properties but can also carry a high risk of immune response. On the other hand, synthetic materials, however, are less immunogenic and more customizable but carry higher risks of toxicity. Furthermore, MSCs can be included in such scaffolds for differentiation to osteogenic lineages and/or implantation for bone defects purposes, although differentiation media are often required. Therefore, HA scaffolds containing MSCs can be used as a combinatorial modality for treating bone disease and degeneration. The combining of stem cells, in particular, MSCs, into the various HA-based scaffolds increases the scaffolds potential use for bone regeneration. Adding the benefits of MSCs immunomodulatory, immune inert, and immune-privileged state to a synthetically or naturally enhanced HA scaffold has demonstrated superior results than the scaffolds alone [125].

In another study, porous chitosan/hydroxyapatite (C/HA) scaffolds were fabricated via freeze-drying with desired pore size. The in-vitro proliferation of Human osteoblasts (hOBs) on the scaffolds were evaluated. Then, these scaffolds were combined with the adenoviral vector encoding vascular endothelial growth factor and green fluorescence protein (Ad-VEGF). In-vivo studies were conducted by subcutaneously implanting inactivated and gene-activated C/HA sponges containing hOBs into the epigastric fasciovascular flaps of Wistar rats. The results show that, in the in-vitro investigation, the adenovirus encoding VEGF gene-activated macroporous C/HA composite scaffold supports transfection of human primary osteoblasts and bone-like tissue formation. In-vivo findings demonstrate that C/HA + AdVEGF + hOBs promote abundant neovascularization during ectopic bone formation, while viral gene therapy has some drawbacks. Generally, all findings support the notion that gene-activated C/HA scaffold could have potential in vascularized bone tissue engineering [126].

Campos et al. studied the synthesis of three-dimensional (3D) scaffolds composed of 50 wt.%HA and 50 wt.% collagen for bone tissue engineering. Self-assembly method with a 0.125% glutaraldehyde solution as cross-linked was used a synthetic route. The in-vitro evaluations are cytotoxicity using MC3T3 cells, proliferation and differentiation. Proliferation and differentiation were tested using STRO-1A human stromal cells for time up to 21 days. The results show that no cytotoxicity was observed in the scaffold by MC3T3 cells. STRO-1A cells were found to adhere, proliferate, and differentiate on the 3-D scaffold, but limited cell penetration was observed [127].

Porous composite bioceramics of hydroxyapatite and dicalcium phosphate dehydrate (HAp/DCPD) were prepared using polyurethane foam. They were designed for application in osteoconductive and osteoinductive scaffolds. In-vitro and in-vivo examinations were performed to evaluate the biological responses of the prepared porous composites. In-vitro studies were performed by immersion of the samples in SBF. The in-vivo test was conducted by inserted porous composite samples into defects in the medial femoral condyle of rabbits. From in vitro and in-vivo studies, it can be concluded that the scaffolds are biocompatible without inflammation. After implementation, necroses or rejection of the tissue was noticed. The application success of the combined HAp and DCPD scaffolds for generating a new bone tissue is attributed to the merge of the biocompatibility property and the formation ability of a favorable 3D matrix for human osteoblast cells to adhere and spread, taking into consideration the advantage of TCP osteoinduction to the superior bioactivity and osteoconduction of HAp. Finally, the prepared scaffolds seem to be a promising biomaterial for low-weight-bearing orthopedic applications [128, 129].

The obtained data from literature indicate that biphasic calcium phosphate is the optimal cell-supporting material. Biphasic calcium phosphate should be recommended as the most suitable matrix for osteogenic cells expansion and differentiation in tissue engineered systems [130]. I.E. biphasic calcium phosphate (BCP) ceramics composed of HA and ß-TCP with varying HA/ß-TCP ratios have been extensively studied in the past two decades, because they combine the excellent biocompatibility and bioactivity of HA and a degradation rate of ß-TCP that matches the growth rate of newly formed bon.

6.3. In-vitro and in-vivo studies of calcium phosphate cement (CPC) scaffolds

It has been reported by Kent et al. in vitro and in vivo measurements of commercial calcium phosphate cement is considerably stronger in vivo compared to in vitro, the cause of this attributed to bone formation within the cement pores and a high degree of osseointegration [131].

Chen et al., 2013 proved that when hUCMSCs and hBMSCs were seeded onto a CPC scaffold. Then, they implanted into the defects, new bone formation increased with time. Compared with CPC control without cells, 88% and 57% increases in new bone were achieved when hBMSCs and hUCMSCs were seeded with CPC, respectively (see **Figure 8**) [132]. These results confirmed by Zeng et al., 2012, whereas he found when h-BMSCs seeding with CPC, the greater new bone was generated greater than in a model without cells in a rabbit maxillary sinus floor elevation model [133].

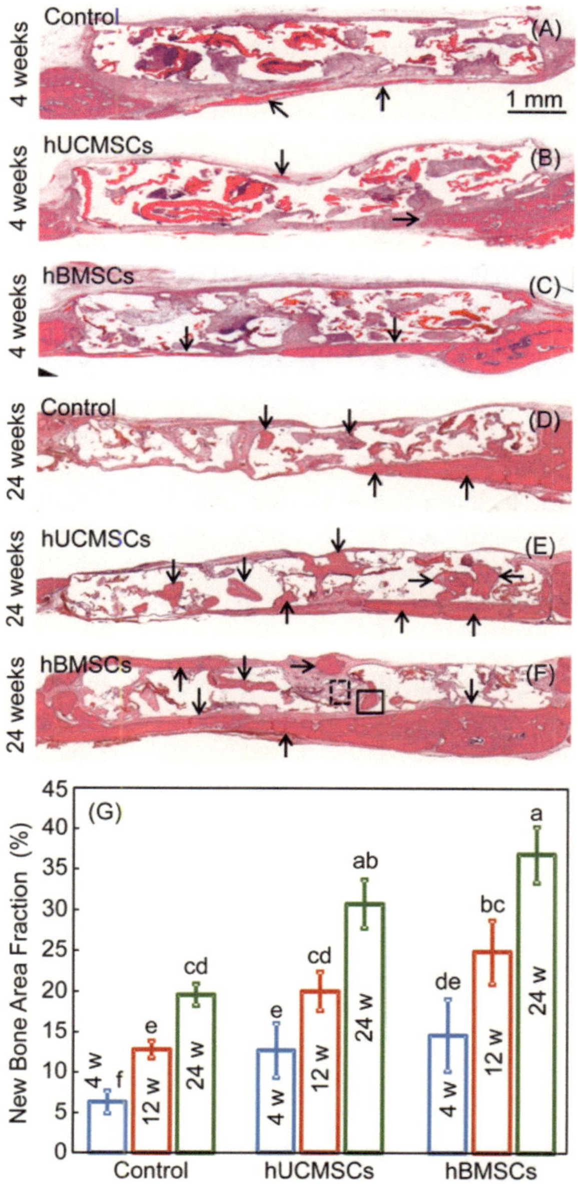

Figure 8. Bone regeneration in critical-sized cranial defects in nude rats with 3 groups (CPC control without cells; CPC with hUCMSCs; CPC with hBMSCs).

7. Scaffolds with extra-functionalities such as drug release ability

Scaffolds are implants or injects, which are used to deliver cells, drugs, and genes into the body. Scaffold matrices can be used to achieve drug delivery with high loading and efficiency to specific sites. This means that the scaffolds must be designed to provide not only the structure integrity required for bone regeneration but also for controlling dose of drug release.

Zhao et al. designed and prepared porous scaffold, which composed of a newly designed polylactone, poly(e-caprolactone)-block-poly(lactic-co-glycolic acid) (b-PLGC) copolymer, and β-tricalcium phosphate (β-TCP). Then, this scaffold was loaded with an antituberculous drug (rifampicin, RFP) to cure serious bone tuberculosis from two points of views (bone regeneration and antituberculous drug therapy). The in-vitro drug release experiment showed that hydrophobic RFP could be released from the b-PLGC/TCP scaffold in a sustained manner for 84 days. Accordingly, RFP concentrations obtained in blood and tissues surrounding the implant could reach a high value in 12 weeks, which was above the effective level needed for the treatment of tuberculosis. The cytological assay proved that the RFP-loaded scaffold has a good cell cytocompatibility. In another trial, it was found that the composite system gave a good regeneration ability for bone. Therefore, the b-PLGC/TCP scaffold can perform a local long-term drug release as well as osteogenesis capability. These achievements are suitable for clinical applications [134].

In another study, Zhu et al. proved that a mesoporous silica nanoparticulate/β-TCP/bioactive glass (BG) composite drug delivery system for osteoarticular tuberculosis therapy much higher antituberculous drugs (rifampicin (INH) and isoniazid (RFP)) loading capacities than pure β-TCP scaffold. The best concentrations of drugs (INH and RFP) for treating tuberculosis (TB) in-vivo can be maintained for an extra-long duration over 42 days without significant long-term lesions to liver and kidney [135].

It has been reported by Kundu et al. that HAp exhibited better drug release than β-TCP when CFS (ceftriaxone sodium and sulbactum sodium in 2:1 w/w ratio) drug was used. HAp and pure β-TCP based porous scaffolds were prepared by applying together starch consolidation and foaming techniques. A bilayered coating was also applied to the pore surfaces of some samples using chitosan and b-lactamase–cephalosporin derivative to assess their effect on sustained drug releasing. The result of bilayered coating of chitosan with CFS provided prolonged release pattern for more than 5 weeks irrespective of the scaffold material, a period that is considered to be sufficient for local drug delivery to combat osteomyelitis [136].

8. Clinical applications

Ceramics include a broad range of inorganic and non-metallic compounds. Although their applications in tissue engineering is recent, they demonstrate good results, whatever they are a single phase or composite. Herein, we will explored the clinical application of the most common examples of bioceramic (HA, TCP, BCP, and bioactive glasses) that are used as in tissue engineering applications.

Hydroxyapatite: By insertion porous HA wedges into the tibias of ten knees in seven patients having high tibial osteotomies, it was found that pores located at the interface, at the time of hardware removal, were completely filled with bone, and the bone depth formed increased consistently with time [137]. In another study, the transverse sections of porous HA implants placed in rabbit tibias demonstrated a new bone growth through the pores. After 8 weeks, a formed concentric lamellae by osteon structure was observed around a single vessel in the pores of sizes 50 and 100 μm of the cylindrical HA implants. Also, similar structures were displayed around the multiple vessels in the pores of sizes 300 and 500 μm of the implants [114]. The examination of 103 patients suffering from cranial defects in which Bone SourceTM was used, a success rate of 97% was recorded [138]. Such high record was achieved by the implant maintenance for 24 months.

Tricalcium phosphate (TCP): TCP has become one of the first calcium phosphates to be used in bioceramics for bone substitution and repair. Thanks to its stability at high temperature and ease of processing as tricalcium phosphate–based ceramics 339 ceramics. β-TCP containing ceramics are one of the major bioresorbable synthetic bone that used daily by orthopedic surgeons and dentists. They are used in the form of porous ceramic pieces and granules to reconstruct all kinds of bone defects, from augmentation of alveolar ridge defects after a tooth extraction and before implant positioning to sinus reconstruction correction of various deformities and bone reconstruction following injury or disease. Recently, α-TCP has also been proposed as ceramic materials for similar applications [139].

Tricalcium phosphate/hydroxyapatite biphasic ceramics (BCP): The BCP concept is based on an optimum balance between the more stable phase (HA) and the more soluble phase (β-TCP). BCP bioceramics of various sizes and shapes are used in maxillofacial surgery, dentistry, ear, nose and throat (ENT) surgeries, and orthopedics. For example, BCP granules with HA/TCP of 60/40 were placed in the alveolar cavity immediately after tooth extraction and followed up radiographically from 0 to 5 years [140].

Bioactive glasses: Bioactive glasses have a wide range of clinical applications in both medicine and dentistry. It is used as bone graft material, a coating materials, and disinfectants.

As bone graft material: Bioglass has been used clinically as a synthetic bone graft material for over 10 years under two different product names: Novabone_ for orthopedics and Perioglass for maxillofacial surgery. The first reported clinical application of bioactive glass was the treatment of conductive hearing loss for the reconstruction of the bony ossicular chain of the middle ear.

As a coating materials: More researchers use bioactive glasses as a coating materials for dental implants. Bioactive silicate glass has also been used for implant coatings, as a bone graft, in dentifrices, and as air-abrasive particles to remove carious enamel and dentin. Goudouri et al. indicated that bioactive glass could be used as a dental material to improve the bonding of the restorative material to dentin [139].

As disinfectants: Bioactive glasses can serve as topical endodontic disinfectants with no effects on dentin stability. Bioactive glass can raise the pH of an aqueous environment to produce its antimicrobial effects. For example, when implanted in areas of periodontal defects, Bioglass can inhibit bacterial colonization at the surgical site by increasing the pH and calcium levels.

As a bone regeneration: Bioactive glass can promote bone regeneration, with osteostimulatory effects in vitro. Similarly, in primate models, bioactive glass filled bony defects by stimulating osteoproduction. Felipe et al. reported that bioactive glass particles were able to treat periodontal defects and triggered the development of mineralized bone in dogs [141, 142].

9. Conclusion

This chapter discussed the bioceramic scaffolds which considered as one-third of the tissue engineering triad. It dealt with the most effective materials that were used in the bioceramic scaffolds. All traditional and advanced techniques for scaffolds manufacturing, their requirements, and challenges taken into consideration the design of the scaffolds were explored. In additions, several examples of the most common bioceramic scaffolds were highlighted from different corners, e.g., their testing in in-vitro and in-vivo, their extra functionalities such as drug release ability and their clinical applications.

Author details

Amira M. M. Amin and Emad M. M. Ewais*

*Address all correspondence to: dr_ewais@hotmail.com

Refractory & Ceramic Materials Division (RCMD), Central Metallurgical R&D Institute (CMRDI), Cairo, Egypt

References

[1] Carrington JL. Aging bone and cartilage: Cross-cutting issues. Biochemical and Biophysical Research Communications. 2005;**328**:700-708

[2] Olshansky SJ, Passaro DJ, Hershow RC, Layden J, Carnes BA, Brody J. Expectancy in the United States in the 21st century. The New England Journal of Medicine. 2005;**352**:1138-1145

[3] Ebnezar J. Textbook of Orthopaedics. London, UK: JP Medical Ltd.; 2010

[4] Cancedda R, Dozin B, Giannoni P, Quarto R. Matrix Biology. 2003;**22**:81-91

[5] Langer R, Vacanti JP. Tissue engineering. Science. 1993;**260**:920-926

[6] Qiang F, Saiz E, Rahaman MN, Tomsia AP. Materials Science and Engineering C. 2011;**31**:1245-1256

[7] Manzano M, Vallet-Regí M, Manzano M, Vallet-Regí M. Progress in Solid State Chemistry. 2012;**40**:17-30

[8] María Vallet-Regí CR. Chimie. 2010;**13**:174-185

[9] Iwasa J, Engebretsen L, Shima Y, Ochi M. Knee Surgery, Sports Traumatology, Arthroscopy. 2009;**17**:561-577

[10] Babensee JE et al. Advanced Drug Delivery Reviews. 1998;**33**:111

http://dx.doi.org/10.5772/intechopen.70194

[11] O'Brien FJ. Materials Today. 2011;**14**(3):88-95

[12] Berger J, Reist M, Mayer JM, Felt O, Peppas NA, Gurny R. European Journal of Pharmaceutics and Biopharmaceutics. 2004;**57**(1):19-34

[13] Chan BP, Leong KW. European Spine. 2008;**17**(4):467-479

[14] Edwards SL et al. AUTEX Research Journal. 2004;**4**:86-94

[15] Lanza R, Langer R, Vacanti JP. Principles of Tissue Engineering. Texas, Austin: R.G. Landes Company and Academic Press Inc.; 1997

[16] http://www.epharmaceuticalnews.com/tissue_engineering.html.

[17] Bell E. Tissue Engineering: Current Perspectives. Boston: Birkhauser, Boston; 1993

[18] Healy KE, Rezania A, Stile RA. Designing biomaterials to direct biological responses. Annals N.Y. Acad. Sci. 1999;**875**:24-35

[19] Service, R.F. Tissue engineers build new bone. Science. 2000;**289**:1498-1500

[20] Peppas NA, Langer R. New challenges in biomaterials. Science. 1994;**263**:1715-1720

[21] Lauffenburger DA. Cell engineering. In: Bronzine JD, editor. The Biomedical Engineering Handbook. Boca Raton: CRC Press; 1994

[22] Sigmund O. Materials with prescribed constitutive parameters – An inverse homogenization problem. Journal of Solids and Structures. 1994;**31**:2513-2529

[23] Lin CY, Hsiao CC, Chen PQ, Hollister SJ. Interbody fusion cage design using integrated global layout and local microstructure topology optimization. Spine. 2004;**29**:1747-1754

[24] Skalak R, Fox CF. Tissue Engineering. 1993;**1988**:26-29

[25] Zippel N, Schulze M, Tobiasch E. Recent Patent on Biotech. 2010;**4**:1-22

[26] Isaksson H, Comas O, van Donkelaar CC, et al. Journal of Biomechanics. 2007;**40**:2002-2011

[27] Wang M. American Journal of Biochemistry & Biotechnology. 2006;**2**(2):80-84

[28] I. Martin et al. Trends Biotechnology. 2004;22:80. nuhs.edu.sg/research/programmatic-research/sm-registered-programms/nustissue-engineering.

[29] Zohora FT, Azim AYMA. European Scientific Journal. 2014;**10**(21):1857-7881

[30] Mekala NK et al. Recent Research in Science and Technology. 2012;**4**(12):5-11

[31] Ryan G et al. Biomaterials. 2006;**27**:2651-2670

[32] Dong-ming XIAO et al. Transactions of the Nonferrous Metals Society of China. 2012;**22**:2554-2561

[33] Bose S, Roy M, Bandyopadhyay A. Trends in Biotechnology. 2012;**30**(10):546-554

[34] Alvarez K et al. Materials. 2009;**2**:790-832

[35] Agrawal M, Koelling KW, Chalmers JJ. Biotechnology Progress. 1998;**14**:517-526

[36] Chu CR, Monosov AZ, Amiel D. Biomaterials. 1995;**16**:1381-1384

[37] Agrawal CM, Ray RB. Journal of Biomedical Materials Research. 2001;**55**:141-150

[38] Grad S, Kupcsik L, Gorna K, Gogolewski S, Alini M. Biomaterials. 2003;**24**:5163-5171

[39] Lucke A, Temmar J, Schnell E, Schmeer G, Gopferich A. Biomaterials. 2000;**21**:2361-2370

[40] Ishaug-Riley SL, Crane-Kruger GM, Yaszemski MJ, Mikos AG. Biomaterials. 1998;**19**: 1405-1412

[41] Trantolo DJ, Sonis ST, Thompson BM, Wise DL, Lewandrowski KU, Hile DD. The International Journal of Oral & Maxillofacial Implants. 2003;**18**:182-188

[42] Cima LG, Langer R. Engineering human tissue. Chemical Engineering Progress. 1993;**89**:46-54

[43] Lu L, Zhu X, Valenzuela RG, Currier BL, Yaszemski MJ. Biodegradable polymer scaffolds for cartilage tissue engineering. Clinical Orthopaedics and Related Research. 2001;**391**:251-270

[44] Tormala P, Vasenius J, Vainionpaa J, Laiho J, Pohjonen T, Rokkanen P. Journal of Biomedical Materials Research. 1991;**25**:1-22

[45] Bostman OM. Current concepts review. Journal of Bone and Joint Surgery. 1991;**73A**:148-153

[46] Mooney DJ, Baldwin DF, Suh NP, Vacanti JP, Langer R. Biomaterials. 1996;**17**:1417-1422

[47] Chim H, Ong JL, Schantz J-T, Hutmacher DW, Mauli Agrawal C. Journal of Biomedical Materials Research. 2003;**65A**(6):327-335

[48] Sikavitsas VI, Bancroft GN, Mikos AG. Materials Research Society Symposium Proceedings. 2001;**662**:LL1.2.1-LL1.2.6

[49] Chen Q et al. Progress in Biomaterials. 2012;**1**:2. DOI: 10.1186/2194-0517-1-2

[50] Shikinami Y, Okuno M. Biomaterials. 1999;**20**:859

[51] Q Chen et al. (Eds. N Ashammakhi et al.) Topics in Tissue Engineering, 2008; Vol 4.

[52] Jarcho M. Clinical Orthopaedics and Related Research. 1981;259-278

[53] Yang S, Leong KF, Du Z, Chua CK. Tissue Engineering. 2001;**7**(6):679-689

[54] Martin RB et al. Biomaterials. 1993;**14**:341-348

[55] Rahaman MN et al. Acta Biomaterialia. 2011;**7**(6):2355-2373

[56] Mieszawska AJ, Kaplan DL. BMC Biology. 2010;**8**:59

[57] Willis RC. ACS Publications. 2004;**7**(9):

[58] Hoppe A, Guldal NS, Boccaccini AR. Biomaterials. 2011;**32**:2757-2774

[59] Gomes ME, Reis RL. International Materials Reviews. 2004;**49**:261-273

[60] Wilson J et al. Journal of Biomedical Materials Research. 1981;**15**:805-817

[61] Chen QZ, Thompson ID, Boccaccini AR. Biomaterials. 2006;**27**(11):2414-2425

[62] Huang W et al. Journal of Materials Science. Materials in Medicine. 2006;**17**:583-596

[63] Yao A et al. Journal of the American Ceramic Society. 2007;**90**:303-306

[64] Wu C, Chang J. Journal of Biomedical Materials Research. Part B, Applied Biomaterials. 2007;**83**(1):153-160

[65] Kokubo T. Biomaterials. 1991;**12**(2):155-163

[66] Sun H, Wu C, Dai K, Chang J, Tang T. Biomaterials. 2006;**27**(33):5651-5657

[67] Liu Q, Cen L, Yin S, Chen L, Liu G, Chang J, et al. Biomaterials. 2008;**29**(36):4792-4799

[68] Wu C, Chang J, Ni S, Wang J. Journal of Biomedical Materials Research. Part A. 2006;**76**(1): 73-80

[69] Xynos ID, Edgar AJ, Buttery LD, Hench LL, Polak JM. Journal of Biomedical Materials Research. 2001;**55**(2):151-157

[70] Valerio P, Pereira MM, Goes AM, Leite MF. Biomaterials. 2004;**25**(15):2941-2948

[71] Huyghe JM, Raats PA, Cowin SC. IUTAM Symposium on Physicochemical and Electro-mechanical, Interactions in Porous Media. Dordrecht, The Netherlands: Springer; 2005 Volume 125

[72] Lv R, Zhou W, Shi K, Yang Y, Wang L, Pan K, Tian C, Ren Z, Fu H. Nanoscale. 2013;**5**:8569-8576

[73] Hammel EC, Ighodaro OL-R, Okoli OI. Ceramics International. 2014;**40**(10):15351-15570

[74] Studart AR, Gonzenbach UT, Tervoort E, Gauckler LJ. Journal of the American Ceramic Society. 2006;**89**(6):1771-1789

[75] Deville S, Saiz E, Tomsia AP. Freeze casting of hydroxyapatite scaffolds for bone tissue engineering. Biomaterials. 2006;**27**:5480-5489

[76] Hollister SJ. Porous scaffold design for tissue engineering. Nature Materials. 2005;**4**: 518-524

[77] Yeong WY, Chua CK, Leong KF, Chandrasekaran M. Trends in Biotechnology. 2004; **22**:643-652

[78] Cesarano J, Segalman R, Calvert P. Ceramic Industry. 1998;**148**:94-102

[79] Hollister SJ. Scaffold design and manufacturing from concept to clinic. Advanced Materials. 2009;**21**:3330-3342

[80] Butscher A, Bohner M, Roth C, et al. Acta Biomaterialia. 2012;**8**:373-385

[81] Butscher A, Bohner M, Hofmann S, Gauckler L, Mueller R. Acta Biomaterialia. 2011;**7**: 907-920

[82] Bergmann C, Lindner M, Zhang W, Koczur K, Kirsten A, Telle R, Fischer H. Journal of the European Ceramic Society. 2010;**30**:2563-2567

[83] Vorndran E, Klarner M, Klammert U, Grover LM, Patel S, Barralet JE, Gbureck U. Advanced Engineering Materials. 2008;**10**:B6 7-BB71

[84] Tarafder S, Davies NM, Bandyopadhyay A, Bose S. Biomaterials Science. 2013;**1**:1250-1259

[85] Tarafder S, Balla VK, Davies NM, Bandyopadhyay A, Bose S. Journal of Tissue Engineering and Regenerative Medicine. 2013;**7**:631-641

[86] Warnke PH, Seitz H, Warnke F, Becker ST, Sivananthan S, Sherry E, Liu Q, Wiltfang J, Douglas T. Journal of Biomedical Materials Research Part B. 2010;**93B**:212-217

[87] Detsch R, Schaefer S, Deisinger U, Ziegler G, Seitz H, Leukers B. Journal of Biomaterials Applications. 2011;**26**:359-380

[88] Vlasea M, Shanjani Y, Bothe A, Kandel R, Toyserkani E. International Journal of Advanced Manufacturing Technology. 2013;**68**:2261-2269

[89] Liu F-H. JMEPEG. 2014;**23**:3762-3769

[90] Chu TMG, Halloran JW, Hollister SJ, Feinberg SE. Journal of Materials Science: Materials in Medicine. 2001;**12**:471-478

[91] Saiz E, Gremillard L, Menendez G, Miranda P, Gryn K, Tomsia AP. Materials Science and Engineering C. 2007;**27**:546-550

[92] Shao H, Liu A, Ke X, Sun M, He Y, Xianyan Yang F, Jianzhong Fu B, Zhang L, Yang G, Liu Y, Xu S, Gou Z. Journal of Materials Chemistry B. 2017;**5**:2941-2951

[93] Butscher A, Bohner M, Hofmann S, et al. Acta Biomaterialia. 2011;**7**:907-920

[94] Sill TJ, Von Recum HA. Biomaterials. 2008;**29**:1989-2006

[95] Yin Z, Chen X, Chen JL, et al. Biomaterials. 2010;**31**:2163-2175

[96] Dalby M, Gadegaard N, Tare RA, et al. Nature Materials. 2010;**6**:997-1003

[97] Li D, Xia YN. Advanced Materials. 2004;**16**:1151-1170

[98] Agarwal S, Wendorff JH, Greiner A. Advanced Materials. 2009;**21**:3343-3351

[99] Barnes CP, Sell SA, Boland ED, et al. Advanced Drug Delivery Reviews. 2007;**59**:1413-1433

[100] Teo WE, Ramakrishna S. Nanotechnology. 2006;**17**:R89-R106

[101] Peng F, Shaw MT, Olson JR, Wei M. Journal of Physical Chemistry C. 2011;**115**:15743-15751

[102] Yu C-C, Chang J-J, Lee Y-H, et al. Materials Letters. 2013;**93**:133-136

[103] Lao L, Wang Y, Zhu Y, et al. Journal of Materials Science. Materials in Medicine. 2011;**22**: 1873-1884

[104] Nie H, Wang C-H. Journal of Controlled Release. 2007;**120**:111-121

[105] Franco PQ, Joao CFC, Borges JP. Materials Letters. 2012;**67**:233-236

[106] Wang HM, Chou YT, Wen ZH, Wang CZ, Chen CH, et al. PloS One. 2013;**8**(11):10.1371

[107] Vivanco J, Aiyangar A, Araneda A, Ploeg H-L. Journal of the Mechanical Behavior of Biomedical Materials. 2012;**9**:137-152

[108] Hutmacher DW. Biomaterials. 2000;**21**:2529-2543

[109] Karageorgiou V, Kaplan D. Biomaterials. 2005;**26**:5474-5491

[110] Boyan BD, Hummert TW, Dean DD, Schwratz Z. Biomaterials. 1996;**17**:137-146

[111] Yoshikawa T, Ohgushi T, Tamai S. Journal of Biomedical Materials Research. 1996;**32**: 481-492

[112] Vallet-Regi M, Gonzalez-Calbet JM. Progress in Solid State Chemistry. 2004;**32**:1-31

[113] Vallet-Regi M, Colilla M, Izquierdo-Barba I. Journal of Biomedical Nanotechnology. 2008;**4**:1-15

[114] Mangano C, Bartolucci EG, Mazzocco C. The International Journal of Oral & Maxillofacial Implants. 2003;**18**:23-30

[115] Fu Q, Rahaman MN, Bal BS, Brown RF, Day DE. Acta Biomaterialia. 2008;**4**:1854-1864

[116] Fu QA, Rahaman MN, Bal BS, Bonewald LF, Kuroki K, Brown RF. Journal of Biomedical Materials Research Part A. 2010;**95A**:172-179

[117] Goodridge RD, Wood DJ, Ohtsuki C, Dalgarno KW. Acta Biomaterialia. 2007;**3**:221-231

[118] Fu QA, Rahaman MN, Bal BS, Kuroki K, Brown RF. Journal of Biomedical Materials Research Part A. 2010;**95A**:235-244

[119] Itala A, Koort J, Ylanen HO, Hupa M, Aro HT. Journal of Biomedical Materials Research Part A. 2003;**67A**:496-503

[120] Lee JH, Lee CK, Chang BS, Ryu HS, Seo JH, Hong S, Kim H. Journal of Biomedical Materials Research. 2006;**77A**:362-369

[121] Yuan HP, de Bruijn JD, Zhang XD, van Blitterswijk CA, de Groot K. Journal of Biomedical Materials Research. 2001;**58**:270-276

[122] Zhao DS, Moritz N, Vedel E, Hupa L, Aro HT. Acta Biomaterialia. 2008;**4**:1118-1122

[123] Poh PSP, Hutmacher DW, Woodruff MA, Stevens MM. Biofabrication. 2014;**6**: 029501 (1pp)

[124] Melchers S, Uesbeck T, Winter O, Eckert H, Eder D. Chemistry of Materials. 2016;**28**(10): 3254-3264

[125] Michel J, Penna M, Kochen J, Cheung H. Stem Cells International Volume. 2015, Article ID: 305217; 13 pages

[126] Koc A, Finkenzeller G, Eser Elcin A, Bjorn Stark G, Murat Elcin Y. Journal of Biomaterials Applications. 2014;**29**(5):748-760

[127] Campos DM, Anselme K, Soares GD d A. Materials Research. 2012;**15**(1):151-158

[128] Baghbani F, Moztarzadeh F, Gafari Nazari A, Razavi Kamran AH, Tondnevis F, Nezafati N, Gholipourmalekabadi M, Mozafari M. Advanced Composites Letters. 2012;**1**(21):

[129] Ebrahimi M, Botelho M. Data in Brief. 2017;**10**:93-97

[130] Joanna Wójtowicz, Joanna Leszczy, nska, Anna Chró, scicka, Anna, Slósarczyk, Zofia Paszkiewicz, Aneta Zima, Krzysztof Rozniatowski, Piotr Jele, Malgorzata Lewandowska-Szumiel. Bio-Medical Materials and Engineering. 2014; 24:1609-1623.

[131] Kent NW, Wilson RM, Blunn G, Coathup M, Karpukhina N, Onwordi L, Davis G, Quak WY, de Godoy RF, Hill R. Journal of Biomedical Materials Research Part B. 2016;**00B**:1-5

[132] Chen W, Liu J, Manuchehrabadi N, Weir MD, Zhu Z, HH X. Biomaterials. 2013;**34**:9917-9925

[133] Zeng D, Xia L, Zhang W, Huang H, Wei B, Huang Q, et al. Tissue Engineering. Part A. 2012;**18**:870-881

[134] Zhao P, Shen H, Li D, Cai Q, Yang F, Decheng W, Ma Y, Wang XYT, Duan S, Wang S. Journal of Materials Chemistry Part B. 2015;**3**:6885-6896

[135] Zhu M, Wang H, Liu J, He H, Hua X, He Q, Ye X, Shi J, Zhang L. Biomaterials. 2011;**32**: 1986-1995

[136] Kundu B, Sen PS, Basu D, Lemos A, Datta S, Soundrapandian C, Ferreira JMF. Journal of Materials Science: Materials in Medicine. 2010;**21**:2955-2969

[137] Freidman CD, Costantino PD, Takagi S, Chow LC. Journal of Biomedical Materials Research (Applied Biomaterials). 1998;**43**:428-432

[138] You CK, Oh SH, Kim JW, Choi TH, Lee SY, Kim SY. Key Engineering Materials. 2003;**240-242**:563-566

[139] Kokubo T. Bioceramics and Their Clinical Applications. 1st ed. Elsevier; 2008

[140] Goudouri OM, Kontonasaki E, Theocharidou A, Kantiranis N, Chatzistavrou X, Koidis P, Paraskevopoulos KM. Bioceramics Development and Applications. 2011;**1**:1-4

[141] Felipe et al. Journal of Periodontology. 2009;**80**(5):808-815

[142] Ali S, Farooq I, Iqbal K. The Saudi Dental Journal. 2014;**26**:1-5

Chapter 4

Graded Cellular Bone Scaffolds

Sakkadech Limmahakhun and Cheng Yan

Additional information is available at the end of the chapter

http://dx.doi.org/10.5772/intechopen.69911

Abstract

Bone scaffolds with graded porosities or graded cellular bone scaffolds are new innovations of bone replacements and biomedical bone implants, especially in cases of long-bone defects, multitissue regenerations, and functional-controlled bone prostheses. The concepts of graded cellular bone scaffolds are based on the complexity of bone characteristics (graded hierarchical structures and heterogeneity), which aims to closer replicate the multifunctions of bone tissues. The designs of graded cellular bone scaffolds are highly fascinating with the relative anatomical, biological, and mechanical similarity to the replaced bones. While it is difficult for the graded designs to replicate the actual bone models, additive manufacturing (AM) techniques with computer-aided designs successfully create well-controlled models with comparable bone properties. Potential advantages of graded cellular bone scaffolds are enormous. Graded pores can direct types of cell regenerations for multitissue regenerations. Furthermore, graded pores promote a greater load-sharing to adjacent bone tissues than conventional scaffolds do, while both mechanical properties are similar. To summarize, bone implants with graded cellular structures can be fabricated using AM techniques, and their mechanical and biological performances can be tailored by modifying the internal architectures.

Keywords: graded cellular structure, bone scaffolds, stress shielding, additive manufacturing techniques

1. Introduction

Bone tissue engineering (TE) scaffold has been considered to be a potential technology for repairing and/or replacing damaged bone tissues and organs. The main function of a scaffold serves as a three-dimensional (3D) template for cell organization and tissue development to eliminate the drawbacks of autologous and allogeneic bone transplantations [1]. The major requirements for the scaffolds are biocompatibility, suitable pore size, and porosity for cell attachment and proliferation, and an adequate mechanical strength under physiological loading conditions. Such scaffolds have been successfully utilized in broad areas including repair of long-bone and osteochondral defects, maxillofacial and spinal surgery, cranial reconstruction, and drug delivery systems.

Although bioceramics and polymers are commonly used to make bone-tissue scaffolds, their mechanical strengths are inadequate to withstand a high loading. Metallic cellular structures, however, have been attractive for application in orthopedic bone implants, since the porous architecture promotes bone anchorage and provides suitable stiffness [2]. On the other hand, dense metals, such as those used for total hip replacement (THR), are often too stiff and can shield the adjacent bone tissue and cause loosening of the implants [3]. The implants based on the concepts of bone TE scaffolds would shift the trend of using porous metallic implants.

A functionally graded structure for bone tissue engineering is a porous biomaterial where the porosity changes with a specific gradient in space [4]. This gradient porosity similarly behaves as a graded structure of bones [5]. Anatomically, graded structures of bone are illustrated with the surface cortical bone toward the inner cancellous bone. The porous structure inside the cancellous bone promotes nutrient and waste transportations for biological functions. Furthermore, the graded structures of bone are also controlled by the mechanical functions required for each area. Many studies have demonstrated that graded cellular bone TE showed the advantages to engineer material with specific structural, morphological, and mechanical properties [6, 7].

2. Fabrications of graded cellular bone scaffold

The scaffolds behave as a structure that promotes tissue formation, therefore the types of cell formations (fibroblast, chondroblast, and osteoblast) are controlled by the pore size, porosity, and surface properties of the scaffolds. The production of implant materials with high porosity allows good and fast bone growth, while the low-porosity materials can withstand early physiological mechanical stress [8].

Pompe et al. [9] reported that functionally graded material could give the implant a suitable strength to withstand the physiological loading, and that the graded porosity structure can optimize the material's response to external loading. A similar feature of graded cellular bones might prove favorable to an artificial bone implant. Bone TE scaffolds with the graded cellular structures like bone structures could be fabricated using conventional and additive manufacturing techniques [10].

2.1. Conventional techniques

The porous materials can be created using porogens as a void spacer. The possible spacers [11] used are various, depending on the particle size and removing techniques, such as sodium chloride, carbomide, poly (methyl methacrylate), magnesium, and so on. The spacers used are subsequently removed by dissolving with solvents or burning-out (**Figure 1**).

http://dx.doi.org/10.5772/intechopen.69911

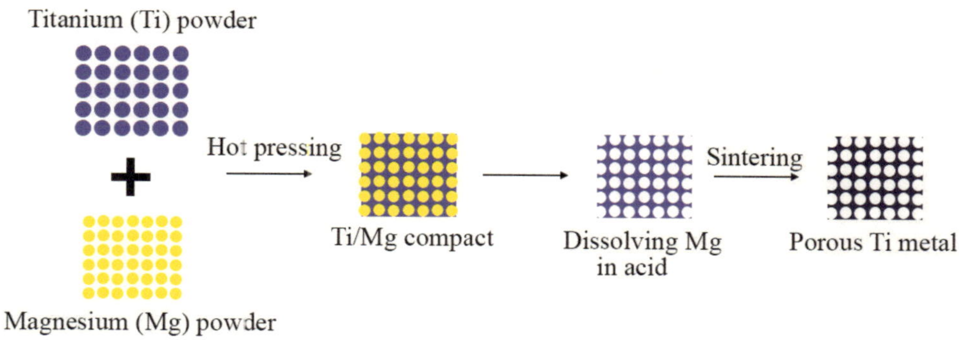

Figure 1. Schematic diagram illustrating fabrication of porous titanium with controlled porous structure and net shape.

Porous materials created with this technique use porogens to create the voids. The pore size is related to the porogen size. There are many types of porogens used as shown in **Table 1**.

Using magnesium as a spacer is suitable for porous metal fabrication during powder compaction and sintering process, since it has a relatively high strength and elastic modulus (45 GPa) and high melting temperature (650°C) compared to polymeric spacers [11]. Graded porosity scaffolds are further created as the composite-laminated materials. Material-porogen mixtures of different porosities are prepared with paste and filled into a mold layer by layer to form laminated multiporous layers (**Figure 2**) [17].

Although the pore sizes are related to sizes of a spacer, an uncontrolled microstructure contributes to internal stress concentrations located around the structural defects [18]. This technique is, therefore, applied to create biomaterials with pores in nano- to micro-scale, and is combined with an additive manufacturing technique to control the micro- to macro-scale architectures.

2.2. Additive manufacturing (AM) techniques

AM techniques have been introduced to TE, recently. The benefits of AM techniques improve the well-defined architectures of the scaffold fabrications according to the designs controlled by computer-based methods. Mimicking the porous structures of bone tissues, an internal

Materials	Porogens	Pore sizes (μm)	References
Mesoporous bioactive glasses	Methylcellulose	100	[12]
Bioglass	Polyurethane	300–600	[13]
Hydroxyapatite/tricalcium phosphate	Polyurethane		[14]
Titanium	Magnesium	100–300	[11]
Hyaluronic acid	Salt	100–600	[15]
Polylactide	Salt	600	[16]

Table 1. Porous materials created with conventional techniques.

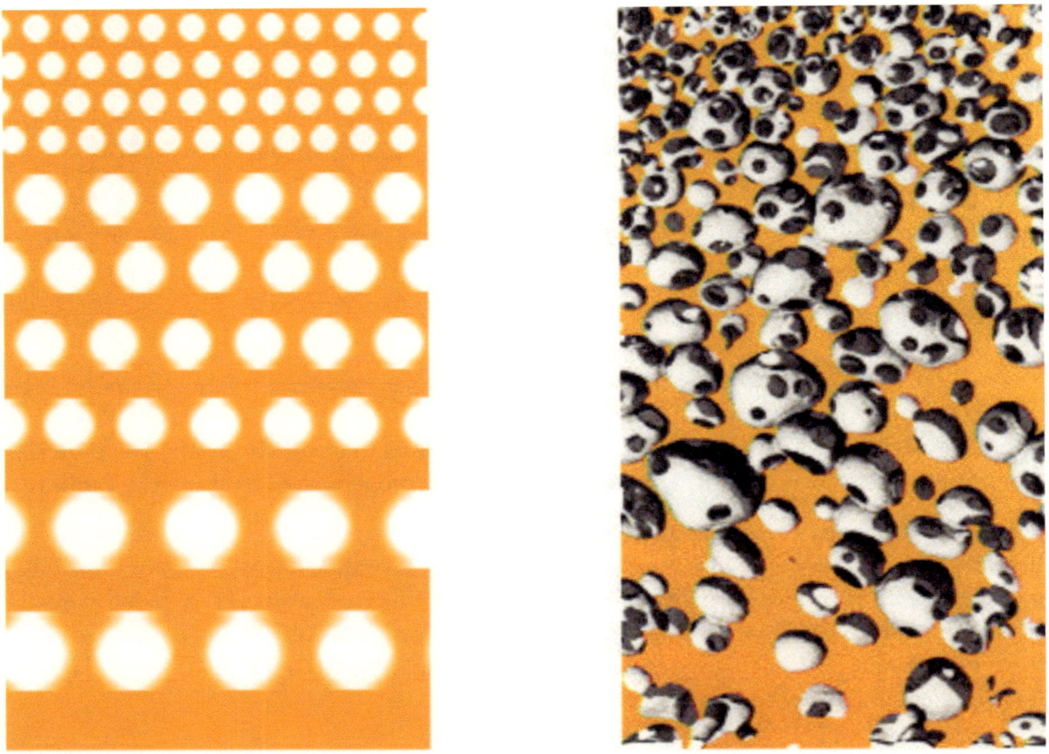

Figure 2. Diagram of material-porogen mixtures of different porosities which are filled into a mold layer by layer to form laminated multiporous layers.

architecture of the scaffolds, significantly affects nutrient diffusion, cell adhesion, and matrix deposition of the regenerated tissues. Scaffolds have to be carefully designed to match specific mechanical, mass transport, and biological requirements; however, customizing the architecture to better suit these requirements remains a challenging issue.

Computer-aided designs (CAD) and computational simulations using finite element analysis (FEA) have played a major role in the reduction of in vitro and in vivo experimental efforts and costs. Furthermore, the accuracy of FEA results is acceptable to predict a design optimization before the manufacturing. An overview of this design paradigm is reported in **Figure 3**.

The images in STL file are directly processed with computer-aided manufacturing (CAM) in order to print the specimens in layer. Selective laser melting (SLM) techniques have shown capability of 3D cellular structure production through printing of layered structures based on CAD (**Figure 4**) [19–21]. During SLM, metallic powder is melted by a scanning laser beam [21], featured by excellent reproducibility and efficiency [22]. The limitations of SLM are the quality of the print which depends on the process setting parameters. Poor setting parameters such as layer thickness, laser-beam spot, laser power, and hatching space contribute to incomplete melting of the metal powders. Furthermore, the minimum scale that can process is limit the lattice thickness and pores in hundreds micrometer. Although, the scale limitation of the print is larger than nano-scale of the real bone, these SLM-processed structures are adequately to allow bone ingrowth (>300 μm).

The methodological approaches for the design of scaffold architectures are classified into the architecture formed by the repetition of unit cells (cellular structures) or that consisting of lattices (lattice domain). This collection of unit cells is also known as computer-aided system for tissue scaffolds (CASTS). Chua et al. [23, 24] and Cheah et al. [25] have developed a parametric library of scaffold structures and an algorithm to automate the entire process of matching desired anatomical shapes in order to reduce the time-consuming processes. **Figure 5** shows examples of the unit cells used in CASTS.

Limmahakhun et al. [26] emphasized that the internal architectures of scaffolds play a crucial role in determining the overall mechanical and biological performances. It is interesting to note that the mechanical response from a cellular structure depends on the loading modes. The cubic structure [0° ± 90°] has much higher compression stiffness but lower shear and torsion stiffness than that in octahedron-type structures [±45°]. The pillar octahedron combined with the strut characteristics featured along 0° and ±45° demonstrates high compression, shear, and torsional stiffness (26.94, 18.93, and 9.52 MPa, respectively). To this end, for axially loaded applications, such as bone implants to fix long-bone defects, the cubic structure

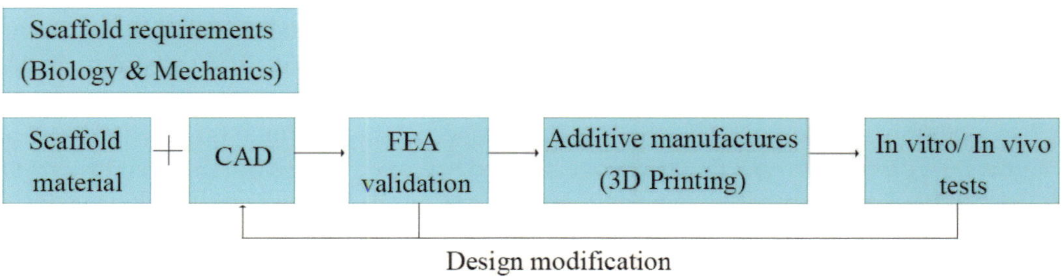

Figure 3. Flowchart steps in the design of tissue-engineering scaffolds.

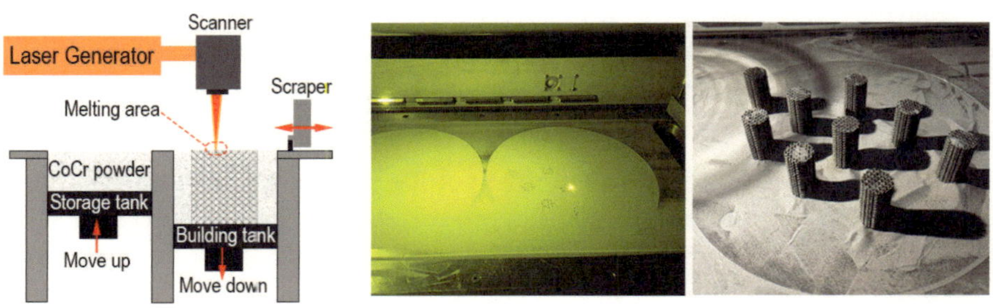

Figure 4. SLM processing techniques. Reprinted from Ref. [31] with permission from Elsevier.

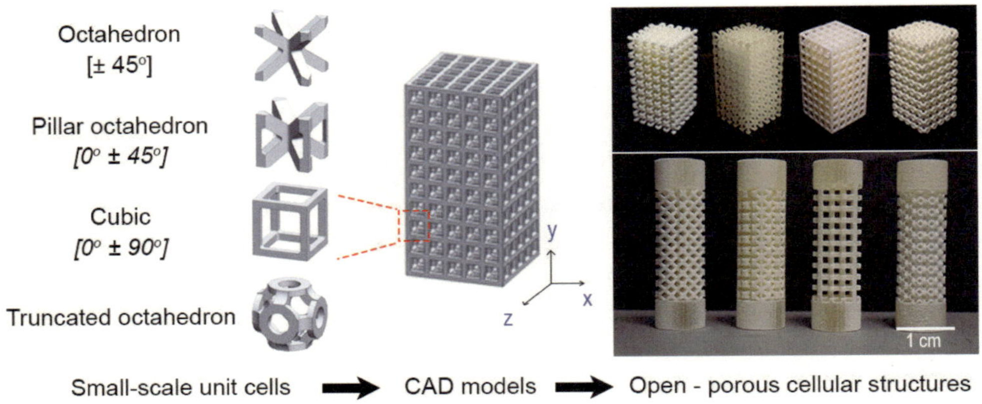

Figure 5. Unit-cell replications of bone tissue engineering scaffolds. Reprinted from Ref. [26] with permission from Elsevier.

is the best choice because of its high stiffness under compressive force. High fatigue life under compression loading was observed in cubic unit cells [19]. On the other hand, when subjected to a combined loading (compression, shear, and torsion), such as femoral hip implants, the pillar octahedron is an appropriate architecture as it provides the greatest shear and torsional stiffness and high compressive stiffness. Note that the pillar octahedral structure also has the greatest accumulated stress during the stress relaxation test [26].

Unit geometry directly controls the tortuosity, a measure of flow distance that travels in the porous materials [27]. The strut intersection in the middle of octahedral types diverts the fluid from the straight direction and slows down the flow rate (**Figure 6b**). The low flow rate when pipetting a cell suspension onto the pillar octahedral shapes would favor cell adherence (**Figure 6d–f**), while the high flow rates of cubic and truncated octahedral shape contribute to cell deposition at the bottom of the well. The greater cell proliferation of the pillar octahedron could be attributed to the greater surface areas of these polyhedral structures that are printed with 3D printing techniques, compared with the cubic and octahedron (**Figure 6a**). Together with its better cell proliferation rate, pillar octahedral structure has demonstrated balanced mechanical and biological properties.

CAD-based methods [28, 29] and implicit surface modeling (ISM) methods [18, 30] are common to create the cellular structures with porosity-graded structures, since the internal architectures are fully controlled with no hanging edge [18, 31]. For the CAD-based methods, graded structures of the pillar octahedral scaffolds could be attained by varying the architectural parameters, such as lattice diameter (t) and unit size (L) (**Figure 7**).

The graded cellular bone scaffolds with desired porosity and pore size on each location could be done using manual or automatic algorithm processes. The primitive unit cell that is recommended by the author's works is the pillar octahedral shape [26]. Our findings proved that this unit has balanced mechanical and biological performances for bone tissue regenerations. **Figure 8** shows an example of pillar octahedral unit cells with different porosities and graded patterns; axially and radially graded patterns, fabricated with cobalt chromium (CoCr) alloys

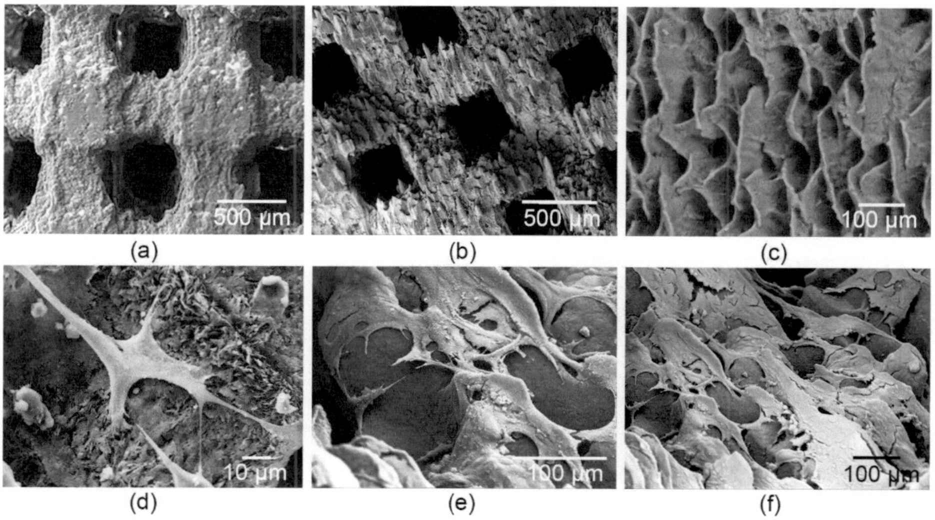

Figure 6. SEM images showing surface roughness of (a) cubic and (b) octahedral unit cells, (c) smaller pores (40 μm) on the surface after printing on the overhanging features, (d) cell adhesion and (e) and (f) cell coverage after 4 and 7 days. Reprinted from Ref. [26] with permission from Elsevier.

using SLM techniques. Manually mating the unit cells inside the graded cellular bone scaffolds could be fully controlled with no error, however it is time-consuming.

On the other hand, ISM allows scaffold architectures to be easily described using a single mathematical equation, with freedom to introduce different pore shapes and architectural

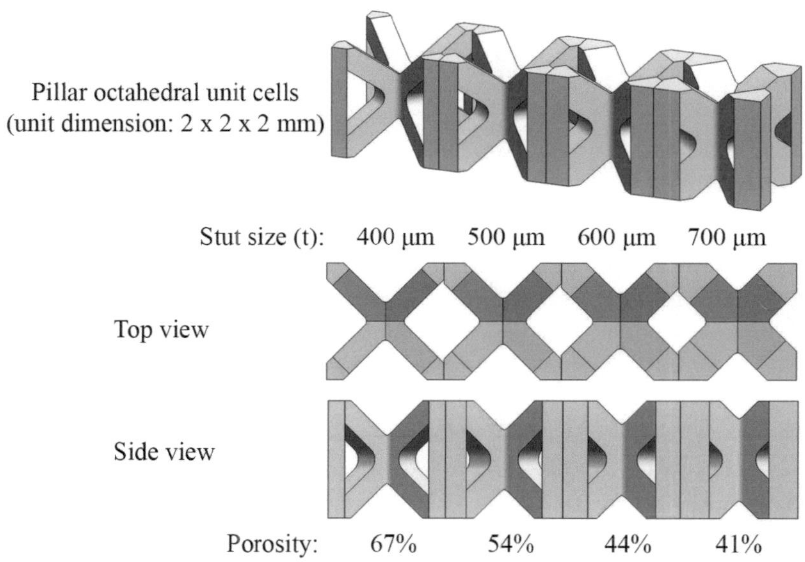

Figure 7. Assembling of different unit porosities.

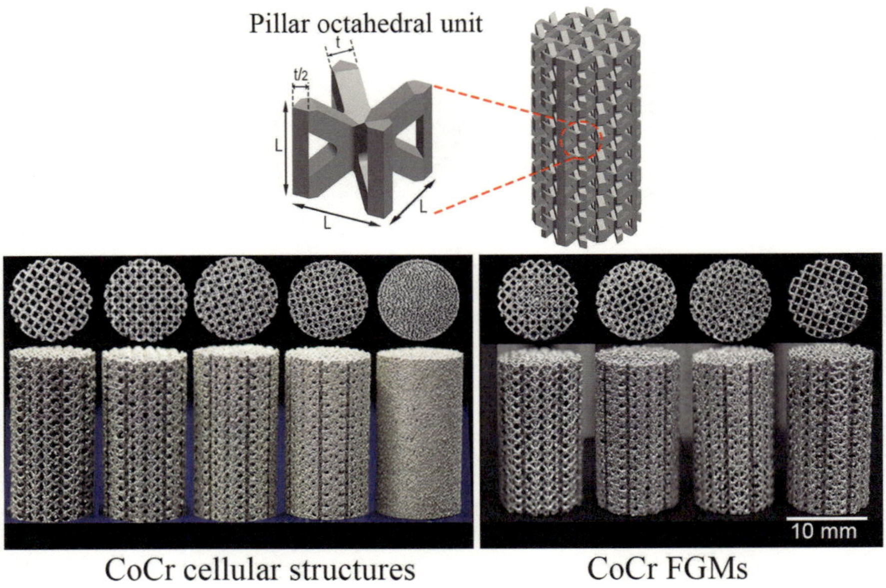

Figure 8. Cellular structures and graded cellular structures (FGMs) built by pillar octahedral unit (L and t are unit and strut size, respectively). Reprinted from [31] with permission from Elsevier.

features, including pore-size gradients. Although, automatically processing time based on the algorithm is more convenient compared to the CAD-based method, the algorithm that could perfectly serve the required graded internal architectures of the scaffolds with no internal defect is more complexity. While many of the ISM are available, both of the Schwarz's Diamond and Schoen's Gyroid shapes are preferable in promoting cell migration and tissue ingrowth [18]. Furthermore, these techniques are suitable to create the graded cellular structures.

3. The effects of graded cellular bone scaffolds

Designing an architecture of the scaffolds that mimics the complex structure of bone tissue is the new generation of bone tissue engineering [32]. Scaffolds with graded cellular structures behave similar to the bone that is replaced. Furthermore, porosity gradation affects the types of cell regeneration and promotes the environment to be more suitable for cell functions. While the biological cell interactions are directly influenced by the porosity gradation, the mechanical performance which is affected by porosity and cellular structures could be modified to match the desired functionalities. The improved tissue regeneration rate of graded cellular structures led the research community to find the possible models of bone tissue engineering. However, such model has not been developed yet.

http://dx.doi.org/10.5772/intechopen.69911

3.1. Biological cell interactions

Different layers of the tissue perform different roles in maintaining the organ functions. Each cell type has individual specific functions and is important for tissue formations. Bone, cartilage, and ligament are formed simultaneously as a bone and joint structure, as a result of tissue formation by osteoblast, chondrocyte, and fibroblast cell-types, respectively. The graded cellular bone scaffolds are critical for multitissue regenerations such as bone and cartilage tissues [33] and ligament-bone interface [34]. These cell types have obviously different environments, thus scaffolds should be tailored with different pore sizes and porosities. For example, fibroblasts (cell size 20–50 μm) can span void spaces up to 200 μm [35]. On the other hand, the preferred pore diameter for osteoblast (cell size 20–30 μm [36]) is 100–350 μm [37].

Beside the pore size, porosity and pore interconnection also facilitate bone ingrowth. The scaffold with porosity >90% and pore interconnection promotes more cell infiltration, proliferation and extracellular matrix deposition, since it has a better flow mechanic for nutrient and waste transportation [38] and allows cellular signals between interconnecting networks [39]. Furthermore, the graded cellular structures also increase the fluid permeability and flow, which enhances the cell diffusion throughout the whole scaffolds [40, 41].

Since different pore sizes affect the type of cell formations, graded cellular scaffolds potentially produce the multi-tissue grafts on the bone-ligament interface by following the phenotypic gradients that exist at the natural ligament-bone interface. In addition to the graded features in architectures, graded features in terms of different material compositions also promote biologic functions like bone materials. Two materials with spatially graded fraction of polymer and hydroxyapatite (HA), fabricated with co-electrospinning techniques, are found to be metabolically active from the study of rat bone marrow stromal cell cultures [34]. Gene expression of bone morphogenic protein-2 and osteopontin was elevated on mineral-containing regions as compared to regions without mineral, which confirmed osteoblastic phenotypic maturation of this polymeric-HA graded scaffolds by day 28.

Computational study has been utilized to optimize the best porosity distribution in functionally graded scaffolds for bone tissue engineering [42]. Based on the porosity distribution law, the graded scaffolds with tri-linear law promote larger amounts of bone formation compared to the models of bi-linear, linear, and constant laws. Alternatively, the more the complexity of porosity distribution laws (i.e., with increasing number of coefficients, Ai), the better the scaffold geometry can be tailored. A larger number of design variables increases the probability that optimizes a geometry to match the specific boundary and loading conditions.

The effect of the loading conditions appears more critical. Boccaccio et al. [42] showed that the loading conditions are essential in determining optimal porosity distribution. For a pure compression loading, it was predicted that the changes of the pore dimension are marginal, and using a graded cellular bone scaffold allows the formation of amounts of bone slightly larger than those obtainable with a homogeneous porosity scaffold. For a pure shear loading, instead, bone formations of graded cellular bone scaffolds are significant compared to a homogeneous porosity scaffold. While increasing the pore diameters leads to an increased value of the scaffold Young's modulus, increasing a porosity distribution law makes a scaffold generate larger amounts of bone formations. Graded porosity characteristics contribute to optimized loading distributions, which enhances sensing signal to maximize osteoblastic cellular activities, as termed as mechanobiologic signal.

3.2. Mechanical performances

A scaffold should have the mechanical properties sufficient to maintain integrity until the new tissue regeneration. Bioceramics (hydroxyapatite, bioglass, etc.) and metals (titanium, tantalum, cobalt chrome, etc.) have been commonly used as a biomaterial for bone tissue regeneration. The limitation of bioceramics is their brittleness and contributes to easily break after replacements. While greater stiffness, endurance, and strength of metals are preferred, the higher stiffness of metals shields the stress distributed to the adjacent bone tissue and leads to bone resorption called "stress shielding." Therefore, the mechanical properties of the scaffolds should match that of the native tissue to both prevent stress shielding and give proper mechanical performances.

Despite the types of materials used for fabrications, the relative modulus of cellular materials (E/E_s) has a power law relation to the relative density (ρ_o/ρ_s), based on Gibson and Ashby model [43]:

$$\frac{E}{Es} = \varphi \left(\frac{\rho o}{\rho s}\right)^n \quad (1)$$

where E and E_s are the moduli of base and cellular materials and ρ_o and ρ_s are the density of base and cellular materials. This relationship depends on the unknown coefficients (φ and n) which are a factor of each cellular unit.

The cellular structures with various relative densities can be created by varying the diameter of the beam thickness (**Figure 7**). The moduli and strength of those cellular scaffolds can be predicted upon the relative density, as shown in an example of cobalt chrome (CoCr) cellular structures (**Figure 9**). It is clear that the elastic modulus (stiffness), yield stress, and ultimate compressive strength of cellular structures increase with decrease in the porosity. The CoCr cellular structures of pillar octahedron have stiffness and compressive strengths between 2.33–3.14 GPa and 113–523 MPa, respectively, which are comparable to those of cortical bone tissues (2.73–17 GPa [44] and 100–150 MPa [45]). In addition, these CoCr cellular structures also demonstrated a greater energy absorption (24.6–116.86 MJ/m^3) than bone tissues.

Unfortunately, the quality of the SLM-built is a major concern that affects the scaffold reproductions. Although FEA could predict the modulus of cellular scaffolds, the accuracy of the prediction is low [46]. Poor correlation between FEA and physical testing results from an incomplete melting of the metal powders. Internal voids of the printed scaffolds increase the stress concentration, and crack may occur, weakening the scaffold constructs (**Figure 10**). Completely selective-melting process in each layer of the manufacture plays a main role to minimize such error of the prediction.

http://dx.doi.org/10.5772/intechopen.69911

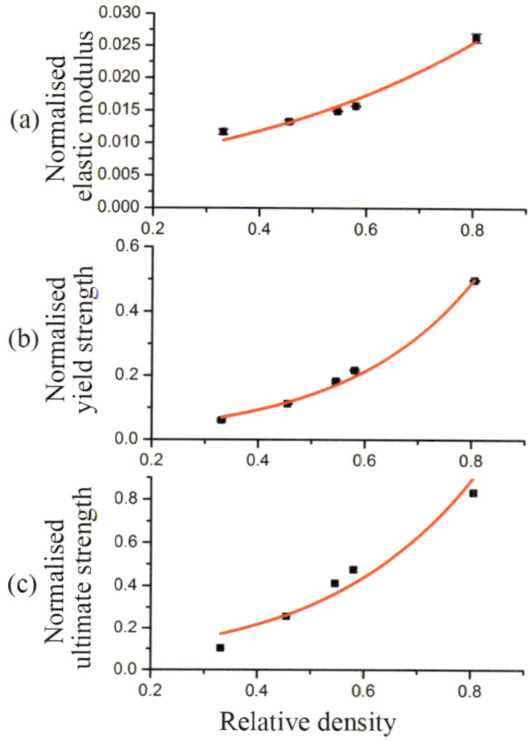

Figure 9. Normalized elastic modulus (a), normalized yield strength (b), normalized ultimate compressive strength (c) of CoCr cellular structures vs. relative density. Reprinted from [31] with permission from Elsevier.

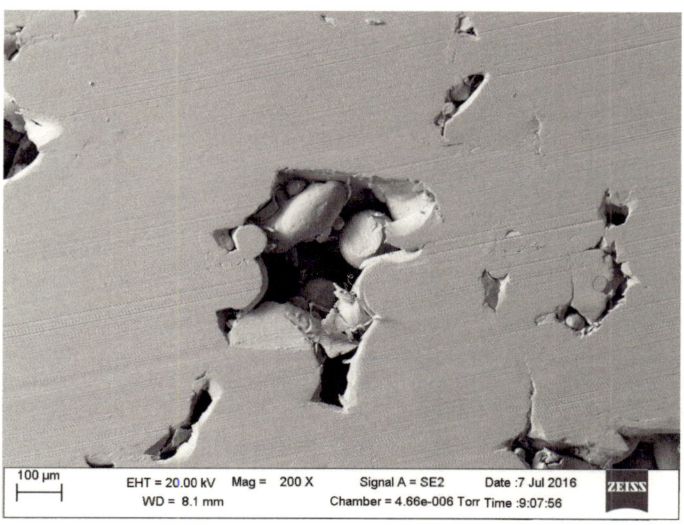

Figure 10. Internal voids of the CoCr laser-melted specimens.

The mechanical properties of titanium alloys fabricated with SLM have been extensively studied under static and fatigue loadings [19, 21, 47–51]. Although titanium cellular structures have similar stiffness as bone tissues, fatigue is still a technical issue, which mainly resulted from poor design of the internal architectures [19, 52]. Naturally, the implant is expected to have similar properties as bone. Structural gradient has been observed in bone tissues, depending on required functions, such as load-bearing capacity and biological properties [5]. However, under axial compression, there is no benefit of graded cellular bone scaffolds over a uniformed bone scaffold with the same relative density [31]. **Figure 11** shows that CoCr graded cellular structures have similar elastic modulus, yield strength, and compressive strength as the uniformed cellular structures. CoCr cellular scaffolds and graded cellular scaffolds with the porosity ranging from 40 to 70% exhibit elastic modulus around 2.7–3.1 GPa, which is in the same level of bone stiffness [44, 53]. Therefore, mechanical properties of the cellular bone scaffolds are mainly related to the relative density rather than the porosity-graded characteristics.

The opposite findings were noticed with the bioceramic graded cellular scaffolds. Wang et al. [17] studied a novel calcium polyphosphate bioceramic scaffold with a graded pore structure similar to the bimodal structure of cortical and cancellous bones (**Figure 12**). The compressive strength of porosity-graded calcium polyphosphate (PG-CPP) scaffolds was better than that of homogeneous calcium polyphosphate (H-CPP) scaffolds, which was significant ($p < 0.05$) at each time point. This fact is also noted that, after 28 days of degradation, the compressive strength of PG-CPP scaffolds was even greater than that of primary H-CPP.

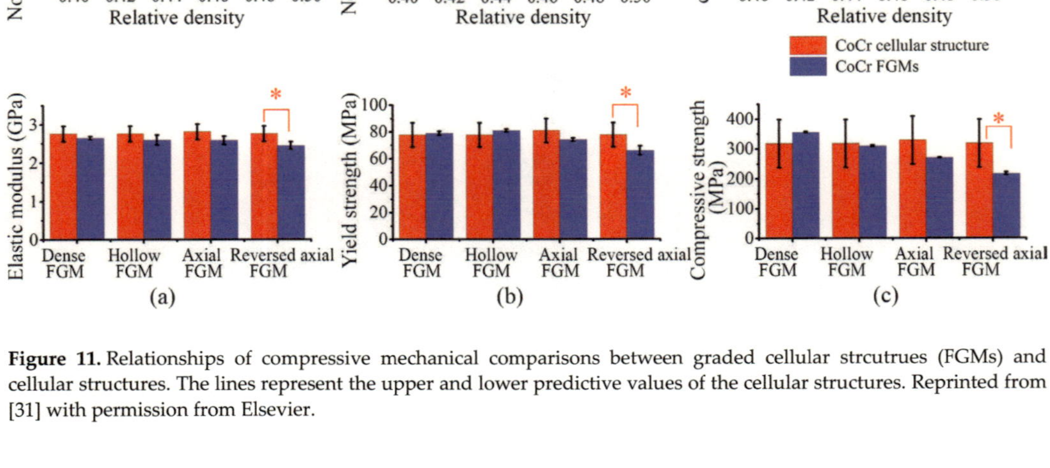

Figure 11. Relationships of compressive mechanical comparisons between graded cellular strcutrues (FGMs) and cellular structures. The lines represent the upper and lower predictive values of the cellular structures. Reprinted from [31] with permission from Elsevier.

http://dx.doi.org/10.5772/intechopen.69911

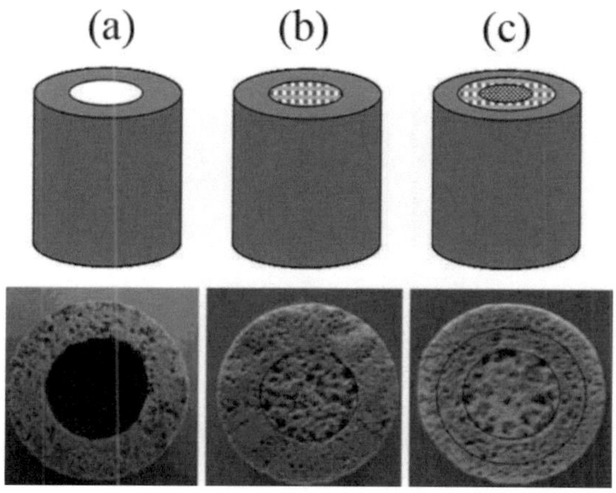

Figure 12. Bioceramic graded cellular scaffolds with different graded porosities: (a) porous scaffold with a hollow center, (b) two graded porous layers, and (c) three graded porous layers [17]. Reprinted with the permission of a creative commons attribution license.

It is still controversial whether the graded cellular scaffolds can improve the mechanical properties of the constructs compared to homogeneous scaffolds. The bioceramic scaffolds with a porosity-graded structure in this study have a much better mechanical property. PG-CPP gives a prolonged deflection besides the elastic deformation which indicates that the porosity-graded scaffolds exhibit a different fracture behavior from that of the homogeneous scaffolds. Furthermore, PG-CPP exhibits nonbrittle fractures whereas the H-CPP fractures catastrophically. Hence, it may be inferred that the mechanical properties of the porosity-graded scaffold can be substantially improved by a graded porous structure [17]. To end this, types of materials used for fabrications, internal structures of the scaffolds, and the design of graded scaffolds play an important role to control the mechanical properties of the graded cellular scaffolds.

Although both homogeneous and porosity-graded metallic bone scaffolds show a reasonable mechanical strength, the benefits of cellular graded structures optimized the functions required for such scaffold. Metallic cellular structures have lighter weight than solid metals, since an introduction of pores inside the materials. The stiffness of cellular structures can be tailored to match that of bone, according to the design and density of the scaffold's architectures. The stiffness of the cellular bone scaffolds is related to the porosity, following a nonlinear relationship as reported by Gibson and Ashby [43]. Therefore, the cellular bone scaffolds share more of the stress to an adjacent bone tissue than the solid counterpart due to the lower stiffness of the scaffolds.

Basically, the cellular structural implants with low stiffness properties reduce the peri-prosthetic bone-stress shielding, yet increase the bone-implant interface failure. **Figure 13** shows that more peri-implant bone-stress shielding occurs with the high-stiffed implant, and finally increases higher risks of the implant loosening. The initial stability of a femoral stem is necessary for biological bony ingrowth, which can be secured by minimizing relative micromotion

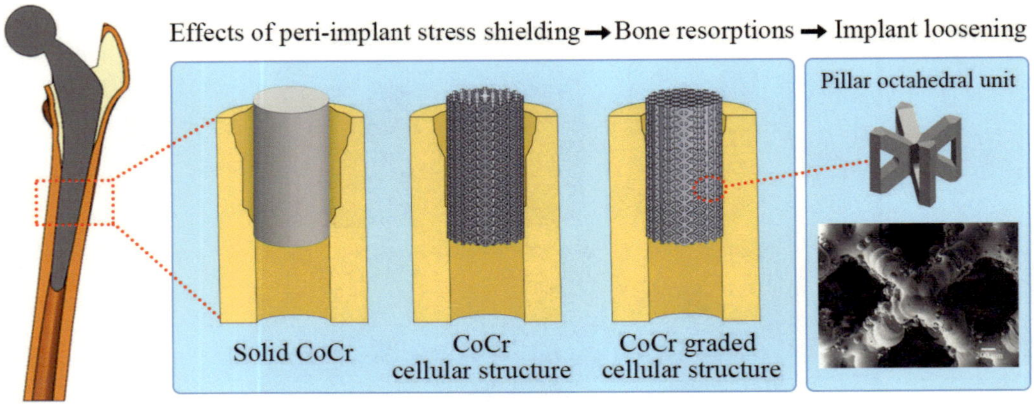

Figure 13. Influences of graded cellular structures on bone resorption. Reprinted from Ref. [31] with permission from Elsevier.

at the bone-implant interfaces. Excessive micromotions (>150 μm) allow fibrous connective tissue to grow, which prevents bone ingrowth between the contact surfaces and leads to aseptic loosening and failure of the implant [54–56]. Although homogeneous cellular implants increase stress sharing to peri-prosthetic bone due to lower construct stiffness, the greater interface failure due to excessive micromotions (>150 μm) adversely causes an initial implant instability and inhibits bone osseointegration.

According to computational study [57], the porous femoral stem with uniform relative density of 50% is approximately three times more flexible than the titanium stem. This implant can qualitatively simulate the behavior of an implant made out of tantalum foam using the well-defined cellular structures. The amount of bone resorption and the interface failure index of this stem are about 34% and 2.87, respectively, and the interface failure is maximum (0.71) at the edge of the proximal region. Compared to the solid titanium implant, the amount of bone resorption decreases by 50%, whereas the maximum interface failure increases about 40%. This shows that a decrease in the implant stiffness with uniform porosity distribution aiming at reducing bone resorption has the undesirable effect of increasing the risk of interface failure at the proximal region.

Kuiper and Huiskes [58] showed that the use of a graded material in an orthopedic stem can lead to a reduction of both stress shielding and bone-implant interface failure. The bone resorption and interface failure of graded cellular stems are 16% and 1.15, respectively [57]. The peak value of the local interface failure is 0.25. Compared to the titanium stem, both the amount of bone resorption and the peak of interface failure were decreased by 76 and 50%, respectively. With respect to the uniformly distributed cellular implant, the decrease in bone resorption and interface failure peak is of 53 and 65%, respectively. A graded cellular implant with optimized relative density distribution is thus capable of reducing concurrently both the conflicting objective functions. In particular, bone resorption reduces as a result of the cellular material which makes the implant more compliant; the interface stress and micromotions, on the other hand, are minimized by the optimized gradients of cellular material.

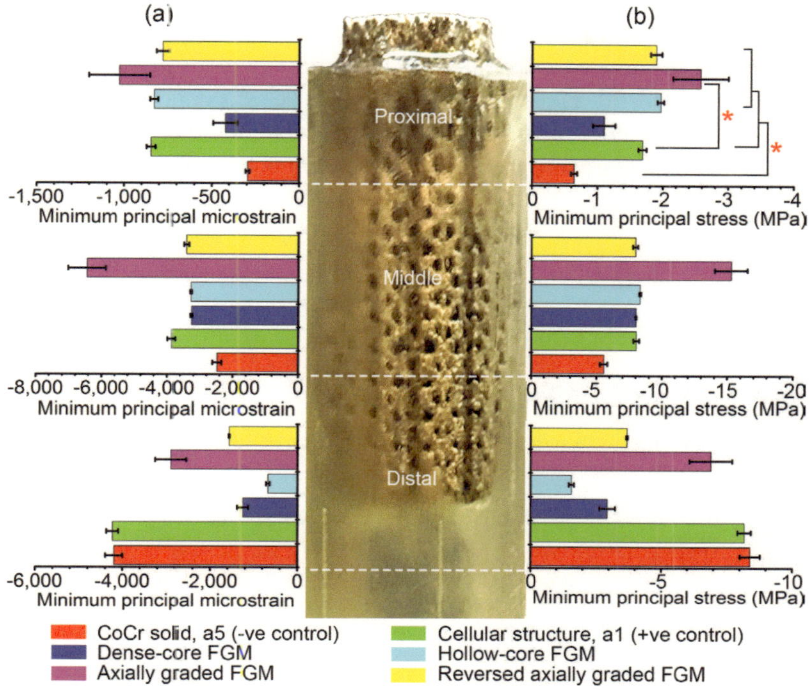

Figure 14. Strain (a) and stress (b) distributions along the proximal, middle, and distal parts of the tube surfaces under compression loads. Reprinted from Ref. [31] with permission from Elsevier.

Furthermore, for designing the scaffolds with functionally graded structures to mimic the graded structures of the host bone, the stress sharing to the adjacent bone is increased around 50% compared to the uniformed cellular bone scaffolds [31]. The degree of proximal bone-stress sharing depends on a porosity-graded orientation. In case of intramedullary implants, such as femoral stems, the scaffolds with porosity gradient along a longitudinal plane present the maximum stress distribution to the proximal bone (**Figure 14**).

Graded cellular implants are functionally designed to match the mechanical and morphological properties of bones. While the graded metallic cellular implants exhibit mechanical properties similar to the uniformed cellular implants, the graded one is more capable of preventing bone-stress shielding and promoting larger amounts of bone ingrowth to the implants. The optimized designs of biomimetic graded cellular implants are complex depending on the functional objectives of the constructs.

4. Conclusion

The graded cellular bone scaffolds show logical concepts for bone tissue engineering. Cellular structures with graded pore sizes and porosities could mimic the graded structure in bones [5]. The advantage of graded cellular structures over the uniformed cellular structures is that the former provide more realistic environment for biological and mechanical functions. For stress-sharing orthopedic applications, the axially graded cellular structure demonstrated balanced mechanical performances and maximized proximal stress transfers around a peri-implant material [31]. Instead of using the uniformed cellular structures, we believe incorporating graded cellular structures in a structure like femoral prosthesis that will improve the load distribution in adjacent bones, greater bone osteointegration, and optimum nutrient permeability of the components [17].

Author details

Sakkadech Limmahakhun[1] and Cheng Yan[2*]

*Address all correspondence to: c2.yan@qut.edu.au

1 Department of Orthopaedic Surgery, Faculty of Medicine, Chiang Mai University, Thailand

2 School of Chemistry, Physics and Mechanical Engineering, Queensland University of Technology, Australia

References

[1] Van Cleynenbreugel T, Van Oosterwyck H, Vander Sloten J, Schrooten J. Trabecular bone scaffolding using a biomimetic approach. Journal of Materials Science: Materials in Medicine. 2002;**13**(12):1245-1249

[2] Mullen L, Stamp RC, Fox P, Jones E, Ngo C, Sutcliffe CJ. Selective laser melting: A unit cell approach for the manufacture of porous, titanium, bone in-growth constructs, suitable for orthopedic applications. II. Randomized structures. Journal of Biomedical Materials Research. Part B: Applied Biomaterials. 2010;**92**(1):178-188

[3] Krishna BV, Bose S, Bandyopadhyay A. Low stiffness porous Ti structures for load-bearing implants. Acta Biomaterialia. 2007;**3**(6):997-1006

[4] Baino F, Fiorilli S, Vitale-Brovarone C. Bioactive glass-based materials with hierarchical porosity for medical applications: Review of recent advances. Acta Biomaterialia. 2016;**42**:18-32

[5] Leong KF, Chua CK, Sudarmadji N, Yeong WY. Engineering functionally graded tissue engineering scaffolds. Journal of the Mechanical Behavior of Biomedical Materials. 2008;**1**(2):140-152

[6] Ajdari A, Canavan P, Nayeb-Hashemi H, Warner G. Mechanical properties of functionally graded 2-D cellular structures: A finite element simulation. Materials Science and Engineering A. 2009;**499**(1-2):434-439

[7] Chua CK, Leong KF, Sudarmadji N, Liu MJJ, Chou SM. Selective laser sintering of functionally graded tissue scaffolds. MRS Bulletin. 2011;**36**(12):1006-1014

[8] Tampieri A, Sprio S, Sandri M, Valentini F. Mimicking natural bio-mineralization processes: A new tool for osteochondral scaffold development. Trends in Biotechnology. 2011;**29**(10):526-535

[9] Pompe W, Worch H, Epple M, Friess W, Gelinsky M, Greil P, Hempel U, Scharnweber D, Schulte K. Functionally graded materials for biomedical applications. Materials Science and Engineering A. 2003;**362**(1-2):40-60

[10] Miao X, Sun D. Graded/gradient porous biomaterials. Materials. 2010;**3**(1):26-47

[11] Kim SW, Jung H-D, Kang M-H, Kim H-E, Koh Y-H, Estrin Y. Fabrication of porous titanium scaffold with controlled porous structure and net-shape using magnesium as spacer. Materials Science & Engineering C: Materials for Biological Applications. 2013;**33**(5):2808-2815

[12] Hsu FY, Lu MR, Weng RC, Lin HM. Hierarchically biomimetic scaffold of a collagen-mesoporous bioactive glass nanofiber composite for bone tissue engineering. Biomedical Materials (Bristol). 2015;**10**(2):2251-2256

[13] Vitale-Brovarone C, Baino F, Verné E. Feasibility and tailoring of bioactive glass-ceramic scaffolds with gradient of porosity for bone grafting. Journal of Biomaterials Applications. 2010;**24**(8):693-712

[14] Hsu YH, Turner IG, Miles AW. Fabrication of porous bioceramics with porosity gradients similar to the bimodal structure of cortical and cancellous bone. Journal of Materials Science: Materials in Medicine. 2007;**18**(12):2251-2256

[15] Karageorgiou V, Kaplan D. Porosity of 3D biomaterial scaffolds and osteogenesis. Biomaterials. 2005;**26**(27):5474-5491

[16] Taboas JM, Maddox RD, Krebsbach PH, Hollister SJ. Indirect solid free form fabrication of local and global porous, biomimetic and composite 3D polymer-ceramic scaffolds. Biomaterials. 2003;**24**(1):181-194

[17] Wang QB, Wang QG, Wan CX. Preparation and evaluation of a biomimetic scaffold with porosity gradients in vitro. Anais da Academia Brasileira de Ciencias. 2012;**84**(1):9-16

[18] Giannitelli SM, Accoto D, Trombetta M, Rainer A. Current trends in the design of scaffolds for computer-aided tissue engineering. Acta Biomaterialia. 2014;**10**(2):580-594

[19] Amin Yavari S, Ahmadi SM, Wauthle R, Pouran B, Schrooten J, Weinans H, Zadpoor AA. Relationship between unit cell type and porosity and the fatigue behavior of selective laser melted meta-biomaterials. Journal of the Mechanical Behavior of Biomedical Materials. 2015;**43**:91-100

[20] Hazlehurst KB, Wang CJ, Stanford M. Evaluation of the stiffness characteristics of square pore CoCrMo cellular structures manufactured using laser melting technology for potential orthopaedic applications. Materials & Design. 2013;**51**:949-955

[21] Mullen L, Stamp RC, Brooks WK, Jones E, Sutcliffe CJ. Selective laser melting: A regular unit cell approach for the manufacture of porous, titanium, bone in-growth constructs,

suitable for orthopedic applications. Journal of Biomedical Materials Research Part B: Applied Biomaterials. 2009;**89**(2):325-334

[22] Mengucci P, Barucca G, Gatto A, Bassoli E, Denti L, Fiori F, Girardin E, Bastianoni P, Rutkowski B, Czyrska-Filemonowicz A. Effects of thermal treatments on microstructure and mechanical properties of a Co-Cr-Mo-W biomedical alloy produced by laser sintering. Journal of the Mechanical Behavior of Biomedical Materials. 2016;**60**:106-117

[23] Chua CK, Leong KF, Cheah CM, Chua SW. Development of a tissue engineering scaffold structure library for rapid prototyping. Part 1: Investigation and classification. International Journal of Advanced Manufacturing Technology. 2003;**21**(4):291-301

[24] Chua CK, Leong KF, Cheah CM, Chua SW. Development of a tissue engineering scaffold structure library for rapid prototyping. Part 2: Parametric library and assembly program. International Journal of Advanced Manufacturing Technology. 2003;**21**(4):302-312

[25] Cheah CM, Chua CK, Leong KF, Cheong CH, Naing MW. Automatic algorithm for generating complex polyhedral scaffold structures for tissue engineering. Tissue Engineering. 2004;**10**(3-4):595-610

[26] Limmahakhun S, Oloyede A, Sitthiseripratip K, Xiao Y, Yan C. 3D-printed cellular structures for bone biomimetic implants. Additive Manufacturing. 2017;**15**:93-101

[27] Eshghinejadfard A, Daróczy L, Janiga G, Thévenin D. Calculation of the permeability in porous media using the lattice Boltzmann method. International Journal of Heat and Fluid Flow. 2016;**13**(29):1-11

[28] Sudarmadji N, Tan JY, Leong KF, Chua CK, Loh YT. Investigation of the mechanical properties and porosity relationships in selective laser-sintered polyhedral for functionally graded scaffolds. Acta Biomaterialia. 2011;**7**(2):530-537

[29] Sudarmadji N, Chua CK, Leong KF. The development of computer-aided system for tissue scaffolds (CASTS) system for functionally graded tissue-engineering scaffolds. In: Liebschner MAK, editor. Computer-Aided Tissue Engineering. Totowa, NJ: Humana Press; 2012. p. 111-123

[30] Yoo D. New paradigms in internal architecture design and freeform fabrication of tissue engineering porous scaffolds. Medical Engineering and Physics. 2012;**34**(6):762-776

[31] Limmahakhun S, Oloyede A, Sitthiseripratip K, Xiao Y, Yan C. Stiffness and strength tailoring of cobalt chromium graded cellular structures for stress-shielding reduction. Materials & Design. 2017;**114**:633-641

[32] Bretcanu O, Samaille C, Boccaccini AR. Simple methods to fabricate Bioglass®-derived glass-ceramic scaffolds exhibiting porosity gradient. Journal of Materials Science. 2008;**43**(12):4127-4134

[33] Sherwood JK, Riley SL, Palazzolo R, Brown SC, Monkhouse DC, Coates M, Griffith LG, Landeen LK, Ratcliffe A. A three-dimensional osteochondral composite scaffold for articular cartilage repair. Biomaterials. 2002;**23**(24):4739-4751

[34] Samavedi S, Guelcher SA, Goldstein AS, Whittington AR. Response of bone marrow stromal cells to graded co-electrospun scaffolds and its implications for engineering the ligament-bone interface. Biomaterials. 2012;**33**(31):7727-7735

[35] Sun T, Norton D, Ryan AJ, MacNeil S, Haycock JW. Investigation of fibroblast and keratinocyte cell-scaffold interactions using a novel 3D cell culture system. Journal of Materials Science: Materials in Medicine. 2007;**18**(2):321-328

[36] Oota Y, Ono K, Miyazima S. 3D modeling for sagittal suture. Physica A: Statistical Mechanics and its Applications. 2006;**359**(1-4):538-546

[37] Yang S, Leong KF, Du Z, Chua CK. The design of scaffolds for use in tissue engineering. Part I. Traditional factors. Tissue Engineering. 2001;**7**(6):679-689

[38] Sun W, Darling A, Starly B, Nam J. Computer-aided tissue engineering: Overview, scope and challenges. Biotechnology and Applied Biochemistry. 2004;**39**(1):29-47

[39] Martin I, Wendt D, Heberer M. The role of bioreactors in tissue engineering. Trends in Biotechnology. 2004;**22**(2):80-86

[40] Melchels FPW, Barradas AMC, van Blitterswijk CA, de Boer J, Feijen J, Grijpma DW. Effects of the architecture of tissue engineering scaffolds on cell seeding and culturing. Acta Biomaterialia. 2010;**6**(11):4208-4217

[41] Sobral JM, Caridade SG, Sousa RA, Mano JF, Reis RL. Three-dimensional plotted scaffolds with controlled pore size gradients: Effect of scaffold geometry on mechanical performance and cell seeding efficiency. Acta Biomaterialia. 2011;**7**(3):1009-1018

[42] Boccaccio A, Uva AE, Fiorentino M, Mori G, Monno G. Geometry design optimization of functionally graded scaffolds for bone tissue engineering: A mechanobiological approach. PLoS One. 2016;**11**(1):1-20

[43] Gibson LJ, Ashby MF. Cellular solids: Structure and properties. 2nd ed. Cambridge (NY): Cambridge University Press; 1997. 1-510 p

[44] Bayraktar HH, Morgan EF, Niebur GL, Morris GE, Wong EK, Keaveny TM. Comparison of the elastic and yield properties of human femoral trabecular and cortical bone tissue. Journal of Biomechanics. 2004;**37**(1):27-35

[45] Roohani-Esfahani SI, Newman P, Zreiqat H. Design and fabrication of 3D printed scaffolds with a mechanical strength comparable to cortical bone to repair large bone defects. Scientific Reports. 2016;**6**:1-8

[46] Chen WM, Xie YM, Imbalzano G, Shen J, Xu S, Lee SJ, Lee PVS. Lattice Ti structures with low rigidity but compatible mechanical strength: Design of implant materials for trabecular bone. International Journal of Precision Engineering and Manufacturing. 2016;**17**(6):793-799

[47] Merkt S, Hinke C, Bültmann J, Brandt M, Xie YM. Mechanical response of TiAl6V4 lattice structures manufactured by selective laser melting in quasistatic and dynamic compression tests. Journal of Laser Applications. 2014;**27**(S1):1-6

[48] Gorny B, Niendorf T, Lackmann J, Thoene M, Troester T, Maier HJ. In situ characterization of the deformation and failure behavior of non-stochastic porous structures processed by selective laser melting. Materials Science and Engineering A. 2011;**528**(27):7962-7967

[49] Wauthle R, Ahmadi SM, Amin Yavari S, Mulier M, Zadpoor AA, Weinans H, Van Humbeeck J, Kruth JP, Schrooten J. Revival of pure titanium for dynamically loaded porous implants using additive manufacturing. Materials Science and Engineering C. 2015;**54**:94-100

[50] Guillén T, Ohrndorf A, Tozzi G, Tong J, Christ HJ. Compressive fatigue behavior of bovine cancellous bone and bone analogous materials under multi-step loading conditions. Advanced Engineering Materials. 2012;**14**(5):B199–B207

[51] Babaee S, Jahromi BH, Ajdari A, Nayeb-Hashemi H, Vaziri A. Mechanical properties of open-cell rhombic dodecahedron cellular structures. Acta Materialia. 2012;**60**(6-7):2873-2885

[52] Weißmann V, Bader R, Hansmann H, Laufer N. Influence of the structural orientation on the mechanical properties of selective laser melted Ti6Al4V open-porous scaffolds. Materials & Design. 2016;**95**:188-197

[53] Öhman C, Baleani M, Perilli E, Dall'Ara E, Tassani S, Baruffaldi F, Viceconti M. Mechanical testing of cancellous bone from the femoral head: Experimental errors due to off-axis measurements. Journal of Biomechanics. 2007;**40**(11):2426-2433

[54] Bieger R, Ignatius A, Decking R, Claes L, Reichel H, Dürselen L. Primary stability and strain distribution of cementless hip stems as a function of implant design. Clinical Biomechanics. 2012;**27**(2):158-164

[55] Lewallen EA, Riester SM, Bonin CA, Kremers HM, Dudakovic A, Kakar S, Cohen RC, Westendorf JJ, Lewallen DG, Van Wijnen AJ. Biological strategies for improved osseointegration and osteoinduction of porous metal orthopedic implants. Tissue Engineering—Part B: Reviews. 2015;**21**(2):218-230

[56] Gortchacow M, Wettstein M, Pioletti DP, Müller-Gerbl M, Terrier A. Simultaneous and multisite measure of micromotion, subsidence and gap to evaluate femoral stem stability. Journal of Biomechanics. 2012;**45**(7):1232-1238

[57] Khanoki SA, Pasini D. Multiscale design and multiobjective optimization of orthopedic hip implants with functionally graded cellular material. Journal of Biomechanical Engineering: Transactions of the ASME. 2012;**134**(3):1-10

[58] Kuiper JH, Huiskes R. Mathematical optimization of elastic properties: Application to cementless hip stem design. Journal of Biomechanical Engineering: Transactions of the ASME. 1997;**119**(2):166-174

Clinical Application of Macroporous Ceramic to Promote Bone Healing in Veterinary Clinical Cases

Pedro Olivério Pinho, José Miguel Campos,
Carla Mendonça, Ana Rita Caseiro,
José Domingos Santos, Ana Colette Maurício and
Luís Miguel Atayde

Additional information is available at the end of the chapter

http://dx.doi.org/10.5772/intechopen.70341

Abstract

Autogenous cancellous bone is the most effective material in promoting rapid healing and still considered the "gold standard" for evaluation of bone graft substitutes. The harvesting process to collect autologous bone is associated with complications and its availability is limited. Allogenic bone is another alternative with osteoconductive properties, and it act as a structural graft when applied in defects of long bones, but some disadvantages are also associated. The development of the bone grafts substitutes has gained tremendous popularity over the last two decades. Osteoconductive materials act as scaffolds were cells from the surrounding tissues with osteogenic capacities can lay new bone, and may be produced using different types of agents, such as bone products, ceramics, bioactive glasses, collagen, polymers, and composites. Bonelike® is produced by the incorporation of P_2O_5–CaO glass-based system within a hydroxyapatite matrix. Bonelike® Poro consists of polygonal granules with 2000–2800 μm and 4000–5600 μm of diameter with pore sizes range from 100 to 400 μm. This chapter will focus on the different techniques were this ceramic synthetic bone substitute was used to promote bone regeneration with special attention in both experimental and clinical cases of veterinary orthopaedics in dogs and cats, horses and ruminants, including results obtained with Bonelike®.

Keywords: bone regeneration, bone graft, bone substitutes, synthetic bone substitutes, orthopaedics, veterinary, clinical cases

1. Introduction

The bone healing process shares many similarities with soft tissues' healing, but, in contrast, the bone is the only tissue that has the capacity of healing without scar formation. Nevertheless, incomplete healing may occur, and the tissue other than the bone may be found at the healing site. Bone healing depends on adequate vascular supply and stability of the bone fragments. Proper bone healing can only occur after restoration of mechanical stability to achieve an ideal biomechanical environment. Many clinical situations may require additional osteosynthesis surgical procedures to acquire the biomechanical stability and immobilization necessary for correct bone regeneration and future functional recovery. Intrinsic mechanisms present unique histological characteristics that appear isolated or in association, depending on the bone fragment mobility [1–4]. The amount of bone *callus* produced depends on the stability of the fracture sides and usually increases with fracture instability. Spontaneous healing of complete fractures often occurs with highly unstable fragment ends and high interfragmentary strain (deformation occurring at the fracture site relative to the size of the gap), to a limit of 2% strain [1, 4]. Fracture healing under restricted motion induces an initial reduced amount of bone *callus* formation. This type of healing relies on fracture configuration and the implant's rigidity and may be achieved by external coaptation of the fracture or after gliding implant fixation with intramedullary pins and nails. With these surgical reconstructive methods, the amount of the *callus* produced is highly variable and dependent on the fracture configuration and the rigidity of the frame used [1, 3]. When the fixation is performed with a bone plate, the amount of *callus* formation will be different if the plate is not applied on the tension side of the bone, if the reduction is not perfect or when the plate lacks rigidity. In stable fractures using a rigid plate for osteosynthesis, there is significantly decreased *callus* formation between the bone fragments [1, 2]. This phenomenon is called 'primary healing', referring to the direct filling of the fracture site with the bone, without formation of significant *callus* (periosteal or endosteal). Healing under these conditions occurs by direct osteonal proliferation with interdigitation of bone fragments providing a very stable union. In opposition when the bone ends are separated by a gap inferior to 0.01 mm and the interfragmentary gap is less than 2%, the primary osteonal reconstruction results in direct formation of lamellar bone, oriented in the normal direction [1, 3]. This is termed contact healing and is initiated by osteoclasts from osteons near the fracture line. Another context for direct bone healing is observed in gaps from 800 μm to 1 mm and interfragmentary strain less than 2%, and gap healing is designated. Here, bone union and Haversian remodellation are separated by sequential process steps. Fracture site is filled directly by intermembranous bone formation, but the newly formed lamellar bone is oriented perpendicular to its long axis, and, later on, it undergoes secondary osteonal reconstruction (remodellation). In this way, the fracture repair depends on two important events: the rate of new bone growth (that replaces the stiffness and strength of the bone) and the strength and duration of the implant (maintenance of its function until it collapses and fails) [1–4]. When new bone formation is compromised and/or a risk of implant failure is expected, promotion of fracture healing is indicated, in order to enhance new bone formation, allowing for corrected bone structure formation and organization [1]. Fractures with impaired bone repair mechanisms or fixation failure will result in a non-union condition. Adequate bone repair depends on four crucial elements: an osteoconductive matrix, an

http://dx.doi.org/10.5772/intechopen.70341

osteoinductive signal, an osteogenic cell acting in response to that signal and an adequate blood supply [1, 3, 4]. There are several clinical situations where it is important to promote and enhance bone healing process in order to restore the original bone structure and function, both in human and veterinary medicine. These include traumatic injuries or tumour resections with substantial and irregular bone loss, gap filling following corrective osteotomy, arthrodesis or arthroplasty, spinal fusion, non-union or delayed bone union, metabolic diseases and local or systemic disease in aged patients. In these situations, bone regeneration is compromised, and the bone defect exceeds the intrinsic biological restoration mechanisms. These clinical cases occasionally result in unsatisfactory outcomes and are a challenging scenario to orthopaedic surgeons. When facing such clinical problems, different treatment strategies can be used to improve new bone formation, avoiding the formation of bone with inferior quality to the original [1, 3, 5–7].

Autogenous cancellous bone grafting (ACBG) is used in veterinary orthopaedic surgery as a bone void filler to improve bone healing in the treatment of bone defects in low-grade fractures in both mechanical and biological assessment score, arthrodesis, delayed unions or non-unions, enhancement of fracture healing, periprosthetic coating, spinal fusion procedures and void filler of bone gaps resulting from fractures, osteotomy, ostectomy, arthrodesis or tumour resection [1, 7]. The 'gold standard' when evaluating bone graft substitutes is still considered to be ACBG. Cancellous bone graft provides osteoconductive properties, acts as a scaffold for osteoprogenitor cells and delivers viable cells without the risk of immune reactions or infectious disease transmission. However, its use is associated with some limitations including the need of an additional surgery for harvesting cancellous bone, donor-site morbidity and limited amount of bone graft. The last point can result in an insufficient amount to completely fill the defect which may implicate the need for harvesting from more than one donor site. In humans, the second surgery for autologous bone graft collections is associated with 25% of morbidity, with major complications occurring in 3–4% of the patients. Complications include pain, sepsis, stress fractures, intraoperative haemorrhage, increased anaesthetic and surgical times and limited supply [8, 9]. Limited supply is more critical in small dogs and toy breeds, in cats or in animals that have been previously submitted to bone graft harvest. In those cases, the harvesting from the humerus may be the best option, because higher amounts of bone can be collected when comparing to the tibia while also presenting accelerated healing and complete restoration cancellous bone. Another drawback is its lack of strength of ACBG, hampering its use as a structural graft [1, 10]. Cortical allograft could be an alternative, which provides structural strength along with osteoinductive and osteoconductive properties. Nevertheless, the transmission of infectious diseases and adverse immune reactions are important risks to consider [11].

In the case of human patients, demineralized bone matrix (DBM) is the most common source of partially purified bone-inducting factors used. During the demineralization process, allogenic bone is chemically sterilized to preserve its osteoinductive properties from the original bone collagen network and molecular signalling. Commercial forms of DBM are also available for canine patients, where immune reactions are reduced by the removal of the periosteum, cartilage and bone marrow and by freezing process [6, 7, 12–14].

In the last decades, the well-known disadvantages of autografts and allografts have encouraged the development of bone synthetic substitutes to be employed as bone grafts, reflecting in the increase of the clinical application of these types of biomaterials. An ideal bone substitute should be biocompatible/safe; should be resorbable, with a similar mechanical resistance as the cortical bone; should have osteoconductive, osteoinductive and osteogenic properties; and should be easily handled and sterilized. The bone substitute should not cause any adverse systemic or local reaction; should provide a favourable environment to be colonized by blood vessels, cells and growth factors; and should be obtained at a low cost [7, 9].

These biomaterials are available for use in veterinary orthopaedics, alone or in combination with other strategies, such as osteoinductive bone morphogenetic proteins (BMPs) or cell-based treatments [6, 7, 11, 14]. They are composed of different materials with osteoconductive properties, such as bone products, ceramics, polymers and composites. The clinical application of each material is defined by its properties, which depend on composition and physical characteristics (e.g. granulometry, shape, pore size and interconnective porosity). The latter is determined by the manufacturing technique used [7, 9]. The synthetic bone substitutes offer an ideal substrate for bone cell colonization and consequent new bone formation [7].

Osteoconductive materials are grouped in two main categories. The first category includes ceramic-based bone substitutes, and the second category includes polymer-based bone graft substitutes (less commonly used). The most popular ceramic-based bone substitutes are calcium sulphate, bioactive glass and calcium phosphate. Calcium sulphate was the first material used as a bone graft substitute in clinical field, where it showed to be user-friendly, inexpensive, readily available and stable for filling-in bone defects, without a negative effect in bone healing. However, it presents fast absorption rates, leading to the loss of mechanical properties before equivalent new bone formation, constituting its main disadvantage and limiting its use in relevant clinical cases. Bioactive glass was designed as a bone graft for dental applications. Its use in orthopaedic surgery seems to be limited by its brittleness, radiopacity (which compromises radiographic evaluation of bone healing) and prolonged resorption times. Compared to bioactive glass, calcium phosphate is less radiopaque with faster reabsorption, but its osteointegration rate is highly variable, depending on its crystal size and stoichiometry. Tricalcium phosphate compounds are available in different presentations, including tricalcium phosphate, hydroxyapatite and a combination of the two.

Regardless of the mineral detail of its composition, the final formulations/presentation forms are of extreme relevance, concerning both its external shape and internal architecture. External shape and granule size will determine its suitability for particular lesion applications, considering effective size and ease of access for implantation. Its architecture is determinant for the resolution of the bone defect since bone ingrowth is dependent on the pore size. According to Ragetly et al., a minimum pore size of 100 μm is required for bone ingrowth and for optimal promotion of ingrowth pores should have a granulometry between 300 and 500 μm [7, 9, 15].

http://dx.doi.org/10.5772/intechopen.70341

2. Preclinical trials

A series of regulatory steps are essential in the development of biomaterials, requiring its physical-chemical characterization; in vitro validation, and at an intermediate stage, in vivo suitability; safety; and performance assessment: biocompatibility, osteoconduction, osteointegration and osteoinduction [16]. In sight of the ethical issues associated to the use of animals for experimental purposes, extensive efforts aim at refining in vitro methods as valid alternatives. Unfortunately, current in vitro models still underachieve in replicating the tissue response of a live animal to a bone substitute [16–18].

The preparation and conduction of preclinical studies are bound to a number of sequential phases. First, the proposed materials ought to be tested in noncritical-sized defects, allowing for the preliminary assessment of multiple material samples' behaviour in defined in vivo conditions. This feature is very important on the initial screening of a biomaterial's in vivo behaviour, enabling the choice of the chemical composition and format of the biomaterial with the most potential for more challenges [19–21]. Furthermore, noncritical-sized defect ensures fast and reliable healing process, allowing for the observation of the various stages of bone healing and biomaterial degradation [21, 22]. Once the ideal composition is chosen and biocompatibility is analyzed, the material may be used directly in clinical trials, or if found necessary, in increasingly critical defects, to confirm the results obtained in the first approach with noncritical defects, and to determine its limits of efficiency and performance [21].

One of the fundamental aspects when choosing the most adequate animal models to test bone replacement materials is the size of the bone defect amenable to assess [23, 24]. A critical-sized bone defect (CSBD) is defined as the smallest bone defect that will not heal spontaneously during the lifetime in a particular bone and species of animal [23, 25]. In a more detailed description, a CSBD has been referred to as a defect that has less than 10% bony regeneration during the lifetime of the animal or duration of the experiment [26, 27]. Although the smallest size that creates a defect designated as 'critical' is not a well-established concept, it has been defined as a segmental bone deficiency of length exceeding 2–2.5 times the diameter of the affected bone [28]. Some animal studies suggest that CSBD in sheep could be approximately three times the diameter of the diaphysis. Therefore, a critical defect in long bone cannot simply be defined by its size, but may also be dependent on the species phylogenetic scale, the location in the skeleton, the surrounding soft tissue envelope and the load bearing on the affected limb.

Bone colonization of macroporous biphasic calcium phosphate (MBCP) ceramics implanted in different sites (femur, tibia and calvaria) on a critical-sized defect in two animal models (rats and rabbits) showed bone ingrowth in all MBCP-implanted sites but with distinct rates. Bone colonization appeared statistically higher in the femur of the rabbits (48.5%) compared to the tibia (12.6%) and calvaria (22.9%) sites. As such, the comparison of results between animals or different bone defect locations is subject to bias, so a well-conducted study and fully validated animal models are essential in the development of new synthetic bone substitutes [29]. Furthermore, the host's age, metabolic and systemic conditions and comorbidities also affect the defect's healing potential [30].

Attempts to repair a CSBD only lead to the formation of fibrous connective tissue rather than the bone [31]. For practical purposes, if there is no more than 30% of the mineralization area after 52 weeks, lifelong incomplete bone healing is to be assumed [32]. The incapacity of natural healing when left untreated represents the negative control, so that the osteogenic potential of the material being tested can be considered unequivocal. Furthermore, CSBD should heal with appropriate treatment, and the autologous cancellous bone grafts are still named as the gold standard or positive control. Any new treatment based on bone tissue engineering should be tested and compared with these two landmarks [25]. Given the clinical targets of such applications (mostly aged or health-impaired patients), animals should be skeletally mature, in order to avoid misleading results deriving from the superior potential of the young animals to regenerate bone defects [23].

Recently, the osteoinductive ability of porous calcium phosphate ceramics was studied in four animal species through the implantation of cylinders of hydroxyapatite/tricalcium phosphate (HA/TCP) (in the proportion of 60 and 40%, respectively) in dorsal muscles in dogs, rabbits, rats and mice. After 1 year, the implants were removed, and histopathology tests were conducted with haematoxylin/eosin staining and Masson's trichrome staining to observe the new bone tissue formation. The study concluded that the material is biocompatible and biologically safe (no tumour or any atypical cells were present) and would be considered as a potential for bone substitute. Apart from the rat's groups, there was new bone and bone marrow tissue development in large amounts. The osteogenic ability of the implant was superior in mice, followed by dog and rabbit [33].

Small rodents, rabbits, dogs and small ruminants are the most popular model species for bone regeneration studies, bracing several of the target species for veterinary clinical applications. Allografts and bone graft substitutes have not been fully evaluated in cats, and for that reason, Dorea et al. compared the efficacy and safety of a Bioglass® with that of autogenous and allogeneic cancellous bone graft in this species. Four defects in the lateral diaphyseal cortex of the femur with a diameter of 4.0 mm were created in each animal. One hole was filled with autogenous CBG, another with allogeneic CBG and a third one with Bioglass®. The fourth defect was left unfilled. The healing process was monitored every 2 weeks by X-ray. After 6 weeks, cats were euthanized, and the resolution of the defects was appreciated. The study indicated an acceptable bone regeneration in all defects. Although with a slower healing rate, Bioglass® showed to be an acceptable alternative to ACBG in cats [34].

3. Clinical cases

The use of synthetic bone grafts, in veterinary medicine, has been increasing in the last years, but even so its application is substantially lower than in human medicine. This is mainly due to the costs involved in the use of biomaterials, but nowadays with the development of society and consequent attention given to the animals, there was an increase in the availability of owners to invest in the treatment of their animals. Besides the sentimental value of certain animals, it is not to be despised the growth of the economic valuation of certain types of animals.

3.1. The use of synthetic bone grafts in small animal clinical cases

Some isolated case reports in small animals using ceramic-based bone graft substitutes have been published in the last decades.

The application of β-tricalcium phosphate (TCP) has already stepped out of preliminary preclinical assays to veterinary patients' applications. Izumisawa et al. successfully solved *pes varus* in two miniature Dachshunds using a wedge of synthetic β-TCP to fill the gaps created by tibial corrective open osteotomies. According to the authors, 2 months after the surgery, the edge of osteotomies was integrated with the bone. The bone plates and screws were removed after 4 months, and, by then, the TCP wedges were completely resorbed and the osteotomy bone was remodelled. In both cases the use of synthetic bone graft TCP avoided the need of a second surgery to harvest autologous cancellous bone graft. In both cases, the postoperative angles were corrected and maintained during the follow-up period, and, morphologically, the body of the tibia in the affected hind limb nearly equalled that found in healthy limbs. Dogs were able to walk a few days postoperatively, and the authors concluded that the corrective transverse-opening osteotomy together with synthetic bone graft substitute β-TCP and veterinary T plate fixation is an effective method for the treatment of *pes varus* in small-breed dogs [35]. A single case of tarsal joint fusion with β-TCP and platelet-rich plasma (PRP) was reported by Hauschild et al., and ACBG, and hence the need of additional surgery to harvest the bone graft, was successfully avoided. The healing process was uneventful and was complete at 4 months after surgery, at which time the dog presented no signs of lameness [36].

Another report on the clinical use of β-TCP as synthetic cancellous bone graft in veterinary orthopaedics was presented by Franch et al. They retrospectively studied 13 clinical cases, where granules of β-TCP, with an irregular form and interconnected porous structure (without dead ends, mean porosity of 60% and mean pore size of 250 μm), were mixed with fresh blood and used as void filler in subcritical-sized bone defects in long bones. The β-TCP used in these clinical series was commercially available in sterile vials containing 2 g of granules with 99% of pure phase. The clinical cases are summarized in **Table 1**. All but one case achieved complete bone union, and radiographic bone ingrowth was at 100% in 10 cases, 90% in 1 case and 75% in another case. The publication reported excellent clinical results confirming the biocompatibility and usefulness of β-TCP as a synthetic bone graft for moderate to large subcritical bone defects with initially expected good biological conditions (blood supply, cellular activity, etc.), on which the main problem is to provide a structural scaffold to allow bone and capillary ingrowth and the healing of the defect [37]. The same author described the treatment of a distal radius atrophic non-union in a 1-year-old male Yorkshire terrier using a 3D-printed β-TCP scaffold with rhBMP-2 (TruScient®) to create a scaffold with the same shape as the defect. After the removal of the bone plate (10 months), load started to transmit along the bone axis, reducing the potential risk of stress protection. Eighteen months after surgery, the scaffold was no longer visible, and complete corticalization of the regenerated bone area was observed on computed tomography (CT) scan evaluation. Given the results, the author suggests that the 3D-printed β-TCP scaffold with rhBMP-2 is an excellent bone substitute, due to good osteoinductive properties given by rhBMP-2 complemented with good osteoconductive potential provided by the open-interconnected macroporosity from the β-TCP scaffold [38].

Type of treatment	Author	Type of macroporous ceramic	No. of cases	Species	Breed	Sex-age-weight	Problem	Reduction device	Bone defect grading	Results			Removal of the implant	Ref.
										Consolidation time (weeks)	Percentage new bone formation	Functional recovery		
β-TCP	Izumisawa et al. 2005	β-TCP wedge shaped	1	Canine	Dachshund	♂-10m	Pes varus and grade 3 lateral patellar luxation	Open-wedge osteotomy in the distal tibia and veterinary 1.5/2.0 T plate	NCSD	About 6	100	Excellent	4 months	[35]
			2	Canine	Dachshund	♂-9m	Pes varus and grade 3 lateral patellar luxation	Open-wedge osteotomy in the distal tibia and veterinary 1.5/2.0 T plate	NCSD	About 6	100	Excellent	4 months	
	Franch et al. 2006	Irregular interconnected granules of β-TCP (60% porosity)	3	Canine	Shiba Inu	♀-7 y-12kg	Traumatic carpal hyperextension	Pancarpal arthrodesis with 2.7 DCP plate	NCSD	12	100	Excellent		[37]
			4	Canine	Shiba Inu	♀-7y-12kg	Traumatic carpal hyperextension	Pancarpal arthrodesis with 2.7 DCP plate	NCSD	10	100	Excellent		
			5	Canine	Crossbreed	♂-12y-10kg	Fractures of the distal radius (cranial bone loss), ulna, III–IV–V metacarpal bones and I phalanx and comminuted carpal bone fracture-luxation	Pancarpal arthrodesis, radial and III metacarpal fractures treated together with 2.7 DCP plate	NCSD	8 (radius) 12 (carpus)	100	Excellent		
			6	Canine	German Shepherd	♂-2y-38kg	Comminuted mid-shaft femoral fracture	Interlocking nail with four cerclage wires	NCSD	9	90	Excellent		
			7	Canine	German Shepherd crossbreed	♂-7y-37kg	Tibial fracture	3.5 DCP plate	NCSD	10	100	Good		
			8	Canine	Belgium Shepherd	♀-8y-34kg	Distal radius fracture	3.5 DCP plate	NCSD	10	100	Excellent		
			9	Canine	English Bulldog	♂-10y-11kg	Tarsometatarsal luxation	Cross pinning tarsometatarsal arthrodesis	NCSD	12	75	Good		

Type of treatment	Author	Type of macroporous ceramic	No. of cases	Species	Breed	Sex-age-weight	Problem	Reduction device	Bone defect grading	Results			Removal of the implant	Ref.
										Consolidation time (weeks)	Percentage new bone formation	Functional recovery		
			10	Canine	Golden Retriever crossbreed	♂-3y-35kg	Radius and ulna comminuted fracture	3.5 DCP plate	NCSD	8	100	Excellent		
			9	Canine	Mastiff crossbreed	♂-2y-48kg	Hypertrophic femoral non-union. Osteopenic bone at surgery due to disuse	Interlocking nail with 4 screws	NCSD	12	100	Excellent		
			11	Canine	Chihuahua	♂-2y-3.5kg	Tibiotarsal fracture/ luxation	Cross pinning tibiotarsal arthrodesis + temporary transarticular external fixator	NCSD	12	100	Good		
			12	Canine	Yorkshire Terrier	♀-7y-3.4kg	II, III, IV and V metacarpal atrophic non-union	Intramedullary pins	CSD	/	0	Poor, the patient needs a permanent splint		
			13	Canine	German Shepherd	♀-5y-44kg	Traumatic carpal hyperextension	Carpal arthrodesis stepped plate	NCSD	8	100	Excellent		
			14	Feline	European crossbreed cat	♀-2y-3.8kg	Highly comminuted distal femoral fracture	2.7 DCP distal femoral prebent plate	NCSD	11	100	Excellent		
β-TCP + PRP 300mg + blood + 0.6 ml PRP	Hauschild et al. 2005		15	Canine	Beagle	♀-10m-?kg	3 weeks old traumatic luxation of the right intertarsal joint and compressive fracture of the fourth tarsal bone	Partial tarsal arthrodesis with lateral 2.0 DCP plate and a K wire plus coaptation bandage for 4 weeks	NCSD	16	100	Excellent, no sign of lameness	7 months later and leaved the K wire in place	[36]
Collagen-TCP sponges + rhBMP-2 + monocortical fenestrated rib	Boudrieau et al. 2004	Collagen-TCP sponges	16	Canine	Golden Retriever	♀-14 month-23.6kg	Mandibular reconstruction after partial mandibulectomy for tumour resection	2.0 miniplate and 2.4 mandibular reconstruction plate in the right side and 2.0 miniplate and 2.4 unilockin plate	2 CSBD (1.5 cm × 1 cm × 2 cm) and (7 cm × 1 cm × 2 cm)	12		Excellent	One exposed plate 1 year later	[47]

Type of treatment	Author	Type of macroporous ceramic	No. of cases	Species	Breed	Sex-age-weight	Problem	Reduction device	Bone defect grading	Results			Removal of the implant	Ref.
										Consolidation time (weeks)	Percentage new bone formation	Functional recovery		
β-TCP +ACBG (1:1) and a titanium mesh cage	Segal and Shani 2010	β-TCP	17	Canine	Crossbreed	♂-7y-25kg	Highly comminuted fractures (femur, radius and ulna) that fail to previous fixation	IM pin through the cage and a locking plate	2 CSBD 71mm segmental femoral bone defect and 27mm segmental radial defect	22–63	70% of femoral cage at 22 weeks and 100% at 63 weeks	Successful long-term clinical outcome	No due to the associated costs	[41]
Contralateral bone widening and transfer and silicate calcium phosphate	Petazzoni 2016	Silicate calcium phosphate	18	Feline	Maine Coon	♀-5y-5kg	Non-union and bone loss	2.5 mm locking plate	CSBD 60mm corresponding to 48% of mechanical axis on the sagittal plane	At 13 satisfactory healing. 22 weeks non-union. 20 weeks post revision surgery with ACBG healing of non-union.		Excellent long-term follow-up	6 weeks–18 weeks good; at 22 weeks implant fails and revision surgery was done; the plate was removed 20 weeks later	[43]
CRM + rhBMP-2	Massie et al. 2017	CRM	19	Canine	Italian Greyhound	3Y-3kg	Non-union of left and right radius/ulna	2.0 mm 14-hole LCP	R 65 mm L 2.5 mm	R 9 and L 10				[46]
			20	Canine	Rat Terrier	1y-7.1kg	Atrophic-oligotrophic non-union of R tibia	2.0 mm 13-hole LCP	32 mm	10				
			21	Canine	Golden Retriever	12y - 24.3kg	Dystrophic non-union of the L tibia	3.5 mm 10-hole LCP	16 mm	10				
			22	Canine	English Pointer	11y - 29.7kg	Atrophic-oligotrophic non-union of R lateral humeral condyle	5.5 mm cortical lag screw + 2.4 mm 8-hole LCP	1 mm	16				
			23	Canine	Schipperke	2y - 7.1kg	Atrophic-oligotrophic non-union of R radius/ulna	2.4 mm 10-hole LCP	5 mm	6				
			24	Canine	Border Collie	9y - 16kg	Atrophic-oligotrophic non-union of L tibia	ESF hybrid	10 mm	10				
			25	Canine	Cocker Spaniel	10y - 12kg	Atrophic-oligotrophic non-union of L lateral humeral condyle	3.5 mm cortical screw and washer + 2.4 mm 4-hole LCP	1 mm	20				
			26	Canine	Miniature Poodle	2y - 3kg	Dystrophic non-union of R an L radius/ulna	R 7-hole LC-DCP and L 8-hole LC-DCP	2 mm	8				

Type of treatment	Author	Type of macroporous ceramic	No. of cases	Species	Breed	Sex-age-weight	Problem	Reduction device	Bone defect grading	Results			Removal of the implant	Ref.
										Consolidation time (weeks)	Percentage new bone formation	Functional recovery		
			27	Canine	Mixed breed	4y - 12kg	Dystrophic non-union of R radius/ulna	2.4 mm 10-hole LCP	10 mm	7				
	Arzi et al. 2015	CRM (collagen sponge with embedded granules of HA and TCP)	28–32	Canine	–	8–9y (mean 8.8Y), 25-37kg (mean 29kg)	Reconstruction of the bone defects after segmental mandibulectomy for tumour resection	Titanium 3.0 mm locking plates	CSD 42–60 mm (mean, 50.5 mm)	8	100	Excellent function, occlusion and quality of life		[49]
	Verstraete et al. 2015	CRM (collagen sponge with embedded granules of HA and TCP)	32–38	Canine	Small-breed dogs	2–11 y (mean, 7.3y), 3.4–6kg (mean, 4.8kg)	Non-union mandibular fractures	2.0 locking titanium miniplate	CSD 5–18 mm (mean, 9.2 mm)	4 to 12		Excellent long-term follow-up		[48]

CRM, compression-resistant matrix; CSD, critical-sized bone defects; DCP, dynamic compression plate; LCP, locking compression plate; LC-DCP, limited contact dynamic compression plate; ESF, external skeletal fixation; L, left; NCSD, noncritical-sized bone defects; R, right.

Table 1. Description of reported clinical cases using ceramic-based bone substitutes in small animals, based on available literature.

Treatment of long bones affected with large segmental defects is challenging in human and veterinary medicine [39]. Different surgical approaches have been developed, but most of these techniques present major complications. Cortical allografts are prone to infection and are rarely fully incorporated [11, 39]. Distraction osteogenesis should only be used in docile animals and cooperative owners because it needs a lot of adjustments and cleaning of the fixator besides having to be kept in place for prolonged periods that could lead to loss of stability and discomfort [40]. Osteoconductive materials and osteoinductive substances, such as canine-demineralized bone matrix, canine autogenous bone grafts and bone morphogenetic proteins, have been used for the treatment of bone defects and comminuted fractures, alone or in combinations with bioceramics [6, 12]. Another alternative treatment that has been used with success is the clinical application of titanium mesh cages [41], nevertheless with extreme variable outcomes [42].

One case report in a dog with large segmental femoral and radial bone defects that failed in the first attempt of surgical stabilization and finally treated with a combined approach with titanium mesh, β-TCP, ACBG into the defects and fractures was stabilized with locking plates. Titanium mesh allowed maintaining bone fragments and grafts within the defect being further, between 22 and 63 weeks, surrounded by a bony bridge. Clinical outcome was successful without visible lameness in pace, but it was visible, while running and reduction in range of motion with crepitation were noticed on the stifle joint. Those limitations could be associated with original trauma or complications of fixation methods. Long-term follow-up (greater than 1 year) shows a satisfactory active mobilization of the limb. In this case, the technique has been simple and should be an alternative for treatment of long segmental bone defects. However, further systematic clinical studies are needed in order to evaluate the efficacy, complications and spectrum of the clinical use of this method [41].

A case report of successful reconstruction of a long segmental tibial defect in a 5-year-old 5 kg-spayed female Maine Coon cat were transverse distraction osteogenesis in the contralateral tibia was used to create free autograft for filling the defect. After fixation of the bone fragments with a locking plate, the autograft was transferred to the defect in the contralateral tibia, and the remnant space was filled with silicate calcium phosphate bone graft substitute. By 27 months, both tibias were healed; implants had been removed with an excellent functional outcome [43].

Combination of BMPs for augmentation of bone regeneration associated with bioceramics in the treatment of non-union fractures of long bones was mentioned in a review as a single case of non-union of the distal radius in a Pomeranian [44] and in a distal radius atrophic non-union in a 1-year old male Yorkshire terrier by Franch [45]. A recent prospective longitudinal cohort study of the use of compression-resistant matrix (CRM) immersed in rhBMP-2 in the treatment of 11 non-union long bone fractures in 9 dogs was evaluated, and treatment was successful with healing time from 7 to 20 weeks (median 10 weeks) and return to full or acceptable function in all dogs [46].

Non-weight-bearing bone lesions can also greatly benefit from the application of synthetic bone substitutes, such as those resulting from dental procedures. Boudrieau et al. published in 2004

http://dx.doi.org/10.5772/intechopen.70341

one case report of treatment of severe mandibular malocclusion after partial mandibulectomy in a 14-month-old golden retriever using fenestrated, monocortical rib grafts with rhBMP-2 in a collagen-TCP sponge that showed new bone formation after 3 months with reconstruction of the defect. One year later, the bone was collected, and the new bone was robust with evidence of continued remodelling. No major complications were noticed, and bony remodelling was evident at 4-year follow-up [47].

According to the authors' knowledge, those are the first reports of the use of rhBMP-2 delivered via adsorption into a CRM for regenerating bone in chronic defect non-union fractures in dogs. Its radiographic appearance seems to have a normal bone density and was integrated to native bone. The combined surgical and regenerative strategies reported achieved predictable, timely reconstruction of defect non-union fractures in small-breed, older dogs. The use of rhBMP-2 should be done with caution due to its very potent effect that is very versatile and with a wide range of functions and dose dependent. Finally, the incorporation of a regenerative strategy into the surgical resolution of non-union fracture defects avoided the morbidity associated with autologous bone grafting and provides fast return to normal function [48].

CRM is made of a collagen sponge with implanted granules of hydroxyapatite and tricalcium phosphate generating a semi-rigid framework that can resist to compressive forces in vivo, and it has been successfully used to fill bone defect in several studies [46, 48, 49]. In a prospective case series published by Arzi et al., CRM was used with rhBMP-2 in immediate reconstruction of segmental mandibulectomies (mandibular defects ≥5 cm) in four dogs for treatment of benign or malignant tumours. After tumour recession, the critical-sized bone defects were stabilized with titanium locking plates, and CRM, soaked with 0.5 mg/ml rhBMP-2 15 min before implantation at a volume equivalent to half of the calculated volume of CRM (with a half to three quarters of the mandibular height and a length 2 mm greater than the defect), was tightly implanted. Radiographic evaluations were made postoperatively at weeks 2, 4, 8 and 12 after surgery. In two clinical cases, a CT scan of the mandible was acquired 3 months after surgery. All dogs had proper occlusion after surgery and in the follow-up evaluation and returned to normal activity. After 2 weeks the entire defect site was covered with gingiva, and at 4 weeks, it was completely solid. There was no recurrence of the tumour or fractures during the controls at 2 and 3 months after surgery, and all owners reported an excellent quality of the life of their dogs [49]. The radiodensity of regenerated mandible increases throughout radiographic controls from postoperative radiographs to 4 weeks after surgery. The CRM scaffold had evidence of new bone formation connecting the adjacent mandible and smooth margins at 4 weeks post-surgery. At 8 weeks the scaffold continued to increase its radiodensity, and a mineralized union with the mandible was noticed. New bone formation and complete integration of CRM material with native mandible tissue were evident on CT image evaluation. The authors concluded that this surgical and regenerative approach achieved a rapid return to normal activity, with normal anatomy and occlusion, bone regeneration and re-established biomechanical function [49].

Verstraete published another case series in six dogs where CRM and rhBMP-2 were used as a regenerative approach to fill non-union mandibular fracture defects stabilized with locking titanium miniplate. All dogs were adopted from a shelter, and, apart from two dogs, it was not possible to know the duration of the non-union. In all cases the non-union defect was debrided and cleaned from fibrous and devitalized tissues and old implant materials. The volume of the defects was estimated with a CT with three-dimensional reconstruction, and an amount of CRM enough to fill a half to three quarters of the mandibular height with a length of 2 mm greater than the mandibular gap was prepared. The CRM was soaked with a volume of rhBMP-2 (0.5 mg/ml) corresponding to 50% of the CRM volume used to fill the defect. As an example, with a CRM piece of 2 cm length, 0.5 cm width and 1 cm height ($2 \times 0.5 \times 1$), the total CRM volume is 1 cm^3, corresponding to 0.5 ml of rhBMP-2 solution with 0.5 mg/ml concentration. An extra-oral approach was used to counter the plate in ventrodorsal position. The wedges of the non-union were debrided to remove sclerotic and devitalized bone and attached soft tissues, and the plate was then fixed to the bone with two to three locking titanium screws in each segment of the mandible and CRM was implanted. Radiograph follow-ups of the mandible were started immediately after surgery and 2, 4, 8 and 12 weeks after surgery. All dogs healed the soft tissues over the defects and with immediate to normal function and correct occlusion, and solid cortical bone formation was noticed within 3 months. There were no recurrence fractures on the follow-up period [48].

Synthetic bone graft (Bonelike®) is produced by the incorporation of P_2O_5–CaO glass-based system within a hydroxyapatite (HA) matrix. Bonelike® macroporous (BL® Poro) consists of polygonal granules with 2000–2800 and 4000–5600 μm of diameter with pore size range from 100 to 400 μm (**Figure 1**). Its osteoinductive and osteoconductive properties have been confirmed in experimental models of bone regeneration in sheep, have been used in clinical orthopaedic application and recently are being used in small animals [50–54].

We used Bonelike® in small granules and Bonelike® presenting macroporous structure (BL® Poro) in vivo in combination with ACBG (**Figure 2(A)**), PRP (**Figure 2(B)**) and rigid internal fixation with bone plates and screws to treat atrophic non-union of long bones in two adult dogs that were referred to the veterinary medicine teaching hospital, Vasco da Gama University, Coimbra, Portugal, for surgical correction of atrophic non-union that fails in at least one previous surgery. After clinical and orthopaedic examination, two orthogonal views of the affected and contralateral bone were obtained with a standard marker to calibrate magnification. Informed consent to use BL Poro® was obtained from all clients. Clinical cases were classified according to Weber-Cech terminology, as defect, dystrophic, necrotic or oligotrophic-atrophic non-union, and were labelled as infected if this was suspected due to either radiographic evaluation or appearance and tissue culture at surgery. For both clinical cases, autologous cancellous bone grafts were harvested from the uppermost proximal humerus as described by Innes [5]. One dog had a large femoral diaphyseal bone defect and is not able to apply loading to the affected pelvic limb. The second dog had an oligotrophic non-union in the proximal radius and two non-unions in the ulnar diaphysis caused by a gunshot trauma. Revision surgery was performed to remove all the failed implants, debridement, ischemic bone fragments and other avascular soft tissues. Sclerotic and atrophic bone ends of the biologically inactive non-unions were osteotomized with a bone. Rongeur until bleeding is seen to expose medullary cavity and improve vascularity in a bone defect. Prior to copious flushing, samples of the removed tissues and deep wound swabs were collected

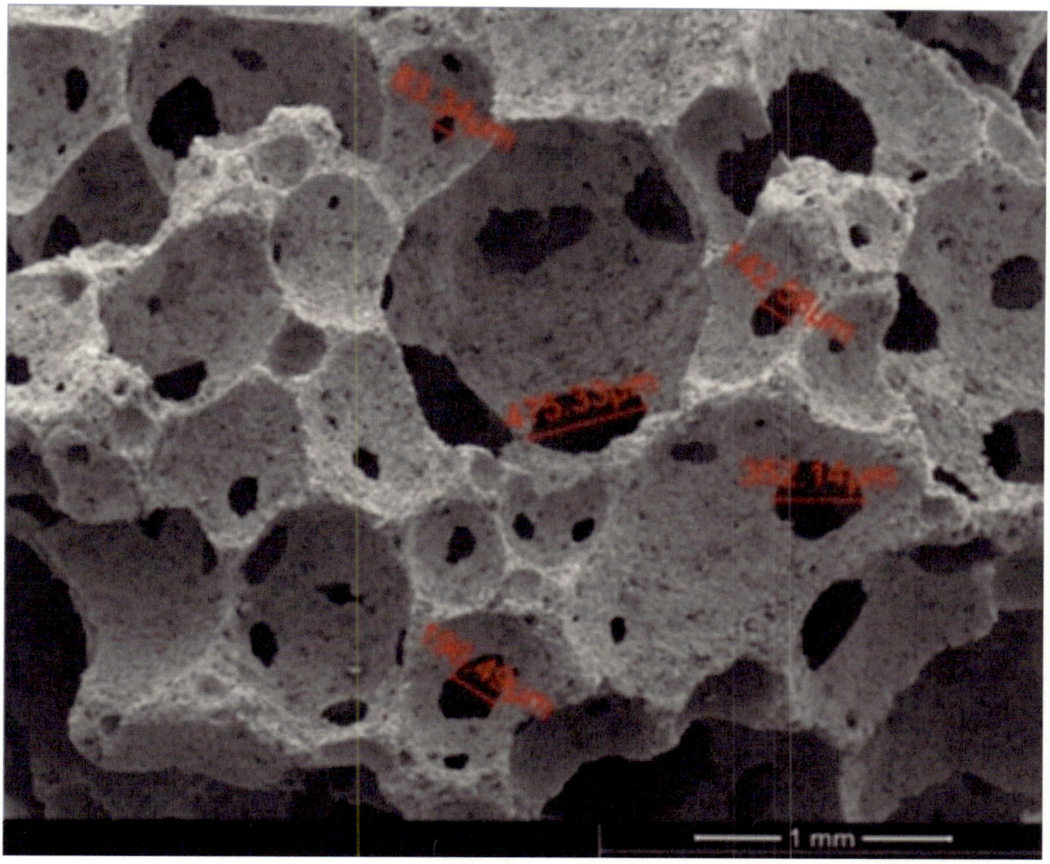

Figure 1. Scanning electron microscope image of a Bonelike® Poro granule showing the interconnected pore architecture with macropores and micropores.

for bacterial culture and sensitivity antibiogram. Joints are aligned, bone segments are pulled apart trying to restore original length of the bone and the fracture was stabilized with a bone plate and screws maintaining a rigid fixation for a prolonged time with minimal discomfort to the animal during postoperative period. The bone defect created after cleaning and the fracture fixation was filled with a homogenous mixture of ACBG and BL® Poro in 1:1 proportion in a PRP gel to maintain the aggregation of the components. The mixture was applied gently avoiding rupture and consequent collapse of the macroporous architecture of BL® Poro (**Figure 3**). Soft tissues were closed routinely and contribute to maintain the bone graft in situ with created soft tissue envelope. All the animals returned to acceptable or good function and decrease lameness grades. All cases achieved bone union between 2 and 8 months without major complications [54].

Our research group also has been using Bonelike® and a BL® Poro mixed with fresh blood to fill voids in noncritical-sized bone defects in dogs and cats with good clinical and radiographical outcomes without any adverse local nor systemic reaction (**Figures 4** and **5**). A

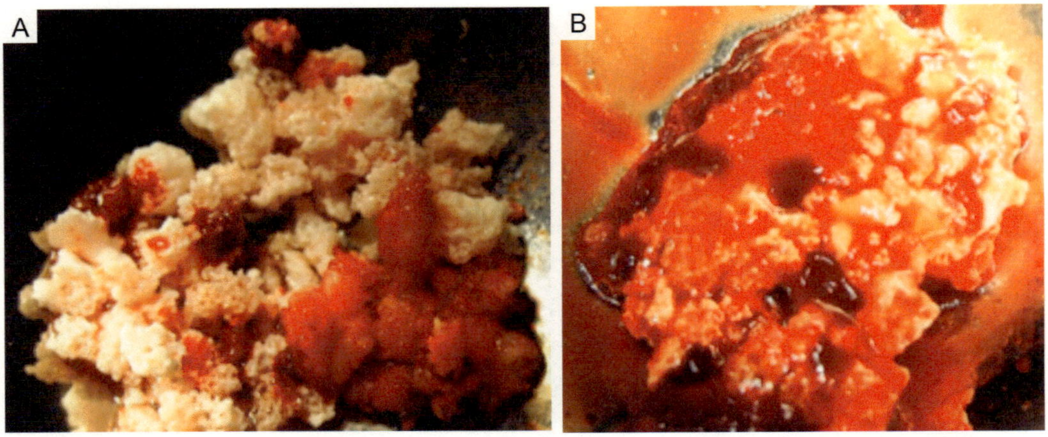

Figure 2. BL® Poro granules (white) mixed with autologous cancellous bone graft (red) (A). Mixture of BL® Poro granules with autologous cancellous bone graft and platelet-rich plasma, ready for implantation (B).

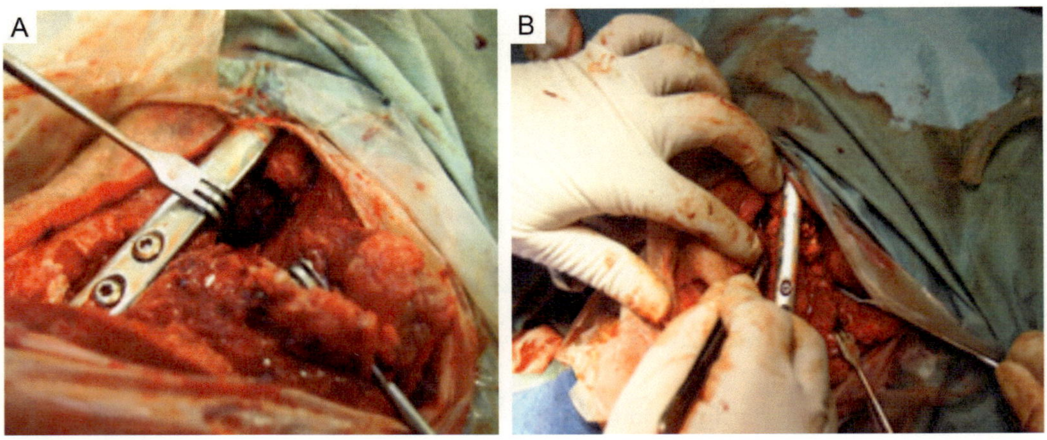

Figure 3. Femoral bone defect after fracture bridging plate fixation (A). Careful filling of defect to avoid the collapse of BL® Poro macroporous architecture (B).

clinical case of an 8-year-old, 21 kg, female Collie with a mandibular fracture along the root of canine teeth. The loose teeth exodontia and improved dental occlusion with an external skeletal fixator where fixation pins were placed along the aboral bone surface avoiding tooth roots were performed. The pins were bended and fixed together with a mouldable-stage application of methyl methacrylate. After hardening of the acrylic, the bone gap was approached directly from the oral mucosa and filled with mixture of Bonelike® (spherical granules of 250–500 μm) and blood. The soft tissues were surgically routinely closed and healed uneventfully. On radiographic evaluation 5 weeks post-surgery, the osteointegration of the biomaterial was visible, and fracture line was no longer visible. The fixator was removed 4 months later, and there was no story of mandible retraction or other types of complications (**Figure 6**) (unpublished data).

http://dx.doi.org/10.5772/intechopen.70341

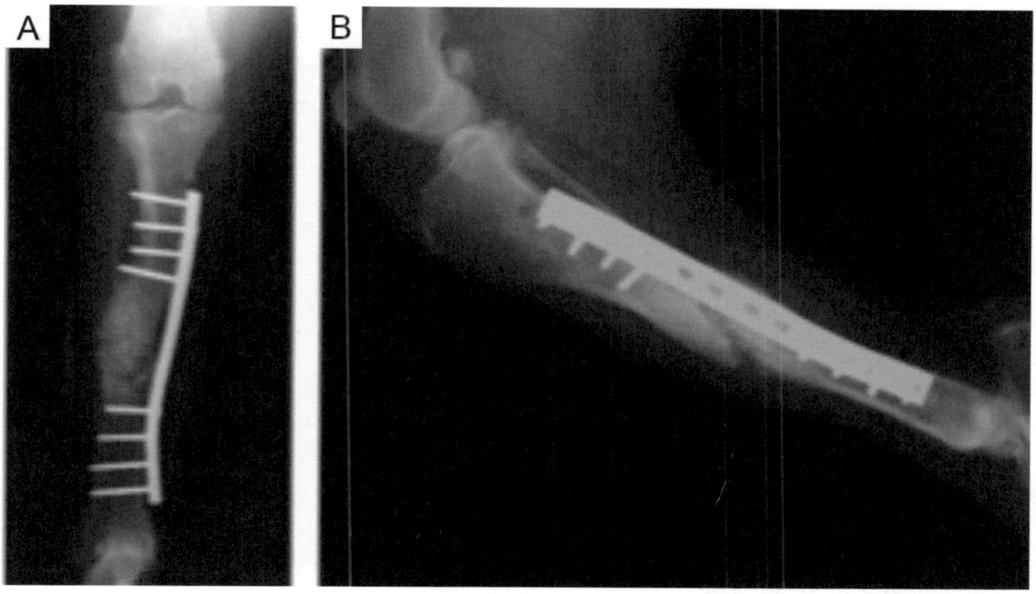

Figure 4. Immediate postoperative radiographs of the tibia of a 7-year-old 35 kg male german shepherd dog with a non-union fracture stabilized with a 3.5 mm 12-hole broad DCP plate. Bone defect was filled with Bonelike® 250–500 μm, autologous cancellous bone graft and platelet-rich plasma: cranio-caudal view (A) and mediolateral view (B). Radiolucent lines were visible within the graft and between the graft and the edges of the bone defect.

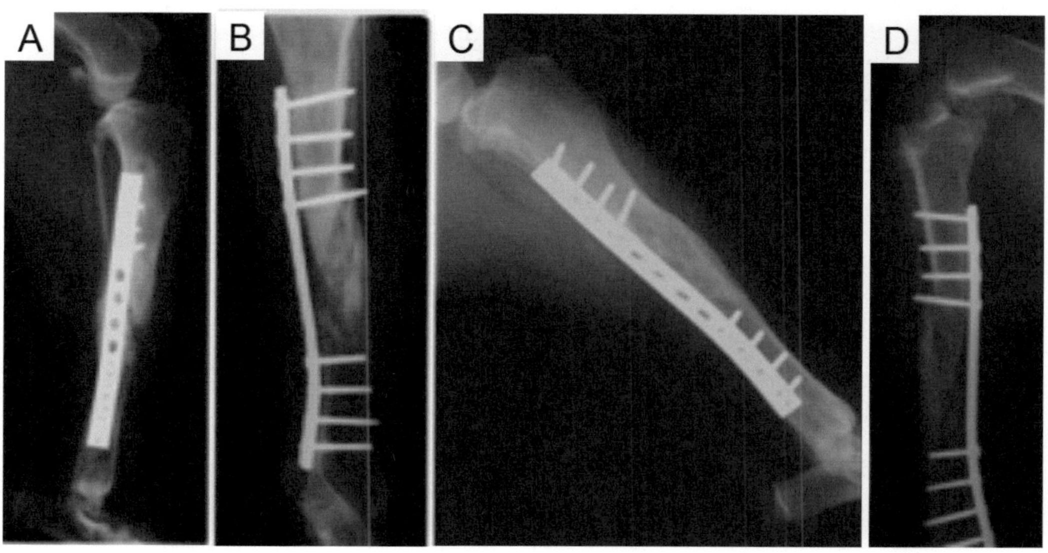

Figure 5. Postoperative radiographic evaluation. Mediolateral view at 2 months of control showing complete bone union on the caudal edge of the defect and not on the cranio-distal edge; here, a radiolucent line was noticed but with a grey colour compatible with new bone formation (A). Cranio-caudal view at 2 months of control, complete bone bridge was observed in lateral and medial cortices with irregular pattern within the bone defect (B). Mediolateral view at 8 months of control (C) and cranio-caudal view at 8 months of control. In both radiographs completed, bone union and remodellation were visible without radiolucent lines in the bone defects (D).

3.2. The use of synthetic bone grafts in equine clinical cases

Besides the majorly emotionally driven small animal clinics, other veterinary medicine fields are growing regarding the clinical application of the biomaterials, as economic implications are added to the scenario. For example, equines are the 'driving force' of an industry that moves enormous financial amounts, with competition prizes and price of some animals reaching astronomical values. With this set of constraints, the use of biomaterials in animals no longer restricts to preclinical trials carried out before their application in humans and is now part of the procedures available to veterinary clinicians. These techniques can be used in a number of clinical situations in veterinary medicine, where there is a bone defect, including fractures such as comminuted fractures, filling of bone cysts and arthrodesis [6, 55–59].

Equine patients are challenging in terms of orthopaedic pathologies and viable treatment options. While most of the small animal cases referred for orthopaedic treatment relate to fractures, and surgical reduction leads to expected full recovery, but even if this is not possible, they can survive with minimal complications with unsupported limb. Contrarily, in result of their weight, equine patients with unsupported limb have a high risk of developed severe secondary pathologies at contralateral limb, like laminitis. Other problems can surge in these equine patients; these animals cannot spend long periods lying down due to complications of myositis, nerve paralysis and decubital scars, so they weight injured limb for prolonged periods. For all those situations, fracture cases are seldom amenable of medical or surgical correction, often resulting in the decision of euthanasia. Other clinical situations are however indicated for grafting and/or biomaterial application, such as arthrodesis procedures.

Arthrodesis techniques were developed for treatment of debilitating osteoarthritis. These are also indicated for treatment of (stable) articular fractures, unstable joint injuries, septic arthritis and osteochondrosis. It is a surgical procedure used to promote the fusion between opposite bones in the joint, resulting in immobilization. Surgical arthrodesis involves the destruction of

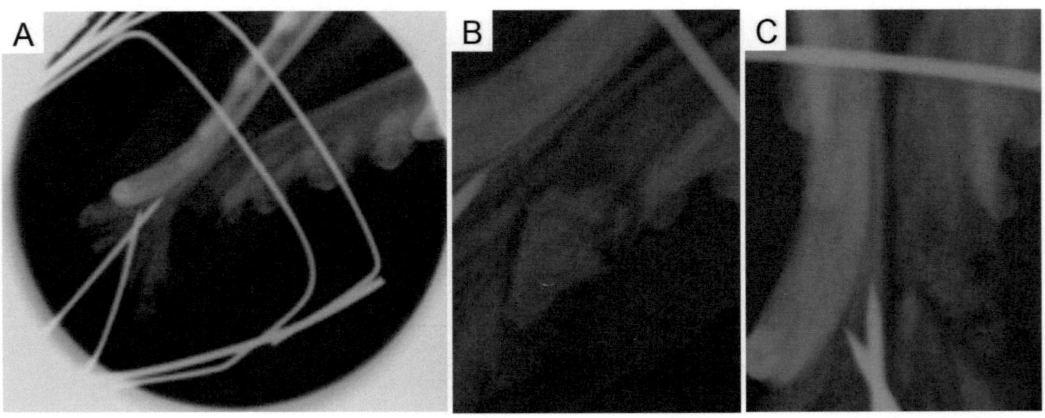

Figure 6. Bone defect resulting from canine teeth exodontia and mandibular fracture along the root of canine stabilized with a methyl methacrylate external skeletal fixator (A); immediate postoperative image of the defect filled with Bonelike® 250–500 and blood. The biomaterial was noticeable due to its multiple radiopaque spherical structures inside the bone defect. Radiolucent lines were visible within the spheres and between the graft and the edges of the bone defect. Fracture line was visible near by the fixation pin (B); 5 weeks of radiographic control, loss of radiopaque appearance of the biomaterial and radiopaque fracture lines was no longer visible due to new bone formation (C).

the articular cartilage of the bones to promote a bone-to-bone contact and subsequent fusion. Then, the joint must be aligned and stabilized into a stable, weight-bearing position [60–63]. Synthetic bone grafts can be used in the arthrodesis technique as void filler, improving the bone contact, helping the joint stabilization and avoiding the autologous bone graft collection. Since synthetic bone grafts deliver important osteoconductive and osteoinductive proprieties, the healing process is improved and shortened [50, 52]. Surgical arthrodesis is used in clinical situations that it is not possible recover the joint. With this technique the joint is stabilized, relieving the pain and improving horse's quality of life. In this species, the proximal interphalangeal, the intertarsal and the tarsometatarsal joints are the joints where more arthrodeses are performed, due to these joints that only have little range of movements, and their permanent immobilization does not affect normal locomotion of the horse. In these cases, the goal of the surgical procedure is to return the horse to its activity. Although the permanent immobilization does not affect normal locomotion, this procedure can weaken the joint region, due to loss of shock absorption capacity of the joint, and fracture risk may be increased [64–66]. Arthrodesis technique is also used in joints that have important movements (like metacarpophalangeal joint), knowing that in these cases the horse will have a mechanical lameness. The use of this technique in these types of joints allows to decrease pain and to increase comfort and the use of the affected limb, thus preventing the appearance of lesions in the contralateral limb overhead [64–66].

As examples of the use of synthetic bone grafts, it will be described in three clinical cases, where Bonelike® was applied in the arthrodesis of metacarpophalangeal joint, in the proximal interphalangeal joint and in the distal intertarsal joint together with tarsometatarsal joint.

The first clinical case is a newborn donkey, born with a severe metacarpophalangeal joint flexural deformity. The flexural deformity is so severe that it is impossible to manipulate and move the metacarpophalangeal joint and the dorsal surface of the fetlock supported the animal's weight. Due to this situation, the dorsal aspect of the fetlock presented an ulceration that communicated with the joint (**Figure 7**). It was decided, as the first approach, the surgical treatment, to quickly as possible resolve the situation and prevent the worsening of the complications [67, 68]. In this type of surgery, to relieve flexor forces, the tightened tendons can be cut [67, 69, 70]. In this clinical case, both the superficial flexor tendon and the deep flexor tendon were involved in the contracture of the joint, as well as the extensor digital tendon, which was displaced caudally, working as a flexor, further forcing the joint to adopt the flexor position (**Figure 8(A)**). Tenotomy of the three involved tendons was undertaken, in order to reposition the joint into a physiological alignment (**Figure 8(B)**). With this aggressive approach, the metacarpophalangeal joint stayed without support; so, a surgical arthrodesis of metacarpophalangeal joint was also performed to stabilize the joint in a physiological position. To do the arthrodesis, first the joint cartilage was removed with a curette to promote a bone-to-bone contact between the metacarpal bone and the first phalange; after that the joint space was filled with Bonelike® spherical granules mixed with autologous blood. A plate with cortical screws was used to stabilize the joint in a physiological position (**Figure 8(C and D)**) [68]. Although arthrodesis is not commonly used on flexural limb deformities, it was decided to apply this technique in the described clinical case, considering the lack of structural support of the joint after the tenotomy of the three tightened tendons. This approach was previously described by [71] that performed arthrodesis in a llama, in a miniature horse and in a

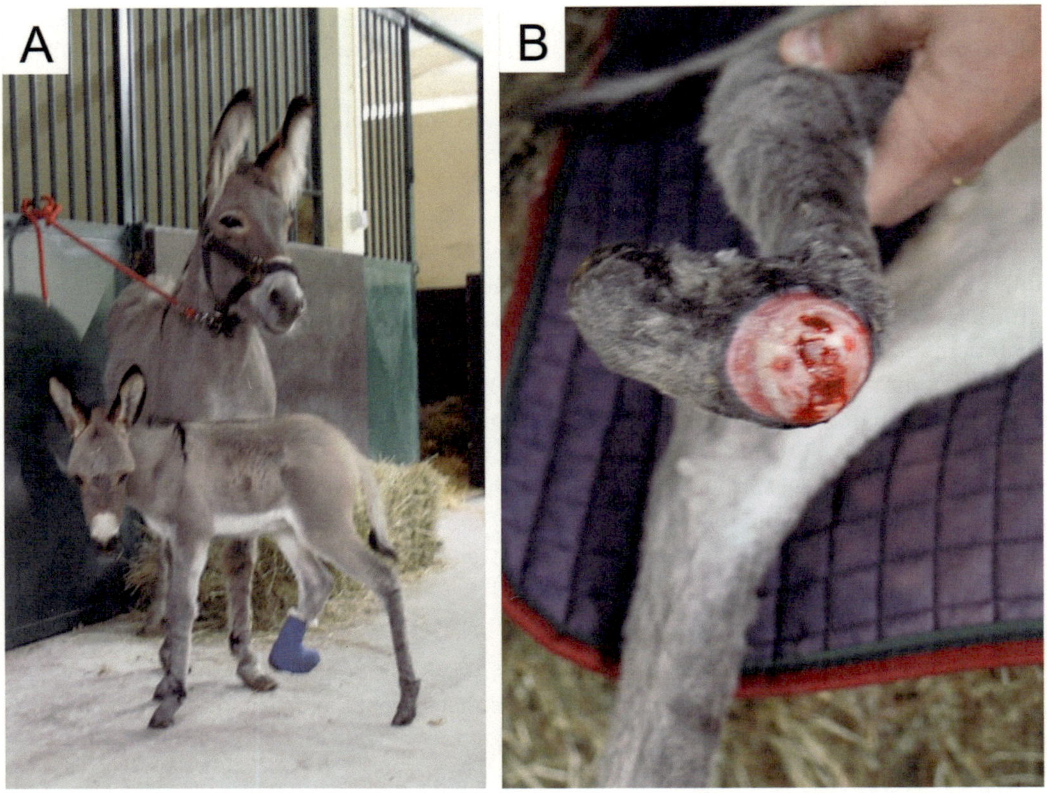

Figure 7. Newborn donkey with a severe metacarpophalangeal joint flexion deformity. The donkey supported the right hind limb with the dorsal aspect of fetlock (A). Large cutaneous ulcer on dorsal aspect of fetlock (B).

miniature donkey, with severe bilateral congenital flexural deformities of the metacarpophalangeal joint. The arthrodesis of the fetlock joint was not aimed at full recovery for future athletic activity but rather having in mind the animal's life quality [68].

The second clinical case was an adult horse that suffered an accident that resulted in the laceration of the medial aspect of the pastern. With this laceration, a medial ligament that causes instability to the pastern joint and consequently lameness was damaged. In clinical examination, a biomechanical instability of this joint was observed, and in X-ray examination, radiologic signs of subluxation and an increase in the medial joint space can be seen. Towards that it was decided to do the arthrodesis of proximal interphalangeal joint [68]. To approach the proximal interphalangeal joint, an I-shaped incision was performed in the skin over the dorsal aspect of the joint, and then a Z-incision was performed over the digital extensor tendon to expose the proximal interphalangeal joint. During the procedure, it was confirmed that this joint was not stable and there was an increased articular gap between the first and the second phalange on the medial aspect of the joint (**Figure 9(B)**). The cartilage from proximal interphalangeal was removed using a curette, and the joint space (**Figure 9(A)**) was filled with Bonelike® (**Figure 9(C)**). Three lag screws 3.2 mm were placed from the first to the second

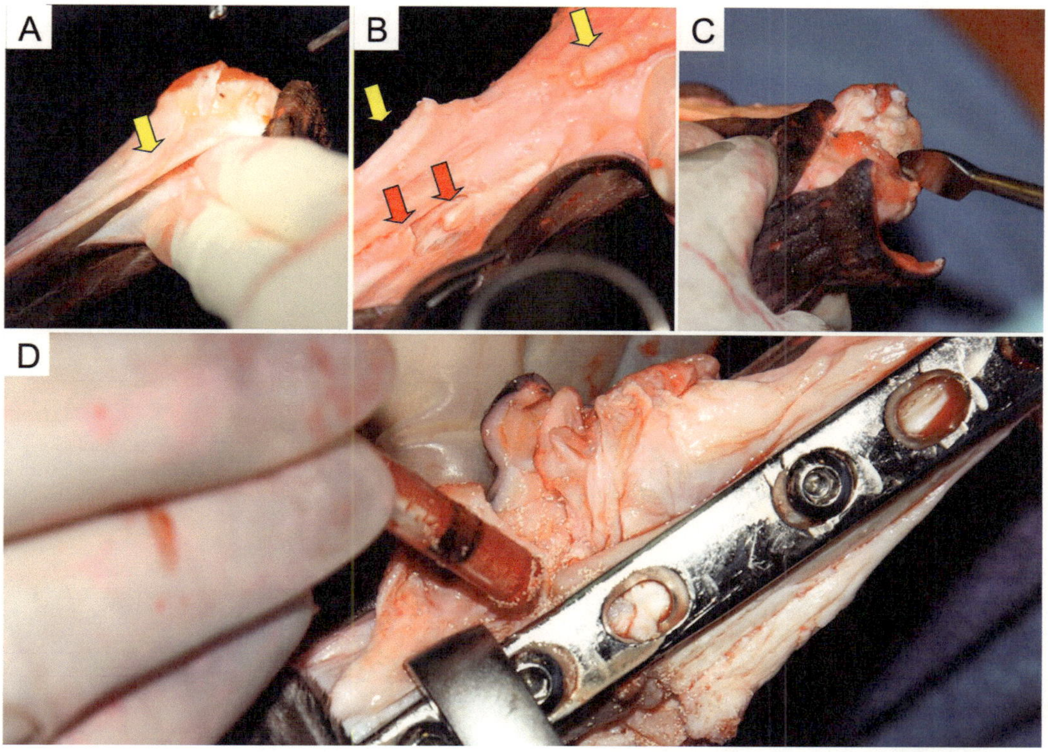

Figure 8. Metacarpophalangeal joint arthrodesis. Extensor digital tendon (yellow arrow) displaced caudally (A). Tenotomy of extensor digital tendon (yellow arrows) and flexor digital tendon (red arrows) (B). Removal of joint cartilage with a curette (C). A plate and screws to stabilize the joint and fill joint space, using a syringe, with Bonelike® spherical granules (D).

phalange crossing the joint space. The head screws stayed in a shelf made previously with an osteotome (**Figure 9(D)**). On the follow-up X-rays performed after the surgery, an important bone proliferation, with evidence of bone fusion and gradual reduction of the joint space, was noticed (**Figure 10**) [68].

The third clinical case was a report of a horse with chronic and intermittent lameness of the right hind limb, with a slight biomechanical instability of the tarsal region during walking, posing clinical and radiological signs of bone spavin. After the confirmation of joint pain by clinical examination with the positive nerve blocks, it was decided to do the distal intertarsal joint together with tarsometatarsal joint arthrodesis. The surgical approach to those joints was made by an incision over the medial aspect of the tarsus perpendicular to the cunean tendon. The cunean tendon was transected, and joints were identified with two needles (**Figure 11(A)** and **(B)**). A 3.2 drill bit was inserted into the joint space of each joint and forced in three different directions, using a single entry point (**Figure 11(C)** and **(D)**) [68]. Drilling paths were filled with Bonelike® (**Figure 11(E)** and **(F)**). The objective of drilling the joint space was to destroy the cartilage and to promote the contact between the bones and subsequent bone fusion. Due to the osteoconductive proprieties of Bonelike®, the application of this bone graft

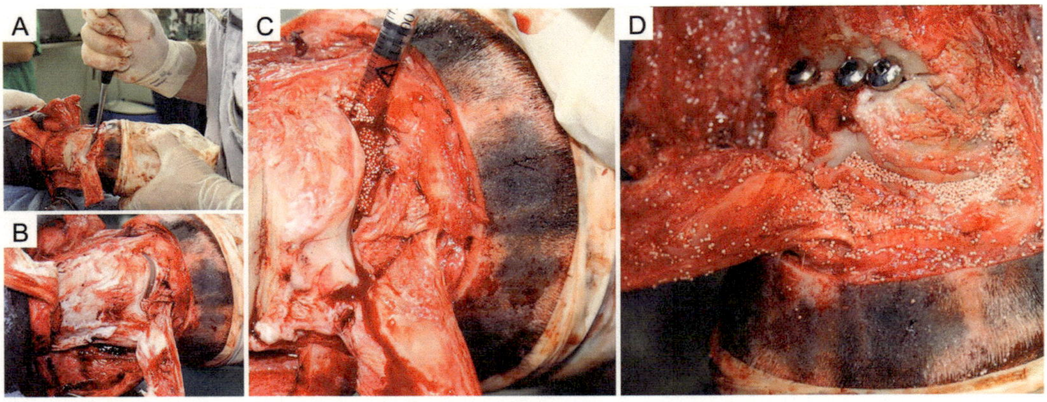

Figure 9. Proximal interphalangeal joint arthrodesis. Removing cartilage from proximal interphalangeal joint using a curette (A). A subluxation gap between the first and the second phalange at medial aspect of the joint (B). Applying with a syringe Bonelike® in proximal interphalangeal joint space (C). Three lag screws were placed from the first to the second phalange crossing the joint space (D).

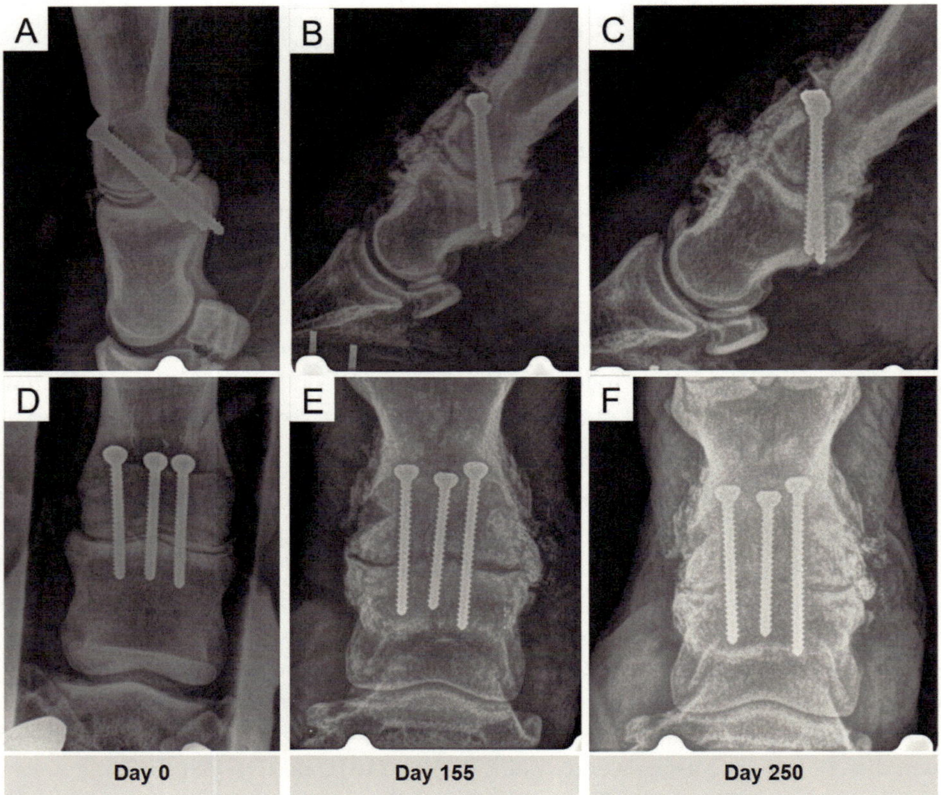

Figure 10. X-ray analysis follow-up of the proximal interphalangeal joint arthrodesis. The X-rays were taken at day 0 (pre-surgery), at day 155 and at day 250, with two projections. Lateral projection (A–C). Dorsoproximal projection (D–F).

http://dx.doi.org/10.5772/intechopen.70341

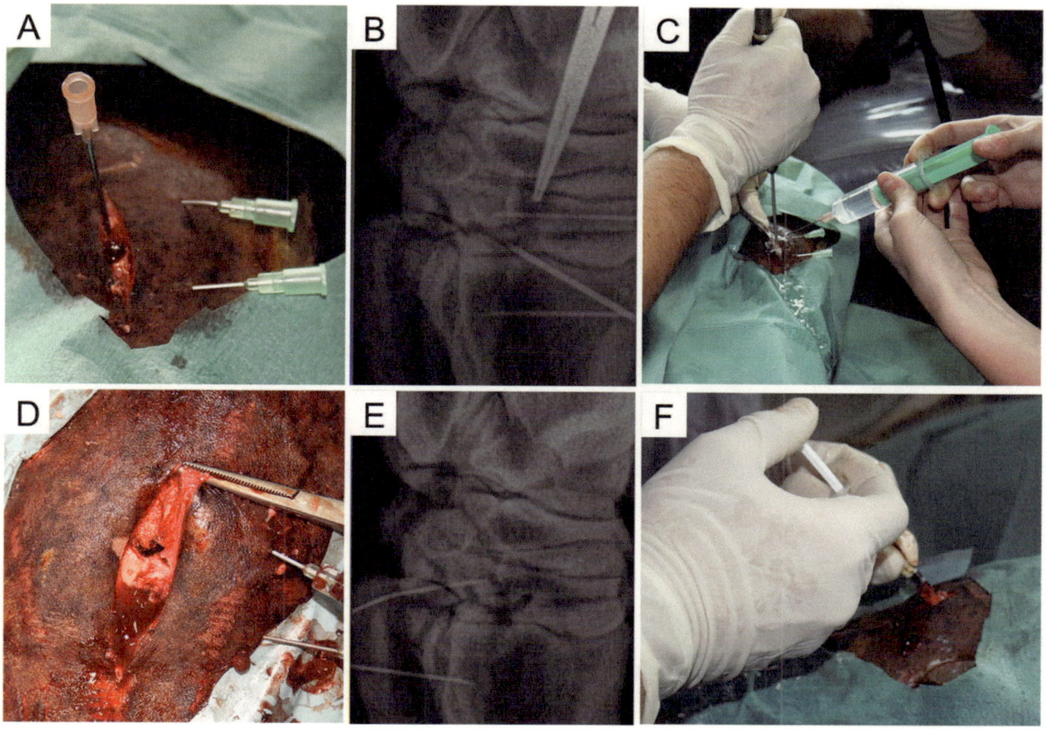

Figure 11. Arthrodesis of distal intertarsal joint together with tarsometatarsal joint. Needle is placed to identify the tarsometatarsal joint (A). Confirmation of correct position of the needle into joint space was made by X-ray (B). Create a hole with 3.2 drill bit in both joints, in three different directions using a single entry point, with a continuous flushing in order to minimize thermal damage (C) and remove any residual bone and cartilage (D). The hole created at the joint after drilling (E). Lateral X-ray of the tarsus showing the created hole (F). Hole filled with Bonelike®.

in the void space of the performed holes improved the biomechanical stability of the distal intertarsal and tarsometatarsal joint, and the bone bridging was accelerated, improving the joints' ankylosis. At the arthrodesis technique, it should be removed as much joint cartilage as possible to allow a greater bone-to-bone contact. But in this specific technique of the distal tarsal joint arthrodesis, as no additional surgical fixation is advised, the excessive drilling causes instability and severe pain [72]. It can be argued that the limited drilling of the distal tarsal joints involved in the three-drill-tract technique does not induce complete arthrodesis, but results in multiple focal areas of arthrodesis and the biomechanical stability of the distal tarsal joints are usually observed, and lameness is eliminated [73]. Using Bonelike® as a bone substitute that will fill the drilled paths, these focal areas of arthrodesis are improved and reinforced [68]. Arthrodesis using Bonelike® as a bone graft promotes bone fusion that permitted the horse to return to the athletic activity and improved the horse quality of life, decreasing the pain and increasing the joint stability. The Bonelike® application can enhance the bone production due to osteoinductive and osteoconductive proprieties, shortening the healing period after the arthrodesis and promoting the joint fusion in a shorter period [68, 74].

These three clinical cases were strongly evidenced that Bonelike® is an appropriate synthetic bone graft substitute with osteoconductive and osteoinductive properties to be used in surgical arthrodesis, as void space filler, associated or not with standard orthopaedic procedures of stabilization of the joints, in order to promote a faster bone fusion without any local or systemic adverse reaction. This procedure permits the horse to return to the athletic performance faster or, at minimum, improves the horse's quality of life, decreasing the pain and increasing the joint stability with positive clinical outcomes [68].

3.3. The use of synthetic bone grafts in ruminant clinical cases

Although the use of synthetic bone grafts in companion animals and horses is becoming a reality, its application in production animals is yet marginal.

Most of the surgery cases in large production animals (ruminants, mostly) are performed in the field settings, and the number of surgical cases referred to specialized veterinary centres is considerably small [75]. Constraints for orthopaedic surgery and bone grafts use in ruminants include the economic (cost of the treatment) and legal issues, the value of the individual in contrast to the group value and the size and weight of the animal [75, 76]. Nevertheless, the high genetic performance and value of some of these animals and the recent improvements in surgical management of production animals, concerning chemical restrain, pain management and surgical techniques, could be accompanied by the use of some biomaterials.

Fractures commonly occur in ruminants. Limbs, digits and skull fractures are often reported in bovine and small ruminants, often subsequent to trauma. The most common limb fractures occur especially in calves, as consequence to incorrect handling of dystocia or injuries due to falls on slippery floors in livestock facilities. Metacarpus and metatarsus fractures represent 50% of the limb fractures cases, tibia fractures represent 12% and radius and ulna almost 7%. Humerus fractures are rare and represent less than 5% of the cases. Still, fractures of the femur, of the pelvis and of the axial skeleton are more uncommon [75, 76].

The selection of the osteosynthesis procedure is influenced by the bone site of fracture, the degree and severity of soft tissue and neurovascular trauma of the status of the environment fractures (close or open), the patient temperament and the surgeon skills. The selected procedure must provide the patient the return to weight-bearing and normal mobility [75–77].

In many cases, it's advisable to do a prompt temporary fracture stabilization with external coaptation (splints or casts), avoiding additional trauma and clinical status worsening (preventing the closed fracture from becoming open or further fracture fragmentation and reducing eburnation of the fracture ends) prior to fracture fixation or during convalescence [75–77]. External coaptation is inadequate in oblique, spiral and comminuted fractures, and with extensive soft tissue trauma. In these cases, surgical procedures are advisable (internal or external fixation) [76].

In ruminants, the most common fractures' surgical procedures include external skeletal fixation (transfixation pin casts and external skeletal fixators) and internal fixation (intramedullary pinning, intramedullary interlocking nail and bone plates). The former procedure can be easily done in the field settings; it's less expensive and allows load bearing in the traumatized limb,

http://dx.doi.org/10.5772/intechopen.70341

and it's indicated in open fractures and wounds, but fracture reduction and reconstruction may fail and may not accomplish ideal fracture stabilization leading to non-union. The latter procedure, although more expensive as it requires specialized equipment and surgeon performance, allows more stability during the osteosynthesis process [76, 78, 79].

Autologous cancellous graft can be collected from the sternum, proximal tibia, proximal humerus and ilium. However, in cattle with chronic non-healing fractures, the lack of movement and the lack of alternating weight support, due to the strong lameness, lead to medullar atrophy, which limits the amount of cancellous bone graft that can be harvested from long bones. To perform the collection of a significant amount of cancellous bone graft from the sternum, general anaesthesia is advised, to avoid thorax trauma [78]. The use of synthetic bone grafts in ruminants' fractures with impaired bone repair mechanisms due to compromised vascular supply after severe trauma, with large long bone defect with septic non-union, infected open fractures, or in fixation failure, will promote and enhance bone healing process in order to restore the original bone structure and function.

The clinical case presented refers to a dwarf goat with sequence of a trauma event, resulting in a tibia open fracture, with soft tissue trauma (**Figure 12(A)** and **(B)**). The patient was 2 months into gestation, in risk of developing complications that could worsen the clinical situation. Surgical treatment, with internal fixation was considered as a first-choice treatment option. Also, it was of importance to restore normal weight-bearing function in a short period of time, as the gestation was close to the end. To stabilize the fracture, an intramedullary pin was applied, and the defect in the fracture ends was filled with Bonelike® spherical granules of 250–500 μm, mixed with autologous blood collected from the jugular vein during the surgery (**Figure 12(C)**).

On the follow-up, X-rays were performed at 2 (**Figure 13(A)**) and 4 (**Figure 13(B)**) weeks after the surgery; bone proliferation and bone fusion were noticed. The intramedullary pin was removed at 5 weeks after surgery (**Figure 13(C)**), presenting complete bone union and good ossification. The intramedullary pin was removed. The patient returned to its normal function. No local or systemic adverse reaction or rejection of the material during the healing process was observed.

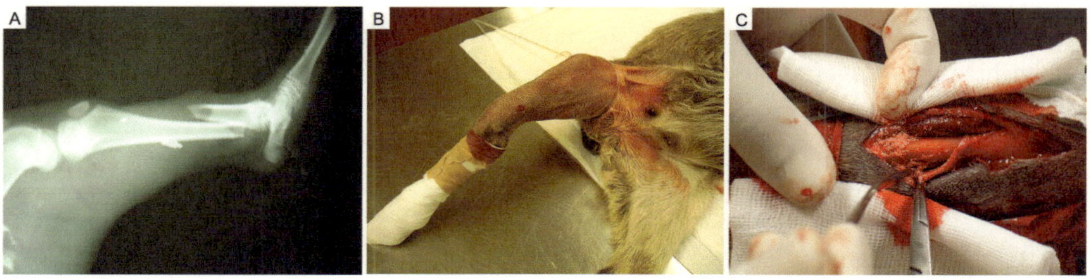

Figure 12. The patient was attended for consultation. The X-ray image showed a spiral fracture in the distal third of tibial bone (A). After hair clipping an open wound was evident. An internal fixation was considered as a first-choice treatment option to avoid additional tissue trauma and further fracture stress (B). An intramedullary pin was applied, and the defect in the fracture ends was filled with Bonelike® spherical granules of 250–500 μm, mixed with autologous blood collected from the jugular vein (C).

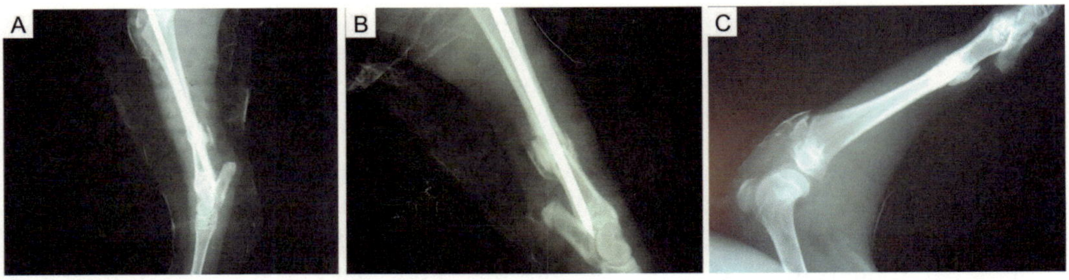

Figure 13. At 2 weeks after surgery, check-up and follow-up X-ray (A). At 4 weeks after surgery, a considerable callus was formed (B). At 5 weeks after surgery, the X-ray showed complete bone union and good ossification (C).

4. Conclusions

Although ceramic-based graft substitute materials and calcium phosphate ceramics such as hydroxyapatite, β-tricalcium phosphate and Bioglass® were used for a long time (since the 1970s in dentistry and 1980s in orthopaedics), its use alone or in combination with osteogenic or osteoinductive strategies still has a place in veterinary surgery. Macroporous granular forms with intrinsic osteoinductive properties could be a practical alternative to a customized 3D-printed scaffold or edges that allow filling of different shapes of bone defects as naturally occurs in the clinical cases. Another presentation of HA and TCP involved in a collagen membrane that resists to compression and could be cut with defects' size before the implantation is also available. Those synthetic bone substitutes offer an adequate alternative allowing the replacement of autologous cancellous bone grafts in management of fractures, vascular non-unions, noncritical-sized bone defects and arthrodesis. Nevertheless, basic surgical principals such as biomechanical stability, vascular preservation and infection control are still vital for providing an ideal environment for bone healing. However, when the local biology is compromised as previously mentioned, the bioceramics scaffold should be complemented with growth factors, cell-based therapies or a combination of those, in order to provide osteoinductive and osteogenic properties to increase the likelihood of bone healing. Experimental and clinical examinations are needed in veterinary surgery, as well as in human field to adequately compare the outcome of the novel treatment options and its combinations, establishing the most appropriate treatment protocols for each clinical presentation.

Acknowledgements

This research was supported by Programa Operacional Regional do Norte (ON.2—O Novo Norte), QREN, FEDER with the project 'iBone Therapies: Terapias inovadoras para a regeneração óssea', ref. NORTE-01-0247-FEDER-003262 and by the programme COMPETE—Programa Operacional Factores de Competitividade, Projects PEst-OE/AGR/UI0211/2011 and PEst-C/EME/UI0285/2013 funding from FCT. This research was also supported by Programa Operacional Competitividade e Internacionalização (P2020), Fundos Europeus Estruturais e de Investimento (FEEI) and FCT with the project 'BioMate—a novel biomanufacturing system to

http://dx.doi.org/10.5772/intechopen.70341

produce bioactive scaffolds for tissue engineering' with reference PTDC/EMS-SIS/7032/2014 and by COMPETE 2020, from ANI—Projectos ID&T Empresas em Copromoção, Programas Operacionais POCI, by the project 'insitu.Biomas—reinvent biomanufacturing systems by using an usability approach for in situ clinic temporary implant fabrication' with the reference POCI-01-0247-FEDER-017771. Ana Rita Caseiro (SFRH/BD/101174/2014) acknowledges FCT, for its financial support.

Author details

Pedro Olivério Pinho[1,2,3], José Miguel Campos[1,2,3], Carla Mendonça[1,2], Ana Rita Caseiro[1,2,4], José Domingos Santos[4], Ana Colette Maurício[1,2] and Luís Miguel Atayde[1,2]*

*Address all correspondence to: ataydelm@gmail.com

1 Centro de Estudos de Ciência Animal (CECA), Instituto de Ciências, Tecnologias e Agroambiente (ICETA) da Universidade do Porto, Praça Gomes Teixeira, Porto, Portugal

2 Departamento de Clínicas Veterinárias, Instituto de Ciências Biomédicas de Abel Salazar (ICBAS), Universidade do Porto (UP), Rua de Jorge Viterbo Ferreira, Porto, Portugal

3 Escola Universitária Vasco da Gama (EUVG), Hospital Veterinário Universitário de Coimbra (HVUC), Coimbra, Portugal

4 Departamento de Engenharia Metalúrgica e Materiais, Faculdade de Engenharia, REQUIMTE/LAQV, Universidade do Porto, Porto, Portugal

References

[1] Griffon DJ. Fracture healing. In: Johnson AL, Houlton JEF, Vannini R, editors. AO Principles of Fracture Management in the dog and cat. New York: AO; 2005. pp. 72-97

[2] Brinker W, Piermattei D, Flo G. Fractures: Classification, Diagnosis and Treatment. Handbook of Small Animal Orthopedics & Fracture Repair; 4th Ed. St Louis, MO: Saunders - Elsevier. 2006. pp. 25-159

[3] Cross A. Fracture Biology and Biomechanics. Veterinary Surgery: Small Animal; 2012. pp. 565-571

[4] Perren S. Basic Aspects of Internal Fixation. Manual of Internal Fixation in Small Animals. editors. Brinker, W.O., Olmstead, M.L., Sumner-Smith, G., Prieur, W.D. (Eds.). Springer; 1998. pp. 1-94

[5] Innes J. Bone grafting in small animal orthopaedic surgery. In Practice. 2014;**36**(4):173-181

[6] Vertenten G, Gasthuys F, Cornelissen M, Schacht E, Vlaminck L. Enhancing bone healing and regeneration: Present and future perspectives in veterinary orthopaedics. Veterinary and Comparative Orthopaedics and Traumatology. 2010;**23**(3):153-162

[7] Ragetly G, Griffon D. The rationale behind novel bone grafting techniques in small animals. Veterinary and Comparative Orthopaedics and Traumatology. 2011;**24**(1):1-8

[8] Conway JD. Autograft and nonunions: Morbidity with intramedullary bone graft versus iliac crest bone graft. Orthopedic Clinics of North America. 2010;**41**(1):75-84; table of contents

[9] Blokhuis TJ, Termaat MF, den Boer FC, Patka P, Bakker FC, Haarman HJ. Properties of calcium phosphate ceramics in relation to their in vivo behavior. The Journal of Trauma. 2000;**48**(1):179-186

[10] Penwick RC, Mosier DA, Clark DM. Healing of canine autogenous cancellous bone graft donor sites. Veterinary Surgery. 1991;**20**(4):229-234

[11] Akagi H, Ochi H, Kannno N, Iwata M, Ichinohe T, Harada Y, et al. Clinical efficacy of autogenous cancellous bone and fibroblast growth factor 2 combined with frozen allografts in femoral nonunion fractures. Veterinary and Comparative Orthopaedics and Traumatology. 2013;**26**(2):123-129

[12] Hoffer MJ, Griffon DJ, Schaeffer DJ, Johnson AL, Thomas MW. Clinical applications of demineralized bone matrix: A retrospective and case-matched study of seventy-five dogs. Veterinary Surgery. 2008;**37**(7):639-647

[13] Southwood LL, Frisbie DD, Kawcak CE, McIlwraith CW. Delivery of growth factors using gene therapy to enhance bone healing. Veterinary Surgery. 2004;**33**(6):565-578

[14] García-Gareta E, Coathup MJ, Blunn GW. Osteoinduction of bone grafting materials for bone repair and regeneration. Bone. 2015;**81**:112-121

[15] Nandi S, Roy S, Mukherjee P, Kundu B, De D, Basu D. Orthopaedic applications of bone graft & graft substitutes: A review. Indian J Med Res. 2010 Jul;**132**:15-30

[16] Pearce AI, Richards RG, Milz S, Schneider E, Pearce SG. Animal models for implant biomaterial research in bone: A review. European Cells & Materials. 2007;**13**:1-10

[17] Martini L, Fini M, Giavaresi G, Giardino R. Sheep model in orthopedic research: A literature review. Comparative Medicine. 2001;**51**(4):292-299

[18] Nahid M, Bottenberg P. Importance of cell cultures in biocompatible dental materials research. Revue belge de medecine dentaire. 2003;**58**(3):189-196

[19] Lu J, Flautre B, Anselme K, Hardouin P, Gallur A, Descamps M, et al. Role of interconnections in porous bioceramics on bone recolonization in vitro and in vivo. Journal of Materials Science: Materials in Medicine. 1999;**10**(2):111-120

[20] Boyd D, Carroll G, Towler M, Freeman C, Farthing P, Brook I. Preliminary investigation of novel bone graft substitutes based on strontium–calcium–zinc–silicate glasses. Journal of Materials Science: Materials in Medicine. 2009;**20**(1):413-420

[21] Atayde LM, Cortez PP, Pereira T, Armada-da-Silva P, Afonso A, Lopes MA, et al. A new sheep model with computing automatized analysis to evaluate the in vivo biomaterial's

behavior on bone tissue regeneration. Journal of Materials Science: Materials in Medicine. 2014;**25**(8):1885-1901

[22] Schopper C, Ziya-Ghazvini F, Goriwoda W, Moser D, Wanschitz F, Spassova E, et al. HA/TCP compounding of a porous CaP biomaterial improves bone formation and scaffold degradation—A long-term histological study. Journal of Biomedical Materials Research Part B, Applied Biomaterials. 2005;**74**(1):458-467

[23] Anderson ML, Dhert WJ, de Bruijn JD, Dalmeijer RA, Leenders H, van Blitterswijk CA, Verbout AJ. Critical size defect in the goat's os ilium. A model to evaluate bone grafts and substitutes. Clinical Orthopaedics and Related Research. 1999;(364):231-239

[24] Salgado AJ, Coutinho OP, Reis RL. Bone tissue engineering: State of the art and future trends. Macromolecular Bioscience. 2004;**4**(8):743-765

[25] Lansdowne JL, Devine D, Eberli U, Emans P, Welting TJ, Odekerken JC, et al. Characterization of an ovine bilateral critical sized bone defect iliac wing model to examine treatment modalities based on bone tissue engineering. Biomed Research International. 2014;**2014**:250958

[26] Hollinger JO, Kleinschmidt JC. The critical size defect as an experimental model to test bone repair materials. Journal of Craniofacial Surgery. 1990;**1**(1):60-68

[27] Bosch C, Melsen B, Vargervik K. Importance of the critical-size bone defect in testing bone-regenerating materials. Journal of Craniofacial Surgery. 1998;**9**(4):310-316

[28] Lindsey RW, Gugala Z, Milne E, Sun M, Gannon FH, Latta LL. The efficacy of cylindrical titanium mesh cage for the reconstruction of a critical-size canine segmental femoral diaphyseal defect. Journal of Orthopaedic Research: Official Publication of the Orthopaedic Research Society. 2006;**24**(7):1438-1453

[29] Le Guehennec L, Goyenvalle E, Aguado E, Houchmand-Cuny M, Enkel B, Pilet P, et al. Small, animal models for testing macroporous ceramic bone substitutes. Journal of Biomedical Materials Research Part B: Applied Biomaterials. 2005;**72**(1):69-78

[30] Reichert JC, Saifzadeh S, Wullschleger ME, Epari DR, Schutz MA, Duda GN, et al. The challenge of establishing preclinical models for segmental bone defect research. Biomaterials. 2009;**30**(12):2149-2163

[31] Schmitz JP, Hollinger JO. The critical size defect as an experimental model for craniomandibulofacial nonunions. Clinical Orthopaedics and Related Research. 1986;(205):299-308

[32] Li Y, Chen S-K, Li L, Qin L, Wang X-L, Lai Y-X. Bone defect animal models for testing efficacy of bone substitute biomaterials. Journal of Orthopaedic Translation. 2015;**3**(3):95-104

[33] Cheng L, Wang T, Zhu J, Cai P. Osteoinduction of calcium phosphate ceramics in four kinds of animals for 1 year: Dog, rabbit, rat, and mouse. Transplantation Proceedings. 2016;**48**(4):1309-1314

[34] Dorea H, McLaughlin R, Cantwell H, Read R, Armbrust L, Pool R, et al. Evaluation of healing in feline femoral defects filled with cancellous autograft, cancellous allograft or Bioglass. VCOT Archive. 2005;**18**(3):157-168

[35] Izumisawa Y, Reona A, Miyoshi K, Maehara S, Wakaiki S, Kushiro T, et al. Axial correction of pes varus by transverse-opening wedge osteotomy and T-plate fixation with beta-tricalcium phosphate (beta-TCP) transplantation in dachshunds. Journal of Veterinary Medical Science. 2005;**67**(4):437-440

[36] Hauschild G, Merten H-A, Bader A, Uhr G, Deivick A, Meyer-Lindenberg A, et al. Bioartificial bone grafting: Tarsal joint fusion in a dog using a bioartificial composite bone graft consisting of beta-tricalciumphosphate and platelet rich plasma—A case report. VCOT Archive. 2005;**18**(1):52-54

[37] Franch J, Diaz-Bertrana C, Lafuente P, Fontecha P, Durall I. Beta-tricalcium phosphate as a synthetic cancellous bone graft in veterinary orthopaedics: A retrospective study of 13 clinical cases. Veterinary and Comparative Orthopaedics and Traumatology. 2006;**19** (4):196-204

[38] Rabillard M, Grand J-G, Dalibert E, Fellah B, Gauthier O, Niebauer G. Effects of autologous platelet rich plasma gel and calcium phosphate biomaterials on bone healing in an ulnar ostectomy model in dogs. Veterinary and Comparative Orthopaedics and Traumatology. 2009;**22**(6):460-466

[39] Johnson KD, August A, Sciadini MF, Smith C. Evaluation of ground cortical autograft as a bone graft material in a new canine bilateral segmental long bone defect model. Journal of Orthopaedic Trauma. 1996;**10**(1):28-36

[40] Degna MT, Ehrhart N, Feretti A, Buracco P. Bone transport osteogenesis for limb salvage following resection of primary bone tumors: Experience with six cases (1991–1996). VCOT Archive. 2000;**13**(1):18-22

[41] Segal U, Shani J. Surgical management of large segmental femoral and radial bone defects in a dog. Veterinary and Comparative Orthopaedics and Traumatology. 2010;**23**(1):66-70

[42] Zoi S, Papadimitriou S, Galatos A, Prassinos N, Psalla D, Dalstra M, et al. Influence of a titanium mesh on the management of segmental long bone defects. Veterinary and Comparative Orthopaedics and Traumatology. 2015;**28**(6):417-424

[43] Petazzoni M. Contralateral bone widening and transfer for limb sparing in a cat. Veterinary and Comparative Orthopaedics and Traumatology. 2016;**29**(2):174-180

[44] Kirker-Head CA, Boudrieau RJ, Kraus KH. Use of bone morphogenetic proteins for augmentation of bone regeneration. Journal of the American Veterinary Medical Association. 2007;**231**(7):1039-1055

[45] Franch J. A preliminary clinical experience in combination with rhBMP-2 in a distal atrophic non-union. 18th ESVOT Congress. London; 2016. pp. 120-121

[46] Massie AM, Kapatkin AS, Fuller MC, Verstraete FJ, Arzi B. Outcome of nonunion fractures in dogs treated with fixation, compression resistant matrix, and recombinant human bone morphogenetic protein-2. Veterinary and Comparative Orthopaedics and Traumatology. 2017;**30**(2):153-159

[47] Boudrieau RJ, Mitchell SL, Seeherman H. Mandibular reconstruction of a partial hemimandibulectomy in a dog with severe malocclusion. Veterinary Surgery. 2004;**33**(2):119-130

[48] Verstraete FJM, Arzi B, Huey DJ, Cissell DD, Athanasiou KA. Regenerating mandibular bone using rhBMP-2: Part 2—Treatment of chronic, defect non-union fractures. Veterinary Surgery. 2015;**44**(4):410-416

[49] Arzi B, Verstraete FJM, Huey DJ, Cissell DD, Athanasiou KA. Regenerating mandibular bone using rhBMP-2: Part 1—Immediate reconstruction of segmental mandibulectomies. Veterinary Surgery. 2015;**44**(4):403-409

[50] Atayde LM, Cortez PP, Afonso A, Santos M, Mauricio AC, Santos JD. Morphology effect of bioglass-reinforced hydroxyapatite (Bonelike®) on osteoregeneration. Journal of Biomedical Materials Research Part B, Applied Biomaterials. 2015;**103**(2):292-304

[51] Gutierres M, Dias A, Lopes M, Hussain NS, Cabral A, Almeida L, et al. Opening wedge high tibial osteotomy using 3D biomodelling Bonelike® macroporous structures: Case report. Journal of Materials Science: Materials in Medicine. 2007;**18**(12):2377-2382

[52] Cortez PP, Silva MA, Santos M, Armada-da-Silva P, Afonso A, Lopes MA, et al. A glass-reinforced hydroxyapatite and surgical-grade calcium sulfate for bone regeneration: In vivo biological behavior in a sheep model. Journal of Biomaterials Applications. 2011(2):201-217

[53] Cortez PP, Atayde LM, Silva MA, Armada-da-Silva P, Fernandes MH, Afonso A, et al. Characterization and preliminary in vivo evaluation of a novel modified hydroxyapatite produced by extrusion and spheronization techniques. Journal of Biomedical Materials Research Part B: Applied Biomaterials. 2011;**99B**(1):170-179

[54] Pinto PO, Campos JM, Caseiro AR, Pereira T, Santos JD, Atayde LM, Maurício AC. Use of a Macroporous Glass-Reinforced Hydroxyapatite Synthetic Bone Substitute in Treatment of Long-Bone Atrophic Non-Union Fracture- Two Clinical Cases in Dogs. London: ESVOT; 2016. pp. 448-449

[55] Schwarz N, Schlag G, Thurnher M, Eschberger J, Dinges HP, Redl H. Fresh autogeneic, frozen allogeneic, and decalcified allogeneic bone grafts in dogs. The Journal of Bone and Joint Surgery British. 1991;**73**(5):787-790

[56] Dowdle S, Spotswood T, Lambrechts N, Duncan N. Aneurysmal bone cyst in the distal radius of a dog: Diagnostic imaging and surgical treatment. Veterinary and Comparative Orthopaedics and Traumatology. 2003;**16**(2):116-121

[57] Duval J, Chambers J, Newell S. Surgical treatment of an aneurysmal bone cyst in a dog. Veterinary and Comparative Orthopaedics and Traumatology: VCOT. 1995;**8**:213-217

[58] Kim CS, Choi SH, Cho KS, Chai JK, Wikesjö UM, Kim CK. Periodontal healing in one-wall intra-bony defects in dogs following implantation of autogenous bone or a coral-derived biomaterial. Journal of Clinical Periodontology. 2005;**32**(6):583-589

[59] Kerwin SC, Lewis DD, Elkins AD, Oliver J, Pechman R, Mccarthy RJ, et al. Deep-frozen allogeneic cancellous bone grafts in 10 dogs: A case series. Veterinary Surgery. 1996;**25** (1):18-28

[60] Lesser AS. Arthrodesis. In: Slatter D, editor. Text Book of Small Animal Surgery. Third ed. Saunders; 2002. pp. 2170-2190

[61] Zubrod CJ, Schneider RK. Arthrodesis techniques in horses. Veterinary Clinics of North America Equine Practice. 2005;**21**(3):691-711, vii

[62] Schoenhaus HD, Lam S, Treaster A, Troiano M. Use of a small bilateral external fixator for ankle fusion. Journal of Foot and Ankle Surgery. 2009;**48**(1):89-92

[63] Sorrel Langley-Hobbs MA, editor. Arthrodesis principles. 3 rd World Veterinary Orthopaedic Congress, ESVOT-VOS 15th ESVOT Congress. Bologna (Italy); 2010

[64] Caron J, Fretz P, Bailey J, Barber S. Proximal interphalangeal arthrodesis in the horse a retrospective study and a modified screw technique. Veterinary Surgery. 1990;**19**(3):196-202

[65] Knox P, Watkins J. Proximal interphalangeal joint arthrodesis using a combination plate-screw technique in 53 horses (1994–2003). Equine Veterinary Journal. 2006;**38**(6):538-542

[66] MacLellan KN, Crawford WH, MacDonald DG. Proximal interphalangeal joint arthrodesis in 34 horses using two parallel 5.5-mm cortical bone screws. Veterinary Surgery. 2001;**30**(5):454-459

[67] Kidd J, Barr A. Flexural deformities in foals. Equine Veterinary Education. 2002;**14**(6):311-321

[68] Atayde LM. Substitutos ósseos para regeneração do tecido ósseo: estudos in vivo e futuras aplicações clínicas e medicina veterinária; 2014

[69] Adams SB, Santschi EM, editors. Management of congenital and acquired flexural limb deformities. Proceedings of the American Association of Equine Practitioners. 2000;**46**: 117-125

[70] Auer JA. Diagnosis and treatment of flexural deformities in foals. Clinical Techniques in Equine Practice. 2006;**5**(4):282-295

[71] Whitehair KJ, Adams SB, Toombs JP, Parker JE, Prostredny JM, Whitehair JG, et al. Arthrodesis for congenital flexural deformity of the metacarpophalangeal and metatarsophalangeal joints. Veterinary Surgery. 1992;**21**(3):228-233

[72] Auer JA. Arthrodesis techniques. In: Stick Aa, editor. Equine Surgery. 1. 3rd ed. Philadelphia: Saunders; 2006. pp. 1073-1086

[73] Dechant JE, Baxter GM, Southwood LL, Crawford WH, Jackman BR, Stashak TS, et al. Use of a three-drill-tract technique for arthrodesis of the distal tarsal joints in horses with distal tarsal osteoarthritis: 54 cases (1990-1999). Journal of the American Veterinary Medical Association. 2003;**223**(12):1800-1805

[74] Lobato JV, Hussain NS, Botelho CM, Maurício AC, Afonso A, Ali N, et al. Assessment of Bonelike® graft with a resorbable matrix using an animal model. Thin Solid Films. 2006;**515**(1):362-367

[75] Anderson DE, Jean GS. Management of fractures in field settings. Veterinary Clinics of North America: Food Animal Practice. 2008;**24**(3):567-582

[76] Pentecost R, Niehaus AJ, Anderson DE. Surgical management of fractures and tendons. Veterinary Clinics of North America: Food Animal Practice. 2016;**32**(3):797-811

[77] Nuss K. Plates, pins, and interlocking nails. Veterinary Clinics of North America: Food Animal Practice. 2014;**30**(1):91-126

[78] Weaver AD, Jean GS, Steiner A. Bovine Surgery and Lameness. John Wiley & Sons; 2013

[79] Vogel SR, Anderson DE. External skeletal fixation of fractures in cattle. Veterinary Clinics of North America: Food Animal Practice. 2014;**30**(1):127-142

TCP-Fluorapatite Composite Scaffolds: Mechanical Characterization and *In Vitro*/*In Vivo* Testing

Achouak Elghazel, Rym Taktak, Jamel Bouaziz,
Slim Charfi and Hassib Keskes

Additional information is available at the end of the chapter

http://dx.doi.org/10.5772/intechopen.69852

Abstract

In the present paper, we investigate the biological performance of the tricalcium phosphate ceramic (β-TCP) bone substitute combined with the fluorapatite (Fap). Porous biocomposites consisting of β-tricalcium phosphate (β-TCP) with 26.5% fluorapatite (Fap) were elaborated and characterized in order to evaluate its potential application in bone graft substitute. Bioactivity was determined with *in vivo* and *in vitro* tests by immersion of samples in simulated fluid body (SBF) for several periods of time. Clinical, radiological, and histological assessments were then carried out to evaluate the biological properties of developed β-TCP-26.5% Fap composites. An *in vivo* investigation revealed the biological properties of the prepared macroporous scaffolds, namely, biocompatibility, bioactivity, biodegradability, and osteoconductivity. The morphological characteristics, granule size, and chemical composition were indeed found to be favorable for osseous cell development. All histological observations of the preliminary *in vivo* study in the tibia of rabbits proved the biocompatibility and the resorption of the investigated bioceramic. In contrast, the implantation period will have to be optimized by further extensive animal experiments.

Keywords: bone substitute, tissue engineering, bioceramic, tricalcium phosphate, *in vivo*, *in vitro*, implant

1. Introduction

Tissue engineering applies methods from materials engineering and life sciences to create artificial constructs for regeneration of new tissue. Even though a range of tissues has been studied, the translation of engineered tissues to clinical applications has been limited [1]. Researchers in bone tissue engineering are working to develop alternatives to allogenic and autologous bone grafts in order to address the growing needs of the population, and much of the research is scaffold based [2–7].

Calcium phosphate ceramics have been extensively used to produce porous scaffolds due to their bone-like chemical composition as well as excellent biological properties, including biocompatibility and osteoconductivity [8–11]. The use of these bioceramics was always restricted because of its fragility and the weak rupture resistance [12]. Hence, there was a need for maximizing mechanical properties of tricalcium phosphate β-TCP suitable for orthopedic applications. In recent investigation, a β-TCP-fluorapatite (Fap) composite has been developed for biomedical applications [13]. These β-TCP-Fap have shown a good combination of compressive strength (95 MPa), flexural strength (15 MPa), and fracture toughness (2.9 MPa $m^{1/2}$) [14]. Pure fluorapatite is known to possess a potential advantage with its high chemical stability and aptitude to delay caries' process without the biocompatibility degradation [15]. In addition, it has a much lower solubility in biological fluid than hydroxyapatite. In several studies, it has been proved that the amount of the released fluoride ions F− affects directly cell attachment, proliferation, morphology, and differentiation of osteoblast cells [16, 17]. It is known that the fluorine ion itself enhances mineralization and crystallization [17]. Since such bioceramic β-TCP-26.5% wt% Fap composition has never been investigated *in vitro* and *in vivo* response as bone substitute. Among various properties, biological response of the developed composites should be evaluated in the light of its potential used in tissue engineering. It worth to note that all newly developed biomaterials are subjected to be approved based on their *in vivo* response results. In fact, *in vitro* experiments cannot produce significant evidences toward the biocompatibility of the investigated materials due to the absence of several hormones and enzymes [18].

In view of the above, the aim of this study is to investigate the bioactivity response of the new biphasic ceramic *in vitro* in simulated fluid body (SBF) and to evaluate its biocompatibility and also the process of bone regeneration in rabbits. The clinical, radiological, and histological assessments were performed.

2. Materials and methods

2.1. Materials

In order to elaborate β-TCP-Fap composites, the used materials are the commercial tricalcium phosphate (Fluka) and synthesized fluorapatite. The Fap powder was synthesized by the precipitation method [19]. In order to improve the biocompatibility of the fluorapatite and the strength of tricalcium phosphate effectively and to search for an approach to produce high performances of the tricalcium phosphate-fluorapatite ceramics. In this study, Fap has been used with a fixed 26.5 wt% amount because the human bone contains 1 wt% of fluorine approximately [20]. Estimated quantities of each powder were milled with absolute ethanol and treated by ultrasound machine for 20 min. The milled powder was dried in a low temperature oven at 80°C to eliminate the ethanol and generate a finely divided powder.

http://dx.doi.org/10.5772/intechopen.69852

The sintered samples used in *in vitro* study [21] were prepared using a molding process to form pellets of 10 mm ϕ × 2 mm of thickness (**Figure 2a**). The details of process optimization, density value, and microstructure could be found elsewhere [22]. All the samples used in the study had almost the same surface area and volume. It is worth noting that the experiments were repeated at least three times. At the end of experiments, all the samples were characterized by various technical tools. Scanning electron microscopy (SEM) was done to reveal the changes in topographical features and to visualize calcium phosphate crystals, the final products for the β-TCP-Fap reaction.

In order to obtain three-dimensional bioceramic for the *in vivo* evaluation, the β-TCP-26.5% Fap samples were prepared using polymeric sponge replication method [23–27]. This technique is based on the reproduction of a polymer foam with open cells by impregnation of a ceramic slurry and removing the support by thermal treatment (**Figure 1b**). This later was optimized based on thermal analysis [28]. Based on previous studies [13], the composite samples were sintered at 1300°C for 90 nm. Helium porosimetry and the Archimedes method were used for investigating density and porosity (both open and closed) of sintered samples.

The mechanical strength of the obtained macroporous biphasic ceramic β-TCP-26.5% Fap was also determined by Lloyd testing machine. The specimens used for compression test were prepared into cylinders (9.0 mm in diameter and 18.0 mm in length) as instructed in ASTM standards C1424-15.

(a) (b)

Figure 1. Photograph of the tested specimens: (a) for *in vitro* tests and (b) for *in vivo* tests.

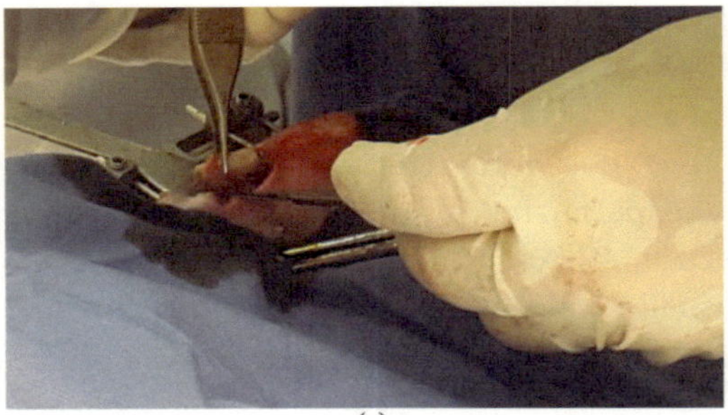

(a)

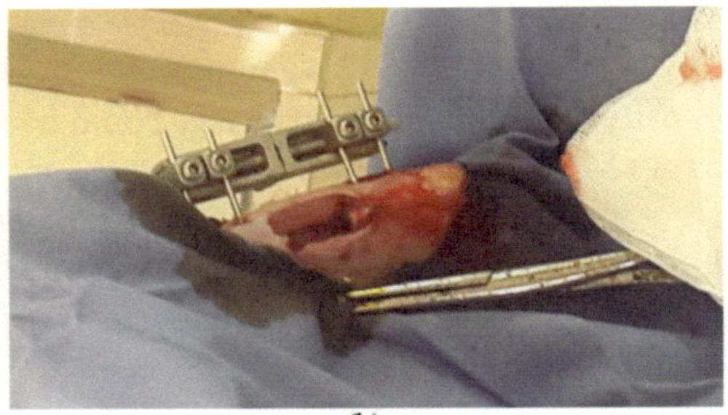

(b)

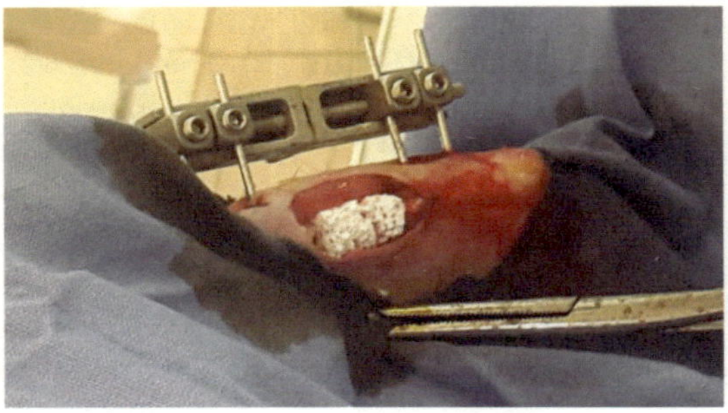

(c)

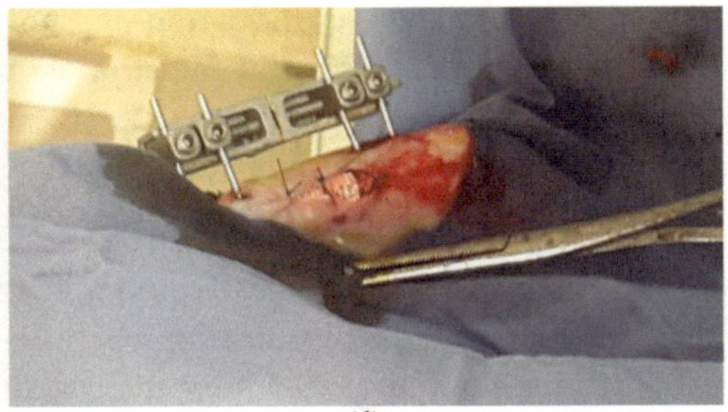

(d)

Figure 2. Digital camera images: (a and b) showing the implantation site and (c and d) β-TCP-26.5% Fap scaffold implanted inside the tibia of rabbit.

2.2. Scanning electron microscopy

The microstructure of the sintered samples was observed by scanning electronic microscopy (SEM) (JSM-5400). The observations were performed on gold-coated specimens and investigated at an acceleration voltage of 15 kV for all samples.

2.3. *In vitro* bioactivity assessment in SBF

Bioactivity was checked through tests of immersion in simulated body fluid (SBF) [21]. In order to study the dissolution behavior of the β-TCP-Fap in physiological environment, the *in vitro* dissolution study was carried out by immersing batches of samples for six different time periods of 4, 24, and 49 h and 6, 12, and 30 days. The initial pH of the solution was kept at 7.42 with tris(hydroxymethyl)-aminomethane and hydrochloric acid. The solution temperature was maintained at body temperature, i.e., 37°C using an incubator. 100 ml of SBF solution was used for each sample. The ion concentration of the SBF is shown in **Table 1**.

2.4. *In vivo* implantation experiments

2.4.1. *Surgical procedures*

Given the closeness of the metabolism of rabbit s to humans, we estimated that the resorption kinetics in rabbits can fairly be assumed to be similar to that in humans.

Ion	Concentration (mmol/l)	
	SBF	Human plasma
Na^+	142.0	142.0
K^+	5.0	5.0
Mg^{2+}	1.5	1.5
Ca^{2+}	2.5	2.5
Cl^-	103.0	147.8
HCO^{3-}	27.0	4.2
HPO_4^{2-}	1.0	1.0
SO_4^{2-}	0.5	0.5
pH	7.2–7.4	7.4

Table 1. Ion concentrations of the human blood plasma and the SBF solution.

In this experiment, two New Zealand white rabbits with an average weight of 2.1 kg were used. The β-TCP-26.5% Fap pellets were sterilized by irradiation from a ^{60}Co gamma irradiation source at a dose of 25 Gy (Equinox, UK) using standard procedures for medical devices. The surgical procedure earned approval from ethical committee of the Tunisian Association of Laboratory Animal Sciences according to the ICLAS ethical guidelines for researchers to committee. All animals were subject to general, backed with local, anesthesia through intramuscular administration of 10 mg/kg ketamine hydrochloride (Ketaminol®, Germany) and 0.1 mg/kg xylazine (Rompun®, France).

In each animal, one lower limb was prepared aseptically for surgery. After shaving the lower limb segment and disinfecting it with Betadine®, a 4 cm skin incision was made medially over the leg, and the tibia was exposed subperiosteally. A 10 mm bony segmental defect was then created by two osteotomies in the middle portion of the tibia. The osteotomy was performed with an oscillating saw at a right angle to the axis of the bone. The defect was then filled with a suitably shaped bioceramic (**Figure 2(a–d)**). To stabilize the tibia and maintain a direct contact between bioceramic and the bone, a mini-stainless steel monoplanar fixator was used. Four pins were inserted through the skin into the bone and then attached to a steel rod outside the limb. Next, the wound was closed for the periosteum and the subcutaneous tissues using a resorbable suture. Finally, a postoperative bandage was made with a sterile compress after local application of Betadine gel®. After the surgical procedure, all the animals were given postoperative care.

Radiographic images were taken at 4 days, 6 weeks, and 12 weeks after implantation to assess the stability of the bioceramic and follow the general structural health of the operated bone; anteroposterior and lateral radiographs of each operated leg were made. Radiographs were obtained using an X-ray machine.

2.4.2. *Histological evaluation*

For microscopic evaluation and to evaluate composite biocompatibility and the biological response of the bone tissue in contact, histological studies were conducted after 12 postoperative weeks. The animals were euthanized, and then the sites of implantations were grossly examined for any evidence of tissue reaction, and tibia bones with the test implant materials were removed and fixed in 10% formaldehyde during 24 h to immobilize the cells for subsequent histological studies. The timing was selected to assess the performance of the biomaterial bone formation before degradation. The samples were then dehydrated in three successive acetone baths. Next, they were placed in methyl methacrylate (MMA) with decalcification. Sections of 4 μm thick were debited along a transverse plane using a sliding microtome (Reichert-Jung). Once colored with hematoxylin eosin, the cut was mounted between slide and slip cover. The colored cuts were observed through a binocular microscope (Olympus® CX-21i), and the observations were photographed with various magnifications using the digital camera.

3. Results and discussion

3.1. Bioactive surface characterization (*in vitro* test): SEM results

To investigate the dissolution behavior of the samples, SEM was performed. It has been seen (**Figure 3b**) that the surface of soaked sample becomes more porous after 4 h immersion, which is a signature that dissolution has occurred. The dissolution affects the strength by

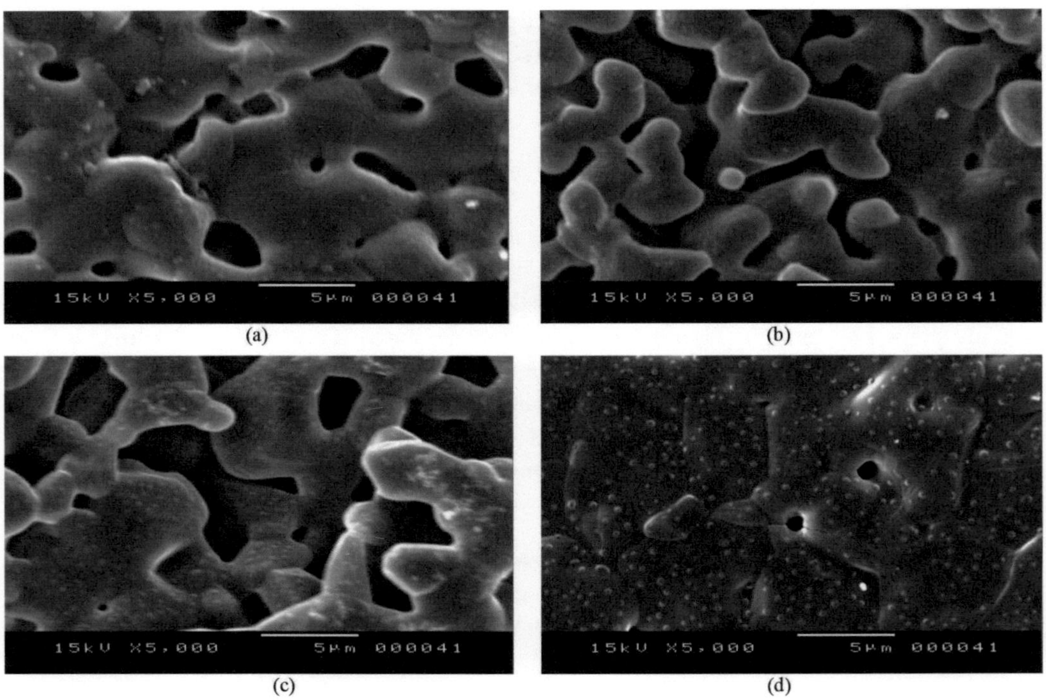

Figure 3. SEM micrographs: (a) sintered β-TCP-26.5% Fap before immersion in SBF, (b) development of porous regions after 4 h of dissolution process, (c) nucleation of apatite crystals after 24 h, and (d) pores filled with formed apatite after 48 h.

increasing the porosity and weakening the grain boundary. After 24 h immersion, the microstructural observation indicates that the surfaces of grains appear mottled.

SEM (**Figure 3c**) indicates precipitation of nodular apatite on the biocomposite grains. With the increase in immersion time, the quantity and size of these apatite particles increase gradually. We note the formation of apatite agglomerate. The apatite formation fills the pores and increases the density and strength of the material (**Figure 3d**). Moreover, EDS analysis also performed on the surface of the soaked sample revealed the presence of calcium and phosphorous elements on the surface, with Ca-P equal to 1.67. This Ca-P ratio is in accordance to stoichiometric biological apatite and indicated that the deposit form during SBF conditioning is apatite layer.

SEM images depict the change in the topographical features recorded for various time intervals during experiment. In fact, the precipitation is more prevalent after immersion showing the formation of bone-like apatite layer due to the dissolution of Ca^{2+} and PO_4^{3-}, followed by deposition of Ca-P (**Figure 4a**). Higher magnification SEM micrograph reveals that particles of apatite grow in a flake-like form and many of these formed agglomerates on the surface of sample (**Figure 4c**). After soaking for 12 days, in the surface of sample, tiny ball apatite particles that formed porous agglomerates were found (**Figure 5b and c**). This outcome has been found by previous studies that proved that the apatite formation occurs in two stages: a formation of globular particles followed by appearance of aggregate which joined together to form a covering layer [29].

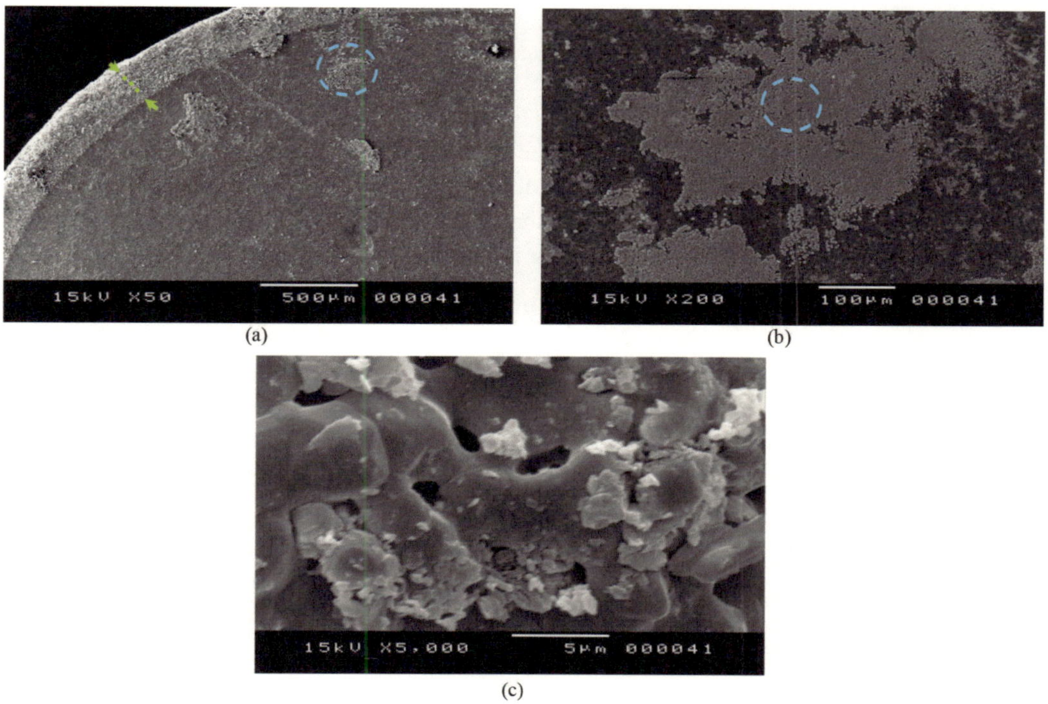

Figure 4. SEM images show the topographic features of β-TCP-26.5% Fap sample after 6 days of dissolution period, (b) is a magnified image of a selected zone from (a), and (c) apatite grows in a flake-like form.

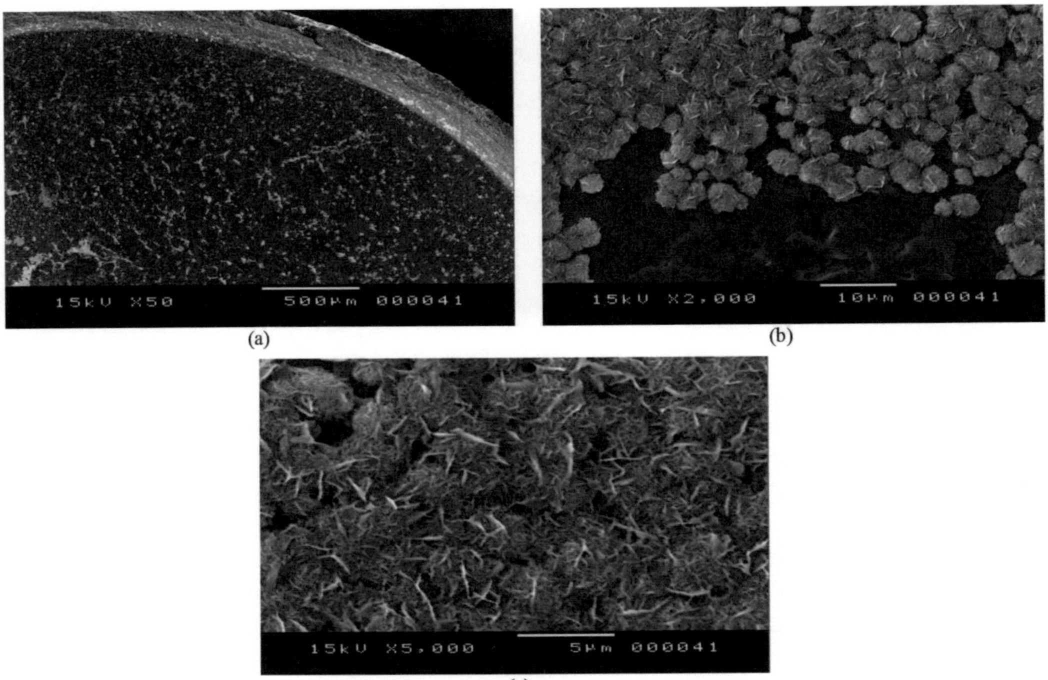

Figure 5. SEM micrographs of sample surface after 12 days of immersion (a) showed the deposit apatite layer, (b) porous agglomerates of apatite particles, and (c) needle-like apatite crystals.

After 30 days of immersion in SBF, a thick and dense mineral layer was formed on the specimen surface, and mineral crystals covered almost the surface of the specimen. **Figure 6b** reveals higher magnification micrograph of the region marked in **Figure 6a** showing typical entanglement of the reprecipitated leaf-like or plate-like crystals (**Figure 6c**).

It is well known that the dissolution of the initial Ca-P compounds has occurred until the oversaturation of the solution, thus inducing the reprecipitation of crystals [10, 30]. The bioactivity of calcium phosphate and other materials has been related to their propensity to nucleate apatite crystals [31]. The presence of bone-like apatite agglomerate on the surface of any implant is always considered as a positive biological response. Therefore, the findings that β-TCP-26.5% Fap shows extensive precipitation of apatite are a clear indication of bioactivity.

3.2 Preliminary *in vivo* evaluation

3.2.1. Microstructure and mechanical characterization of bioceramic scaffolds

The SEM analysis for β-TCP-26.5% Fap scaffolds presented highly porous and interconnective pore architecture with porosity of 75% which has an open porosity 71%. As can be observed in **Figure 7**, the ceramic foams appeared to be macroporous with different pore sizes. The pore size distribution seemed to be uniform. Pore sizes were roughly about 300 μm, which is above the critical value of 100 μm to allow bone [32].

http://dx.doi.org/10.5772/intechopen.69852

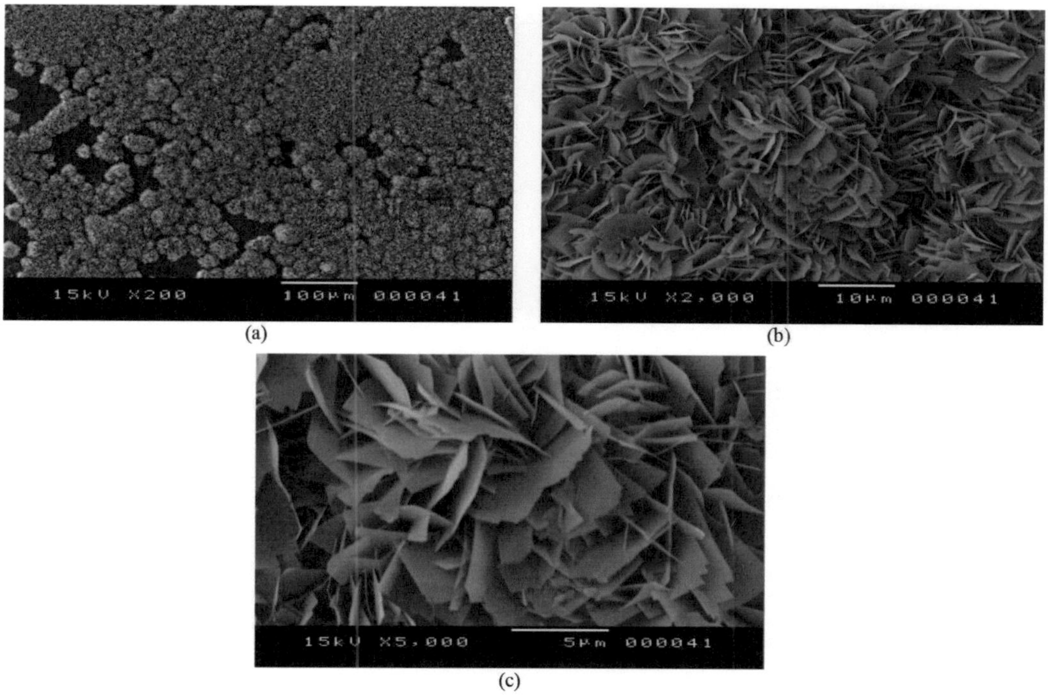

(a) (b) (c)

Figure 6. SEM images of β-TCP-26.5% Fap sample after soaking in SBF for 30 days: (a) joined aggregate forms a covering layer, (b) showing needle-like and plate-like crystals, and (c) is a magnified image of selected zone from (b).

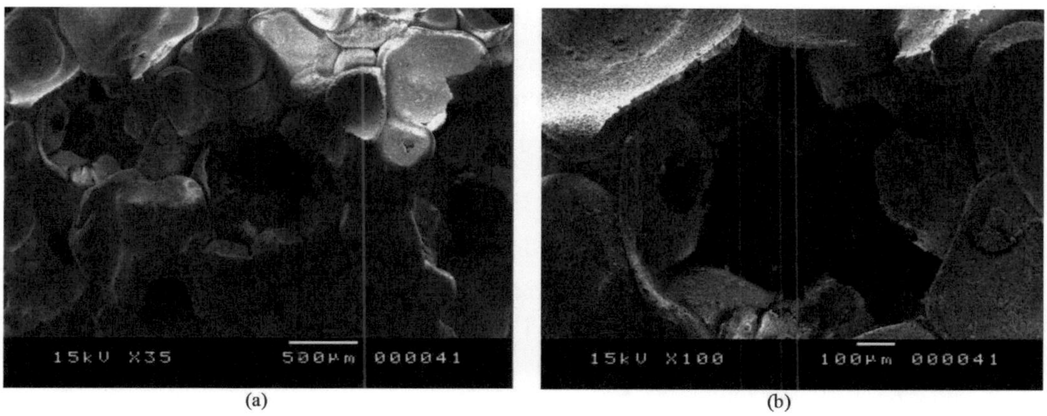

(a) (b)

Figure 7. (a) SEM images of porous β-TCP-26.5% Fap and (b) zoom.

The mechanical strength of the β-TCP-26.5% sintered foam was plotted as function of porosity (**Figure 8**). At least six samples were tested under compression test condition. Average strength values were calculated. The maximum error obtained was found to be less than 5%. As expected, the compressive strength decreased with increasing porosity. Moreover, the compressive strength presented a sharp decrease as soon as macropores are introduced.

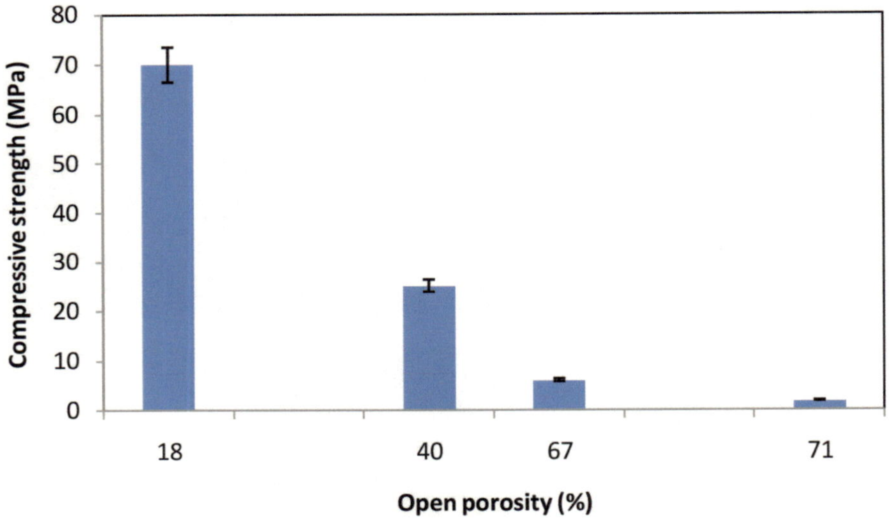

Figure 8. Compressive strength versus open porosity of β-TCP-26.5% bioceramics.

3.2.2 Clinical control and radiological findings

During the postoperative, the daily clinical evaluation exhibited no sign of surgical site infection (i.e., swelling, redness, or wound disunion) on the operated rabbits, which had overcome the tibia osteotomy. After 6 weeks postimplantation, the radiographs showed healing of the host bone and mild degeneration of the implanted bioceramic. The bioresorption is partly revealed by the change in implant dimension (**Figure 9b**).

At 12 postoperative weeks, the radiograph showed complete healing of the host bone and disappearance of the implanted biomaterial (**Figure 9c**). In fact, bone consolidation was noted in the two rabbits.

3.2.3. Organ harvesting and general observations

After completion of the expected implantation time (12 postoperative weeks), the operated rabbits were sacrificed, and the samples were extracted immediately and then stored in an ethanol or BBS solution prior to the histological study. The explants were then transferred to the laboratory of histology and embryology for inclusion in resin and histological study. Given the small amount of data currently being processed, we cannot conclude on the osteoconductive potential of our implant. However, it should be noted that no major problems occurred during the operation or during the implementation periods. In addition, observation of the explants also showed almost complete dissolution of the elaborated implant and the formation of a new neoformed bone (**Figure 10**). This observation will be confirmed by the histological study.

3.2.4. Histological analysis

Although tricalcium phosphate and fluorapatite are reported to be biocompatible *in vivo*, the combination of these phases is not yet tested *in vivo*. Therefore, the biocompatibility

http://dx.doi.org/10.5772/intechopen.69852

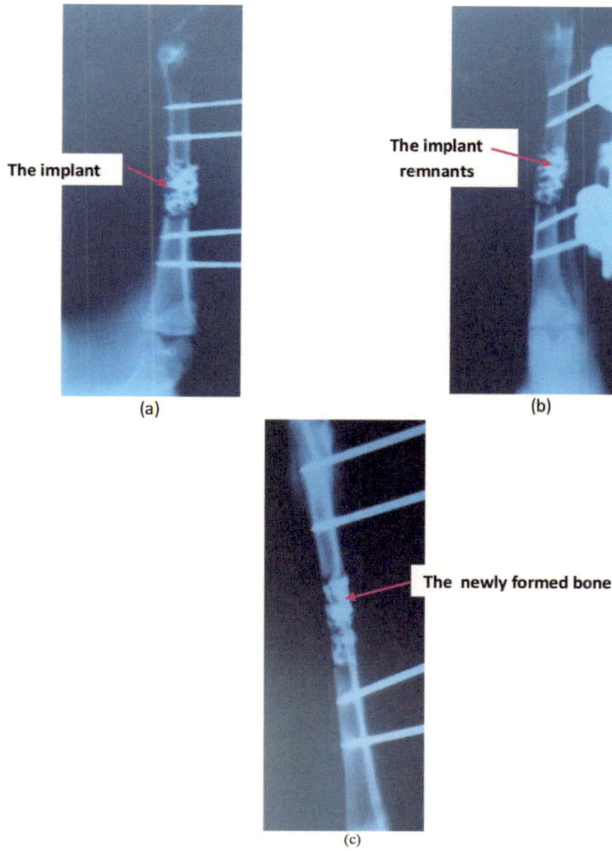

Figure 9. X-ray radiographs of postimplant rabbit's tibia: (a) after 4 days, (b) after 6 postoperative weeks, and (c) after 12 postoperative weeks.

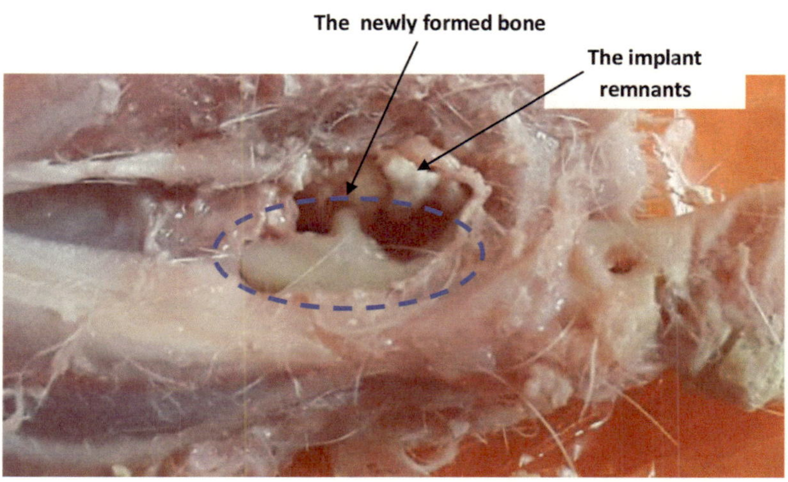

Figure 10. Extraction of the implant after 12 postoperative weeks.

evaluation of this biphasic material would be of great interest. It is to be noted here that in the investigated composite, the major phase is β-TCP, which is well known for its bioresorbability. However, in our case, the resorption is limited due to the presence of second-phase fluorapatite.

Figure 11 illustrates the representative optical microscopy images of the histological sections of bone implant interface. A cellular infiltrate, predominantly composed of macrophages, was observed. Although the short-term implantation of 12 weeks is being carried out in this investigation, the experimental observations should provide clear indication about the biocompatibility *in vivo*. These findings revealed that β-TCP-26.5% Fap is biocompatible and appears to induce acceptable *in vivo* response. However, it appeared that the resorption rate of this bioceramic was far too low to induce neobone formation after 12 weeks postimplantation. Despite these results, the proposed material presented markedly better mechanical properties than monophasic β-TCP [14]. Hence, the incorporated fluorapatite can be used to modulate the biological and the mechanical properties. Thus, the challenge is to find a balance between the biological performance and the changes in mechanical behavior provoked by both the macroporosity and the composite effects. Further encouragement from these results is that the novel bioceramic should have an attracted potential as bioactive coating for orthopedic. Such application required long-term and sufficient mechanical stability when under physiological stresses associated with locomotion to not detach from the implant surface [33].

Considering the importance of the time required for integration and osteointegration, which is determined by the newly formed bone at the host bone-material interface, we should investigate the biological response according to ISO-10993 guidelines. Additionally, the implantation period will have to be optimized by further extensive animal experiments. An *in vitro cell* culture model can be also an *alternative* to study the cytocompatibility of the novel bioceramic as well as other research [34–36].

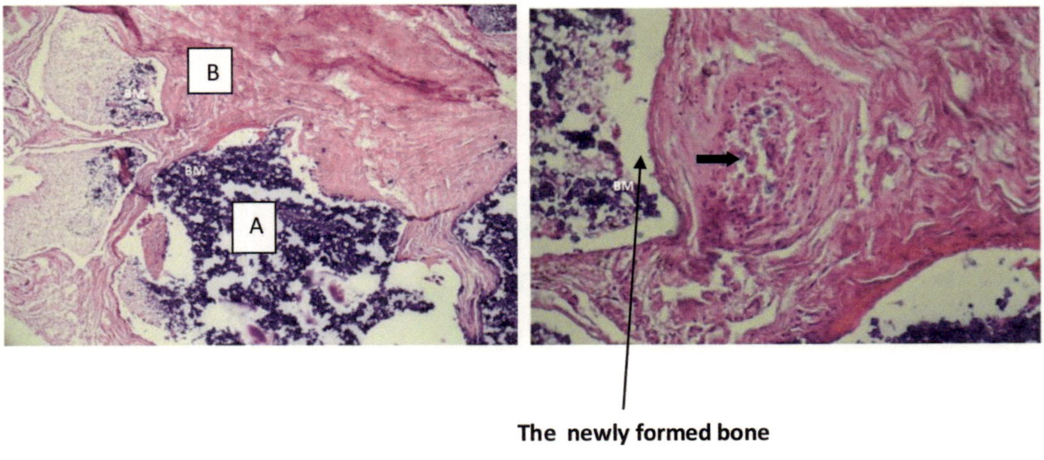

Figure 11. Histological sections of β-TCP-26.5% Fap scaffold after 12 weeks of implantation in rabbit. (A) Fibrous tissue surrounding the biomaterial. (B) Inflammatory infiltrates around degradated biomaterial (➡).

http://dx.doi.org/10.5772/intechopen.69852

4. Conclusion

In this study, a novel biphasic material that consisted of two major phases of tricalcium phosphate and fluorapatite was investigated. During bioactivity experiments, the bone substitute material exhibited to be able to produce porous apatite layer *in vitro*. The combination of dissolution and precipitation is found to be the mechanism of apatite formation. The results showed that fluoride released ion decreased the dissolution rate by improving the crystallization.

A preliminary *in vivo* investigation revealed that the new bioceramic scaffold with highly porous architecture exhibits biocompatibility. It can be concluded that β-TCP-26.5% Fap may take few more weeks for deposition of neobone. Besides, there are many aspects related to *in vivo* behavior that need further *in vivo* results in order to confirm or improve the formulation of the porous bioceramic that should promote both osteoconductive and resorption processes.

Author details

Achouak Elghazel[1]*, Rym Taktak[2], Jamel Bouaziz[1], Slim Charfi[3] and Hassib Keskes[4]

*Address all correspondence to: elgazelachwak@live.fr

1 Laboratory of Industrial Chemistry, National School of Engineers of Sfax, Sfax, Tunisia

2 Materials Engineering and Environment Laboratory, National School of Engineers of Sfax, Sfax, Tunisia

3 Department of Pathology CHU Habib Bourguiba, Sfax, Tunisia

4 Orthopedics-Traumatology Research Unit, Faculty of Medicine of Sfax, Tunisia

References

[1] Lanza R. Principles of Tissue Engineering, 4th ed. USA: Elsevier; 2014

[2] Gómez S, Vlad MD, López J, Fernández E. Design and properties of 3D scaffolds for bone tissue engineering. Acta Biomaterialia. 2016;**42**:341-350

[3] Reno CO, Lima BFAS, Sousa E, Bertran CA. Scaffolds of calcium phosphate cement containing chitosan and gelatin. Materials Research. 2013;**16**:1362-1365

[4] Hea D, Zhuanga C, Xub S, Kec X, Yanga X, Zhangc L, Yangc G, Chend X, Moud X, Liue A, Goua Z. 3D printing of Mg-substituted wollastonite reinforcing diopside porous bioceramics with enhanced mechanical and biological performances. Bioactive Materials. 2016;**1**:85-92

[5] Stevens MM. Biomaterials for bones tissue engineering. Materials Today. 2008;**11**:18-25

[6] Bohner M, Galea L, Doebelin N. Calcium phosphate bone graft substitutes: Failures and hopes. Journal of the European Ceramic Society. 2012;**32**(Special Issue: E Cer S XII, 12th Conference of the European Ceramic Society):2663-2671

[7] Yea X , Caia S, Xub G, Doua Y, Hub H, Yeb X. Preparation and in vitro evaluation of mesoporous hydroxyapatite coated β-TCP porous scaffolds. Materials Science and Engineering. 2013;**33**:5001-5007

[8] Szubert M, Adamska K, Szybowicz M, Jesionowski T, Buchwald T, Voelkel A. The increase of apatite layer formation by the poly(3-hydroxybutyrate) surface modification of hydroxyapatite and β-tricalcium phosphate. Materials Science and Engineering. 2014;**34**:236-244

[9] Azevedo AS, Sá MJC, Fook MVL, Nóbrega Neto PI, Sousa OB, Azevedo SS, Teixeira MW, Costa FS, Araújo AL. Use of chitosan and β-tricalcium phosphate, alone and in combination, for bone healing in rabbits. Journal of Materials Science Materials in Medicine. 2014;**25**:481-486

[10] Zhang Y, Li S, Wu C. The in vitro and in vivo cementogenesis of CaMgSiO bioceramic scaffolds. Journal of Biomedical Materials Research. Part A. 2014;**102**:105-116

[11] Levent Aktuğa S, Durdu S, Yalçınc E, Çavuşoğluc K, Usta M. Bioactivity and biocompatibility of hydroxyapatite-based bioceramic coatings on zirconium by plasma electrolytic oxidation. Materials Science and Engineering. 2017;**71**:1020-1027

[12] Boslama N, Chevalier Y, Bouaziz J, Ben Ayed F. Influence of the sintering temperature on Young's modulus and the shear modulus of tricalcium phosphate—Fluorapatite composites evaluated by ultrasound techniques. Materials Chemistry and Physics. 2013;**141**:289-297

[13] Elghazel A, Taktak R, Bouaziz J. Determination of elastic modulus, tensile strength and fracture toughness of bioceramics using the flattened Brazilian disc specimen: Analytical and numerical results. Ceramics International. 2015;**41**:12340-12348

[14] Elghazel A, Taktak R, Bouaziz J. Combined numerical and experimental mechanical characterization of a calcium phosphate ceramic using modified Brazilian disc and SCB specimen. Materials Science and Engineering A. 2016;**670**:240-251

[15] Ben Ayed F, Bouaziz J. Sintering of tricalcium phosphate–fluorapatite composites by addition of alumina. Ceramics International. 2008;**34**:1885-1892

[16] Tredwin CJ, Young AM, Abou Neel EA, Georgiou G, Knowles JC. Hydroxyapatite fluor-hydroxyapatite and fluorapatite produced via the sol-gel method: dissolution behaviour and biological properties after crystallization. Journal of Materials Science Materials in Medicine. 2014;**25**:47-53

[17] Nathanaela J, Mangalarajc D, Hongb SI, Masudad Y, Rheee YH, Kime HW. Influence of fluorine substitution on the morphology and structure of hydroxyapatite nanocrystals prepared by hydrothermal method. Materials Chemistry and Physics. 2013;**137**:967-976

[18] Nath S, Basu B, Mohanty M, Mohanan PV. In vivo response of novel calcium phosphate-mullite composites: results up to 12 weeks of implantation. Journal of Biomedical Materials Research Part B: Applied Biomaterials. 2009;**90**:547-557

[19] Ben Ayed F, Bouaziz J, Bouzouita K. Pressureless sintering of fluorapatite under oxygen atmosphere. Journal of the European Ceramic Society. 2000;**20**:1069-1076

[20] Moreno EC, Kresak M, Zahradink RT. Nature. 1974;**247**:64-65

[21] Kokubo T, Kushitani H, Sakka S, Kitsugi T, Yamamwo T. Solutions able to reproduce in vivo surface structure changes in bioactive glass-ceramic A-W. Journal of Biomedical Materials Research 1989;**24**:721-734

[22] Bouslama N, Ben Ayed F, Bouaziz J. Effect of fluorapatite additive on densification and mechanical properties of tricalcium phosphate. Journal of the Mechanical Behavior of B iomedical Materials 2010;**3**:2-13

[23] Lee JH, Kim HE, Kohr YH. Highly porous titanium (Ti) scaffolds with bioactive microporous hydroxyapatite/TiO_2 hybrid coating layer. Materials Letters. 2009;**63**:1995-1998

[24] Li J, Li S, Van Blitterswijk C, De Groot K. A novel porous Ti6Al4V: Characterization and cell attachment. Journal of Biomedical Materials Research. Part A. 2005;**73**:223-233

[25] Lee JH, Kim HE, Shin KH, Koh YH. Improving the strength and biocompatibility of porous titanium scaffolds by creating elongated pores coated with a bioactive, nanoporous TiO_2 layer. Materials Letters. 2010;**64**:2526-2529

[26] Hsu HC, Hsu SK, Wu SC, Wang PH, Ho WF. Design and characterization of highly porous titanium foams with bioactive surface sintering in air. Journal of Alloys and Compounds. 2013;**575**:326-332

[27] Wang C, Chen H, Zhu X, Xiao Z, Zhang K, Zhang X. An improved polymeric sponge replication method for biomedical porous titanium scaffolds. Materials Science and Eng ineering C. 2017;**70**:1192-1199

[28] Chaari K, Ben Ayed F, Bouaziz J, Bouzouita K. Elaboration and characterization of fluorapatite ceramic with controlled porosity. Materials Chemistry and Physics. 2009;**113**:219-226

[29] Martíneza M, Meseguer-Olmob L, Bernabeu Esclapezb A, Velásqueza PA, De Aza PN. In vitro behavior of α-tricalcium phosphate doped with dicalcium silicate in the system Ca_2SiO_4–$Ca_3(PO_4)_2$ I. Materials Characterization. 2012;**63**:47-55

[30] Hua H, Liua X, Dinga C. Preparation and in vitro evaluation of nanostructured TiO_2/TCP composite coating by plasma electrolytic oxidation. Journal of Alloys and Compounds. 2010;**498**:172-178

[31] Priya A, Nath S, Biswas K, Basu B. In vitro dissolution of calcium phosphate-mullite composite in simulated body fluid. Journal of Materials Science Materials in Medicine. 2010;**21**:1817-1828

[32] Hench LL. Bioceramics. Journal of the American Ceramic Society. 1998;**81**:1705-1728

[33] Zhang BG, Myers DE, Wallace GG, Brandt M, Choong PF. Bioactive coatings for orthopaedic implants-recent trends in development of implant coatings. International Journal of Molecular Sciences. 2014;**15**:11878-11921

[34] Liu H, Li H, Cheng W, Yang Y, Zhu M, Zhou C. Novel injectable calcium phosphate/chitosan composites for bone substitute materials. Acta Biomaterialia. 2006;**2**:557-565

[35] Ulum MF, Arafat A, Noviana D, Yusopa AH, Nasutiona AK, Abdul Kadir MR, Hermawan H. In vitro and in vivo degradation evaluation of novel iron-bioceramic composites for bone implant applications. Materials Science and Engineering C. 2014;**36**:336-344

[36] Liu W, Zhai D, Huan Z, Wu C, Chang J. Novel tricalcium silicate/magnesium phosphate composite bone cement having high compressive strength, in vitro bioactivity and cytocompatibility. Acta Biomaterialia. 2015;**21**:217-227

Section 2

Scaffolds for Soft Tissue Engineering and Organ Repair

Chapter 7

Mechanical Stimulation of Cells Through Scaffold Design for Tissue Engineering

Carolina Oliver Urrutia,
Ma. Victoria Dominguez-García,
Jaime Flores-Estrada, Antonio Laguna-Camacho,
Julieta Castillo-Cadena and Miriam V. Flores-Merino

Additional information is available at the end of the chapter

http://dx.doi.org/10.5772/intechopen.69925

Abstract

Tissue engineering scaffolds attempt to mimic the stem cell environment by creating different biophysical and chemical signals. On the other hand, stem cells are able to sense these characteristics and change their destiny. Scientists try to explain these phenomena through scaffold design and *in vitro* experiments, but the mechanisms implicated remain unclear. Moreover, environment-cell interactions are a key process to get organs and tissues arrangement. Therefore, this chapter deals with the mechanical signals and mechanism involved in cell behaviour through scaffolds as a strategy in tissue engineering.

Keywords: stem cells, extracellular matrix, tissue engineering, scaffold, mechanical cues

1. Introduction

It has been more than 300 years since Robert Hooke first observed a cell and more than 150 years that the cell theory was postulated. Although all living organisms are made up of cells, not all cells are the same. There is a great variety in their shape and most importantly in their function [1, 3]. Different aspects have been revealed about how cells communicate, differentiate and respond to certain stimuli. Nevertheless, the answers remain incomplete and cell responses can be catalogued as dynamic and complex.

Individual cells sense and respond to the environment at different levels (micro- and nanoscale). In multicellular organisms, cells interpret signals of the micro-environment and neighbouring cells. All these signals end up coordinating the growth and development of organs and tissues [1]. According to Bissell et al. [2], the function of a tissue is regulated reciprocally and dynamically. The micro-environment is formed by the extracellular matrix (ECM) and is able to send signals to the cells. These signals can reach the nucleus and direct the cellular behaviour. The above is supported by the theory called 'dynamic reciprocity'. These concepts are important for tissue engineering, which is defined as a multidisciplinary science that applies principles of engineering and life sciences to the development of biological substitutes that restore, maintain and improve the functions of tissues. The concept that cellular behaviour can be directed by modifying cell micro-environments implies that biomaterials need to build and imitate the ECM. Since each cell resides in a different micro-environment, the biomaterials or scaffolds should be available with precise characteristics [3].

In particular, mechanical forces in cell behaviour have only recently begun to receive attention. For example, mechanical overloading can induce deformation and remodelling of cells, which significantly affects the cellular function. Also, living cells support or create forces; mechanical loading induces deformation and remodelling, which influence many aspects of human health and disease.

Therefore, more importance has been given to stress in cell behaviour [4]. Modelling the constitutive behaviour of cells through biophysical signals poses a challenge. The stimuli reside *in vivo*, but the challenge is mimicking the properties *in vitro* [5]. Imitating stem cell biophysical niches with biomaterials could facilitate the production of large numbers of stem cells needed for *in vitro* regenerative medicine. In recent years, researchers have tried to evaluate the significance of physical cues that influence stem cells; such as stiffness of cell culture substrates and other applied mechanical forces [6]. Several studies explore the regulation of stem cells via fluid shear stress, hydrostatic pressure, ECM elasticity, substrate topography and tension [5]. However, how cells can sense mechanical forces or deformation and convert them into signals is not well understood. Also, the mechanism and the communication pathways remain unclear.

2. Mechanical properties of natural extracellular matrix

ECM is a macromolecular aggregate where the cells reside, proliferate and perform different functions. Their components are normally produced by cells or provided by bloodstream [7]. ECM can also be defined as secreted molecules (including grown factors, cytokines and cell adhesion molecules) that are immobilized outside the cell. Macromolecules of ECM are collagen, elastic fibres and proteoglycans; they are mainly responsible for tissue-type specific extracellular architecture [7, 8].

Collagen and elastic fibre system constitute the architecture of ECM [7].Collagen is a large family of molecules having the ability to aggregate making a supramolecular structure. It is composed of three polypeptide chains forming a triple helix. Elastic fibres are assembled by elastin, an insoluble polymer. Additionally, glycoproteins act as adhesion molecules of intercellular substrates, which are very important in cell-cell and matrix-cell interactions.

Basically, proteoglycans are macromolecules covalently associated between polypeptide chains and glycosaminoglycans [8]. They contribute to cellular adhesion through interaction between matrix components and cellular surface [8, 9]. Principal multifunctional glycoproteins include: laminin, fibronectin, tenascins and thrombospondin. They interact with other molecules, such as integrins, cadherins, immunoglobulins and selectins, and serve as a union between ECM and cytoskeleton [8].

ECM has an important role in regulating the development, function and homeostasis of all eukaryotic cells. This matrix provides physical support for cells and participates in establishment and maintenance of differentiated tissues and organs. Also, it regulates the presence of growth factors and receptors, the level of hydration and pH of the local environment [8, 9]. Interactions between cells and the environment (i.e. ECM) are important in processes such as development, homeostasis and pathogenesis [9]. ECM composition and topography are generated through a dynamic biochemical and biophysical interplay between the various cells in each tissue. The mature ECM can also undergo dynamic remodelling in response to environmental stimuli, such as applied force injury, which enables the tissue to maintain homeostasis [10].

Biophysical considerations of native ECM include the mechanical properties, which vary depending on the tissue. For example, animal connective tissues (tendons and the dermis of the skin) can be rough and flexible, or hard and dense like bone. The range of elasticity in tissues is very wide (~0.1 kPa to 20 GPa) (**Table 1**). For instance, the variation in stiffness can have deep effect in cells (spreading, migration, signalling, differentiation and tumour formation) [7, 8]. Differences in the mechanical properties of tissues may depend on the presence of a disease process or the age. For example, while normal breast tissue has an elastic modulus on the order of 1.2 kPa, breast tumours are significantly stiffer (2.4–4.8 kPa). Another example is the progression of carcinomas, where matrix stiffness increases due to an increased deposition of collagen [8].

The theory of tensegrity states that there is a balance of compression and tension. In this context, the elements resist compression and bring the system into a self-sustained state that maintains size and form. The cytoskeleton is a complex structure that supports and responds to mechanical forces and changes depending on the extracellular forces or conformational alterations in the membrane. Forces can be transmitted, due to modifications, directly to the

Tissue	Elastic modulus (kPa)
Pre calcified bone [7]	~80
Trabecular bone [7]	2×10^7
Muscle [7, 9]	~10–13
Brain [7, 9]	~0.2–1
Adipose [9, 11]	~2–4

Table 1. Elastic modulus of different tissues in human body.

cell nucleus and alter shape, rearrange centromeres and modify gene transcription. Therefore, structural alterations and remodelling of the cytoskeleton in response to mechanical forces might be essential in mechanosensing and cell behaviour. Moreover, stem cells can change from quiescence into differentiated cells in response to biophysical signals such as mechanical forces [5, 6].

Force application to a single element results in the distribution of forces and rearrangement of elements that can span across long distances and different scales. Actomyosin filaments can generate tensions, which are driven by molecular motors that convert the chemical energy of adenosine triphosphate into mechanical forces [5]. Different cytoskeletal filament systems are interconnected with each other. Therefore, a tensile stress is generated in cytoskeleton through a balance between opposing forces. Depending on the level of the tensile stress, cell stiffness increase in a proportional manner, this is called prestress. The prestress in cells can be elevated internally by stimulating actomyosin-based contraction, disrupting microtubule compression struts, or externally increasing the ability of the ECM or other cells to resist contractile forces.

Living cytoskeleton is stabilized by a tensile prestress that is generated and maintained through a force balance between contractile actomyosin filaments. Actin cytoskeleton has a prestress transmitted by traction forces that act at cell-anchoring points. There is a coupling between the cytoskeletal contractile actin network and microtubules analogue to tension-compression coupling in tensegrity structures. Stiffening in living cells is mainly due to geometrical rearrangements, bending or buckling of the structures of the cytoskeleton [6, 12].

Overall, mechanical forces play a central role in understanding how biological patterns and morphologies emerge and vary along evolution. In multicellular organisms, tensional forces applied by cells to the ECM are balanced by equal and opposite forces. Stress is defined as force per unit area. Several studies explore the regulation of stem cells via fluid shear stress, hydrostatic pressure, extracellular matrix (ECM) elasticity, substrate topography and tension [5, 11].

There is a challenge in the characterization of the mechanical properties of natural ECM that arise from their complexity and dynamic nature. For instance, the heterogeneous characteristics of ECM complicate the task. Also, the variability of a biological structure depends on several factors (i.e. tissue type, age, etc.). Simple methods used to measure mechanical properties are those based on the analysis of deformations without association with actual forces. More sophisticated methods include the use of tools such as atomic force microscopy (AFM) [13]. For example, Wu et al. [14] described a protocol to measure the membrane plasticity and mechanical dynamics of individual hippocampus neurons in a murine epilepsy model with AFM.

3. Cell matrix interactions

Cells are surrounded by ECM and are responsible for its composition, structure and mechanical properties. For example, fibroblasts build the ECM in soft connective tissues. At the same time, ECM is fundamental in many cellular processes (spreading, migration, proliferation and differentiation) and tissue functions [15]. Therefore, a type of communication is generated between cells and ECM. In the tissues, cells adhere to ECM or to nearby cells through the formation of (cell-matrix and cell-cell). These junctions allow a transmission of signals (i.e. mechanical) among the different biological structures (**Table 2**) [16, 12].

Cell-ECM interactions are mediated by an integrin family of migration-promoting receptors that interact with the actin cytoskeleton in the cell. The integrins are heterodimeric receptors consisting of α and β chains with large ligand-binding extracellular domains and short cytoplasmic domains. Humans have at least 24 different kinds of integrins, which recognize different extracellular structures and have distinct functions depending on the type of cell in which they reside. Recent studies have identified a set of integrin-associated cytoplasmic proteins such as talin, vinculin and p130Cas [13, 18, 19]. Integrins themselves do not have catalytic activity and do not initiate signalling cascades; therefore, signals are transmitted through direct and indirect interactions with several integrins. Nowadays, there are many questions related to the molecular mechanisms mediated by integrins, although it is clear that integrin-mediated adhesions can sense the properties and characteristics of the ECM [14].

On the other hand, actin is an abundant protein, which can be found in globular (G-actin) or filamentous (F-actin) form. Actin proteins are well known as an essential component of the cytoskeleton. Cells have an actin layer coating the plasmatic membrane that has a critical role in controlling changes in cell morphology [6, 14]. Integrins and actin are separated by a high focal adhesion core-region consisting of specific protein layers. It is suggested that the first section includes a signalling layer consisting of cytoplasmic tails, focal adhesion kinase (FAK) and paxillin. The second layer is an intermediate stratum related to force transduction it contains talin and vinculin. The third layer is composed of an actin-regulatory surface containing vasodilator-stimulated phosphoprotein, zyxin and α-actinin [6, 20].

Focal adhesions and actin proteins have important functions in various cell-signalling pathways and cell fate. The signalling and mechanosensory system of the adhesions are organized in a nanoscale manner. Focal adhesions are flat and elongated structures often located near the periphery of cells [21].Recent studies have revealed a set of proteins responsible for

Cell-cell	**Associated proteins**
Tight junction	Transmembrane proteins, actin
Adherens junction	Actin micro-filaments
Desmosome	Intermediate filaments
Gap junction	Connexins
Hemidesmosome	Intermediate filaments
Cell-matrix	
Focal adhesion	Actin micro-filaments, integrin
Hemidesmosome	Intermediate filaments, integrin

Table 2. Cell junctions and associated proteins [12, 13].

sensing mechanical force and regulating cell-ECM and cell-cell adhesions. They are a part of the linkage between cytoskeleton and cell adhesions and are subject to tensile forces produced by actomyosin contraction [22]. Biomaterials (synthetic or natural) can modulate the effects of these soluble factors by temporally or spatially controlling their delivery. They promote the organization of focal adhesions. For example, Sequeira et al. [23] investigated the influence of scaffolds surfaces in cell attachment, tissue morphology and formation of focal adhesion complexes. In this study, they used an adult mouse submandibular salivary gland ductal epithelial cell line. It was a relationship between the focal adhesions complexes formed and the type of substrate used. Moreover, cells seeded in nanofibre scaffolds showed the fewest focal adhesion complexes; meanwhile in polymer films were abundant. Focal adhesions complexes are mechanisms for scaffold-cell communication. Therefore, they are important to sense biomaterial cues that can direct their fate.

Several authors have discussed and trying to explain the communication pathways between the cells and the micro-environment. Bissell et al. [24] have shown in previous studies that some pathways can be turned on and change gene or protein expression indicating a dialogue between the components of the tissues.

4. Understanding mechanical stimulation through scaffold design

Mechanotransduction is the process by which physical cues are translated into biochemical signals. This route is mediated by focal adhesions. There are two types of forces that the cells can experience, those applied from the environment and those that the cell generates itself. In response to external forces or other stimuli, cells can produce internal forces either by extending membranes or by rearranging their actin cytoskeleton. In this way, they produce endogenous contractile forces [25]. It has been suggested that mechanical forces applied to proteins may perturb the conformations and expose the hidden binding sites, resulting in mechanical signalling processes [12, 13].

Externally applied forces are detected by numerous cell-surface adhesion receptors, such as integrins and cadherins. The ability of these receptors to respond to external forces directs cell behaviour and tissue homeostasis. The force that is applied to integrins is sensed and supported by cytoplasmic components, which at the same time are capable of generating a response [6, 16]. Forces applied trigger actin cytoskeletal rearrangements, activating the small GTPase RhoA and enhancing the activity of myosin II. Subsequently, contraction forces are generated through actin and myosin II filaments. These events create a response through the association of adhesion complexes and the establishment of an internal force. This process is known as reinforcement or cell stiffening [16].

An important theory has been introduced, *Extracellular Matrix Tethering Hypothesis*. In this case, the cells do not directly sense the bulk stiffness of the underlying substrate; instead, respond to the mechanical feedback presented by covalently anchored ECM molecules such as collagen. The exact sequence of events and molecular mechanisms remain to be unrevealed [9]. Despite this, it has been well established that the mechanical properties of materials regulate cell behaviour. Although it is unclear how physical properties (i.e. stiffness and viscoelasticity) are capable of controlling different functions in cells, biophysical aspects of the ECM are associated with different cellular actions *in vitro* and *in vivo* [26].

Since biophysical and biochemical properties of native ECMs are difficult to control, synthetic materials are important to recreate mechanical characteristics of ECM. Several studies have analysed cell behaviour depending on different mechanical properties controlled through synthetic substrates. For example, Chaudhuri et al. [19] investigated the influence of hydrogel viscoelasticity and stress relaxation on spreading, proliferation and differentiation of mesenchymal stem cell (MSC). MSC differentiation depended strongly on the initial elastic modulus of 3D hydrogel matrices, with osteogenesis occurring only when the initial elastic modulus was 17 kPa. In this work, an approach to modulate stress relaxation properties in alginate hydrogels was showed and demonstrated that substrate stress relaxation influences cell behaviour [19, 25].

Also, Baker et al. [27] explain mechanisms of how cells interpret ECM stiffness in fibrous networks, which are synthesised by electrospinning and soft lithography and coupled with RGD peptides. They found that fibrillar topography had a stronger influence on cell morphology than the biochemical nature of these interactions. Moreover, Huebsch et al. [28] studied the response of mouse mesenchymal stem cells (mMSC) seeded on injectable void-forming hydrogels. The morphology of mMSC was initially similar in standard and void-forming hydrogels. But, after void formation, cells neighbouring to pores exhibited extended, spread morphology, whereas cells in standard hydrogels maintained a rounded morphology. Furthermore, Fusco et al. [17] studied the existence of a relationship between substrate stiffness and characteristics of focal adhesions with mouse embryo fibroblast NIH/3T3. They developed two different materials: polydimethylsiloxane and polyacrylamide. Their results suggested that focal adhesions are sensitive to elastic properties of the materials while cell spreading is dependent of substrate viscoelasticity.

Other studies have been focused on techniques to stimulate cultured cells with mechanical cues. Special attention has been given to the bone cell lineage since skeleton is responsible for withstanding load bearing. Techniques such as mechanical compressive forces have been shown a variety of cell responses *in vitro* that include cell proliferation and differentiation. Compressive loading of stem cells, in particular, mesenchymal stem cells, has been studied and related to chondrogenic differentiation. For example, Steinmetz et al. [29] demonstrated that in a dynamic *in vitro* environment, human mesenchymal stem cells (hMSC)seeded on hydrogels showed high expression of collagen (I, II and X). This study suggested that mechanical stimulation has a positive impact on hMSC differentiation. Also, some studies have been focused on the effects of mechanical stimulation on diseases states of cells. For instance, Tse et al. [30] suggested that compressive stress accumulated during tumour growth could enable migration of cancer cells, therefore promoting cancer cell invasion.

The mechanical stimulation *in vitro* has been studied with the addition of molecules that are able to induce expression of genes involved in differentiation processes. Also, some studies demonstrated that mechanical loading is able to induce differentiation of cells without the help of biochemical molecules. For instance, a recent study showed the effects of mechanical strain in mesenchymal stem cells seeded on silicon substrates. In this study, the mechanical stimulation was the main variable. There was not any addition of chemicals to promote proliferation or differentiation. Their finding suggested that mechanical strain enhances the proliferation of MSCs [31].

Despite several studies have been documented the effects of mechanical properties on cells with the help of synthetic materials, several questions concerning the mechanisms remain unclear. For example, it is not known the specific pathways that regulate the switching between homeostatic and disease states. Moreover, these states are related to the progression from soft to stiff characteristics in tissues.

5. Tissue engineering and scaffold mechanical properties

Tissue engineering is an interesting approach aimed to reconstruct or create new tissues. However, building new tissues is an enormous challenge, for instance, several tissues are composed of different cell populations [32]. An advantage is the self-repair ability of cells that can be used in favour of tissue engineering scientist. However, the poor understanding of cell repair mechanisms and the additional challenges of biomaterial design have been slowed the progress in this area. When some circumstance (age, wound size, inflammation or chronic disease) inhibits the natural repair process, an alternative method to healing is required. Tissue engineering considers the use of biomaterials and cells from autologous or external sources. Basically, biomaterials or scaffolds are aimed to help cells in the proliferation and differentiation processes. Then, biomaterials and cells are the beginning formulae to create a new organ or tissue. However, we have to remember that cells need to get the right instructions to start the process of self-repair. These instructions are delivered through physical or chemical cues included in biomaterials [3, 23].

The inspiration of biomaterial design is the ECM, which properties are crucial for cell behaviour as discussed before. All cells receive signals from ECM, so scientists have attempted to mimic the physical and chemical characteristics [23]. Some scientists have been focused on imitating patterns, forms, textures and specific characteristics such as mechanical resistance and chemical structure. For example, Zhang et al. [33] constructed a three-dimensional system to create tissue architecture. The scaffold systems were synthesised with an elastic modulus similar to brain tissue. Additionally, they encapsulated a laminin protein, which is a neural ECM component. A rapid maturation of neurons from human induced pluripotent stem cells was associated with the physical properties of the scaffold systems, which are similar to the mechanical properties of the natural extracellular matrix in the brain.

On the other hand, stem cells are widely employed in the tissue engineering area due to their potential to give rise to different cell types. Also, stem cell differentiation through biomaterial mechanical properties remains a critical goal [34]. For example, changes in the bulk stiffness of ECM-coated hydrogels elicit different cell responses. In the case of mesenchymal stem cells, bone differentiation is favoured by stiffer substrates, whereas adipocyte differentiation is promoted by softer substrates. The influence of mechanical properties on stem cell differentiation has been demonstrated on a range of substrates, including collagen and hyaluronic acid gels, Poly(D-lactide-co-glycolide acid) electrospun nanofibres and polydimethylsiloxane, among other biomaterials [35]. For example, Shih et al. [31]

studied the mechanisms of osteogenic differentiation from bone marrow mesenchymal stem cells. Polyacrylamide substrates with different young's modulus were synthesised to analyse secretion of molecules involved in cell differentiation. They found that production of collagen type I increased in cells seeded in stiffer substrates. Also, they demonstrated a higher level of mineralization and a higher FAK and RAK activation (mechanoresponsive elements) when stiffer matrices were used. The expression of integrin was also different depending on the elastic modulus of the biomaterial. For instance, integrin expression per cell was statistically higher on stiffer matrices.

In the same way, Banerjee et al. [36] examined the behaviour of neural stem cells encapsulated in three-dimensional scaffold (alginate hydrogels) with a variable elastic modulus. They analysed the differentiation of cells with neural marker β-tubulin III. Proliferation of cells increased significantly with a decrease in the elastic modulus of hydrogels. The maximum intensity of β-tubulin III staining was observed in cells grown in the hydrogel with the lowest modulus. The modulus ~180 Pa promotes neuronal differentiation which is related to the elasticity of brain tissues. Overall, these results demonstrated the influence of the mechanical characteristics of the biomaterials on cellular behaviour *in vitro*. Moreover, there is a diversity of biomaterial systems that can be used to investigate cellular behaviours to mimic native ECM. Stem cells in these different biomaterials had a diverse behaviour related with mechanical properties of the scaffold and confirm that stiffness is an important factor [37].

In the 1980s, the main function of a biomaterial was limited to support cells. However, as stated above, biomaterials can influence different cellular processes depending of the physical characteristics such as young's modulus [32]. As an example, biomaterial stiffness has been found to affect the transcriptional process [38]. In this context, studies have been shown that certain cell lines develop larger focal adhesions on stiffer surfaces. Also, cell migration speed had showed a dependence on mechanical properties. Other studies had demonstrated that cells migrate preferentially to stiffer surfaces. However, the influence of substrate mechanical properties on cell phenotype also depends on the cell type [23, 38].

The cellular behaviour and mechanical stimuli *in vivo* models have also been analysed. For example, Moshayedi et al. [37] developed a hydrogel material to control neural cells fate *in vivo*. In this study, an injectable hydrogel was designed and showed to promote survival and differentiation towards immature states of human neural progenitor cells. Another study employed magnetically responsive ferrofluid microdroplets to measure local mechanical properties in developing embryos. Their results suggested that tissue mechanics might play a critical role in morphogenesis [39].

6. Additional considerations about cell biomechanics: the case of the adipocyte

As shown in this chapter, cells respond to external environmental forces. Such understanding about cell behaviour would also benefit from studying how cells react to biomechanical disturbances inside them. In this section, a source of physical strain within the cell is presented. The case is made for the fat overload in adipose cells, a common condition in people with obesity.

Adipocytes are cells specialized in storing triglycerides in the form of lipid droplets [40]. The adipocyte can form one giant lipid droplet as large as 100 μm, this constitutes the most efficient cellular packaging of energy per volume, which is a favourable trait to conserve energy that could be used when energy supply decreases [41]. However, there are factors in modern life such as frequent intake of energy-dense food that contribute to adipocyte hypertrophy [42].

A question is why the adipocyte continues accumulating fat even when it is unhealthy [43]. One possibility is that triglyceride accumulation may be part of adaptive mechanisms that prevent toxicity induced by high levels of lipids [44].

Despite the high resilience of adipose cells to fat overload, excessive accumulation of triglycerides within the adipocyte impairs its cellular functions [45]. For instance, a negative effect of excessive packing of fat by the adipocyte includes induction of cellular hypoxia through inhibition of effective oxygen supply from the circulation [46]. Another negative intracellular effect of adipocyte hypertrophy is a mechanical stress on the endoplasmic reticulum, condition that impairs protein folding [47]. Indeed, the adipocyte displays a potent inflammation as effect of the high storage of fat [48]. The impact of hypertrophy can be so adverse as to trigger adipocyte apoptosis [49]. Nevertheless, before such ultimate death phase occurs, the adipocyte enacts a series of responses to improve its own functioning as fatness accumulation increases in its intracellular space.

It has been proposed that adipocytes contribute to sense the levels of body energy (fat content) and are able to signal such state to the central nervous system that in turn modulates individual's intake and expenditure [41]. Although the somatic influence on appetite seems to be not as strong as needed to reduce overeating behaviour [50], to deal locally with lipid accumulation, the adipocyte increases its metabolic pathway for fat oxidation [51, 52] In addition, the adipocyte signals immune cells that phagocyte and oxide fat [53].

There is on-going research showing promising findings that adipocytes are a ready body source of cells that could be used for tissue-engineering reconstruction [54]. For instance adipose stem cells (ASCs) are a great promise for regenerative medicine applications. The use of de-cellularized human adipose tissue ECM combined with ASCs is a strategy that can be employed in the tissue engineering area [55]. Kim and collaborators designed a free-cell scaffold for adipose tissue regeneration; the aim was creating a specific scaffold to recruit cells into a desire cell type [56].Hence, research on adipocyte biomechanics has potential for evidence that could be applied to the development of methods for tissue construct. Indeed, a high proportion of reconstructive procedures involve repairing adipose tissue.

This case of adipocyte behaviour in the face of overload of fat illustrates how cells of the body are highly specialized systems that display impressive responses against mechanical forces outside and inside them. Thereby, any engineered treatment related to biomaterials for cells and tissues should rely on proper understanding of cell behaviour under unfavourable stimuli. In particular, biomaterials characteristics should aim to act in synergy with the natural cell systems in order to improve the conditions in which healing of cells and tissues can occur.

7. Future directions

Since tissue engineering appeared in the 1990s, research on biomaterials has increased and advanced greatly. Now these materials have specific characteristics depending on the tissue in which they want to be applied. Moreover, the physical characteristics (i.e. mechanical) of living systems are important in order to create artificial scaffolds. It is possible to reprogram cells through mechanical cues and synthetic constructs. However, the challenges consist of controlling such properties according to certain outcomes in cell behaviour. Also, the integration of more than one mechanical characteristic (i.e. external dynamic stimuli and matrix stiffness) imitating the *in vivo* conditions is required. Finally, further studies of mechanisms that direct cells to create new tissues are important to understand the way cells behave and respond to external and internal mechanical forces.

Author details

Carolina Oliver Urrutia[1,2], Ma. Victoria Dominguez-García[1], Jaime Flores-Estrada[3], Antonio Laguna-Camacho[1], Julieta Castillo-Cadena[1] and Miriam V. Flores-Merino[1]*

*Address all correspondence to: mvfloresm@uaemex.mx

1 Centro de Investigacion en Ciencias Medicas (CICMED), Universidad Autonoma del Estado de Mexico (UAEMex), Toluca, State of Mexico, Mexico

2 Facultad de Enfermeria, UAEMex, Toluca, State of Mexico, Mexico

3 Facultad de Quimica, UAEMex, Toluca, State of Mexico, Mexico

References

[1] Bray A, Johnson H, Raff L, Walter R. Essential Cell Biology. 3rd ed. New York:Garland Science; 2010. p. 731

[2] Inman J, Robertson C, Mott J, Bissell M. Mammary gland development: Cell fate specification, stem cells and the microenvironment. Development. 2015;**142**:1028-1042. DOI: 10.1242/dev.087643

[3] Dimmeler S, Ding S, Rando T, Trounson A. Translational strategies and challenges in regenerative medicine. Nature Medicine. 2014;**20**(8):814-821. DOI: 10.1038/nm.3627

[4] Bao G, Suresh S. Cell and molecular mechanics of biological materials. Nature Materials. 2003;**2**:715-725. DOI: 10.1038/nmat1001

[5] Govey PM, Loiselle AE, Donahue HJ. Biophysical regulation of stem cell differentiation. Current Osteoporosis Reports. 2013;**11**(2):83-91. DOI: 10.1039/c4bm00128a

[6] Dalby MJ, Gadegaard N, Oreffo ROC. Harnessing nanotopography and integrin-matrix interactions to influence stem cell fate. Nature Materials. 2014;**13**(6):558-569. DOI: 10.1038/nmat3980

[7] Wade RJ, Burdick JA. Engineering ECM signals into biomaterials. Materials Today. 2012;**15**(10):454-459. DOI: 10.1016/S1369-7021(12)70197-9

[8] Keely P, Nain A. Capturing relevant extracellular matrices for investigating cell migration. F1000Research. 2015:1-14. DOI: 10.12688/f1000research.6623.1

[9] Das RK, Zouani OF. A review of the effects of the cell environment physicochemical nanoarchitecture on stem cell commitment. Biomaterials. 2014;**35**(20):5278-5293. DOI:0.1016/j.biomaterials.2014.03.044

[10] Mouw JK, Ou G, Weaver VM. Extracellular matrix assembly: A multiscale deconstruction. Nature Reviews Molecular Cell Biology. 2014;**15**(12):771-785. DOI: 10.1038/nrm3902

[11] Higuchi A, Ling QD, Chang Y, Hsu ST, Umezawa A. Physical cues of biomaterials guide stem cell differentiation fate. Chemical Reviews. 2013;**113**(5):3297-3328. DOI:0.1021/cr300426x

[12] Ridley AJ. Cell migration: Integrating signals from front to back. Science. 2003;**302**(5651):1704-1709. DOI: 10.1126/science.1092053

[13] Polacheck W, Chen C. Measuring cell-generated forces: A guide to the available tools. Nature Methods. 2016;**13**(5):415-423. DOI: 10.1038/nmeth.3834

[14] Wu X, Muthuchamy M, Reddy DS. Atomic force microscopy protocol for measurement of membrane plasticity and extracellular interactions in single neurons in epilepsy. Frontiers in Aging Neuroscience. 2016;**8**:1-12. DOI: 10.3389/fnagi.2016.00088

[15] Fusco S, Panzetta V, Embrione V, Netti PA. Crosstalk between focal adhesions and material mechanical properties governs cell mechanics and functions. Acta Biomaterialia. 2015;**23**:63-71. DOI: 10.1016/j.actbio.2015.05.008

[16] Levy JR, Holzbaur ELF, Parsons JT, Horwitz AR, Schwartz MA, Larson DR, et al. Cell adhesion: Integrating cytoskeletal dynamics and cellular tension. Molecular Cell. 2010;**11**(9):633-643. DOI: 10.1038/nrm2957

[17] Khalil AS, Xie AW, Murphy WL. Context clues: The importance of stem cell material interactions. ACS Chemical Biology. 2014;**9**(1):45-56. DOI: 10.1021/cb400801m

[18] DeMali KA, Sun X, Bui GA. Force transmission at cell-cell and cell-matrix adhesions. Biochemistry. 2014;**53**(49):7706-7717. DOI: 10.1021/bi501181p

[19] Chaudhuri O, Gu L, Klumpers D, Darnell M, Bencherif SA, Weaver JC, et al. Hydrogels with tunable stress relaxation regulate stem cell fate and activity. Nature Materials. 2015;**15**:326-333. DOI: 10.1038/nmat4489

[20] Sergé A. The molecular architecture of cell adhesion: Dynamic remodeling revealed by videonanoscopy. Frontiers in Cell and Developmental Biology. 2016;**4**(36):1-9. DOI: 10.3389/fcell.2016.00036

[21] Geiger B, Bershadsky A, Pankov R, Yamada KM. Transmembrane crosstalk between the extracellular matrix-cytoskeleton crosstalk. Nature Reviews Molecular Cell Biology. 2001;**2**(11):793-805. DOI: 10.1038/35099066

[22] Yan J, Yao M, Goult BT, Sheetz MP. Talin dependent mechanosensitivity of cell focal adhesions. Cellular and Molecular Bioengineering. 2015;**8**(1):151-159. DOI: 10.1007/s12195-014-0364-5

[23] Sequeira SJ, Soscia DA, Oztan B, Mosier AP, JeanGilles R, et al. The regulation of focal adhesion complex formation and salivary gland epithelial cell organization by nanofibrous PLGA scaffolds. Biomaterials. 2012;**33**(11):3175-3186. DOI: 10.1016/j.biomaterials.2012.01.010

[24] Bissell MJ, Kenny PA, Radisky DC. Microenvironmental regulators of tissue structure and function also regulate tumor induction and progression: The role of extracellular matrix and its degrading enzymes. Cold Spring Harbor Symposia on Quantitative Biology. 2005;**70**:343-356. DOI: 10.1101/sqb.2005.70.013

[25] Humphrey JD, Dufresne ER, Schwartz MA. Mechanotransduction and extracellular matrix homeostasis. Nature Reviews Molecular Cell Biology. 2014;**15**(12):802-812. DOI: 10.1038/nrm3896

[26] Elosegui-Artola A, Oria R, Chen Y, Kosmalska A, Pérez-González C, Castro N, et al. Mechanical regulation of a molecular clutch defines force transmission and transduction in response to matrix rigidity. Nature Cell Biology. 2016;**18**(5):540-548. DOI: 10.1038/ncb3336

[27] Baker BM, Trappmann B, Wang WY, Sakar MS, Kim IL, Shenoy VB, et al. Cell-mediated fibre recruitment drives extracellular matrix mechanosensing in engineered fibrillar microenvironments. Nature Materials. 2015;**14**(12):1262-1268. DOI: 10.1038/nmat4444

[28] Huebsch N, Lippens E, Lee K, Mehta M, Koshy ST, Darnell MC, et al. Matrix elasticity of void-forming hydrogels controls transplanted stem cell-mediated bone. Nature Materials. 2015;**14**:1-19. DOI: 10.1038/nmat4407

[29] Steinmetz NJ, Aisenbrey EA, Westbrook KK, Qi HJ, Bryant SJ. Mechanical loading regulates human MSC differentiation in a multi-layer hydrogel for osteochondral tissue engineering. Acta Biomaterialia. 2015;**21**:142-153. DOI: 10.1016/j.actbio.2015.04.015

[30] Tse JM, Cheng G, Tyrrell JA, Wilcox-Adelman SA, Boucher Y, et al. Mechanical compression drives cancer cells toward invasive phenotype. Proceedings of the National Academy of Sciences. 2011;**28**:911-916.DOI: 10.1073/pnas.1118910109

[31] Shih YR V, Tseng KF, Lai HY, Lin CH, Lee OK. Matrix stiffness regulation of integrin-mediated mechanotransduction during osteogenic differentiation of human mesenchymal stem cells. Journal of Bone and Mineral Research. 2011;**26**(4):730-738. DOI: 10.1002/jbmr.278

[32] Place ES, Evans ND, Stevens MM. Complexity in biomaterials for tissue engineering. Nature Materials. Eff Br mindfulness Interv acute pain Exp An Exam Individ Differ. 2009;**8**(6):457-470. DOI: 10.1038/nmat2441

[33] Zhang ZN, Freitas BC, Qiand H, Luxa J, Acab A, Trujillo CA, et al. Layered hydrogels accelerate iPSC-derived neuronal maturation and reveal migration defects caused by MeCP2 dysfunction. Proceedings of the National Academy of Sciences. 2016;**22**:3185-3190. DOI: 10. 1073/pnas.1521255113

[34] Watt FM, Huck WTS. Role of the extracellular matrix in regulating stem cell fate. Nature Reviews Molecular Cell Biology. 2013;**14**(8):467-473. DOI: 10.1038/nrm3620

[35] Matthys OB, Hookway TA, McDevitt TC. Design principles for engineering of tissues from human pluripotent stem cells. Current Stem Cell Reports. 2016;**2**(1):43-51. DOI: 10.1007/s40778-016-0030-z

[36] Banerjee A, Arha M, Choudhary S, Ashton RS, Bhatia SR, Schaffer DV, et al. The influence of hydrogel modulus on the proliferation and differentiation of encapsulated neural stem cells. Biomaterials. 2009;**30**(27):4695-4699. DOI: 10.1016/j.biomaterials.2009.05.050

[37] Moshayedi P, Nih LR, Llorente IL, Berg AR, Cinkornpumin J, Lowry WE, et al. Systematic optimization of an engineered hydrogel allows for selective control of human neural stem cell survival and differentiation after transplantation in the stroke brain. Biomaterials. 2016;**105**:145-155. DOI: 10.l016/j.biomaterials.2016.07.028

[38] Murphy WL, McDevitt TC, Engler AJ. Materials as stem cell regulators. Nature Materials. 2014;**13**(6):547-557. DOI: 10.1038/nmat3937

[39] SerwaneF, MongeraA, RowghanianP, KealhoferDA, LucioAA, HockenberyZM, et al. In vivo quantification of spatially varying mechanical properties in developing tissues. Nature Methods. 2017;**14**:181-186. DOI:10.1038/nmeth.4101

[40] Krahmer N, Farase RV, Walther TC. Balancing the fat: Lipid droplets and human disease. EMBO Molecular Medicine. 2013;**5**(7):905-915.DOI: 10.1002/emmm.201100671

[41] Laguna-Camacho A. Obesidad y control de peso. México: Trillas; 2009

[42] Swinburn B, Sacks G, Ravussin E. Increased food energy supply is more than sufficient to explain the US epidemic of obesity. The American Journal of Clinical Nutrition. 2009;**90**:1453-1456. DOI: 10.3945/ajcn.2009.28595

[43] Heymsfield SB, Wadden TA. Mechanisms, pathophysiology and management of obesity. The New England Journal of Medicine. 2017;**376**:256-266. DOI: 10.1056/NEJMc1701944

[44] Schaffer JA. Lipotoxicity: When tissues overeat. Current Opinion in Lipidology. 2003,**14**(3): 281-287.DOI: 10.1097/01.mol.0000073508.41685.7f

[45] Blüher M. Adipose tissue dysfunction in obesity. Experimental and Clinical Endocrinology & Diabetes. 2009;**117**(6):241-250. DOI: 10.1055/s-0029-1192044

[46] Wood IS, de Heredia FP, Wang B, Trayhurn P. Cellular hypoxia and adipose tissue dysfunction in obesity. The Proceedings of the Nutrition Society. 2009;**68**:370-377. DOI:10.1017/S0029665109990206

[47] Sharma NK, Das SK, Mondal AK, et al. Endoplasmic reticulum stress markers are associated with obesity in nondiabetic subjects. The Journal of Clinical Endocrinology and Metabolism. 2008;**93**:4532-4541. DOI:10.1210/jc.2008-1001

[48] Gregor MF, Hotamisligil GS. Inflammatory mechanisms in obesity. Annual Review of Immunology. 2011;**29**:415-445. DOI: 10.1146/annurev-immunol-031210-101322

[49] Alkhouri N, Gornicka A, Berk MP, et al. Adipocyte apoptosis, a link between obesity, insulin resistance, and hepatic steatosis. Journal of Biological Chemistry. 2010;**285**:3428-3438. DOI: 10.1074/jbc.M109.074252

[50] Park HK, Ahima RS. Physiology of leptin: Energy homeostasis, neuroendocrine function and metabolism. Metabolism. 2015;**64**(1):24-34.0020. DOI: 10.1016/j.metabol.2014.08.004

[51] Lass A, Zimmermann R, Oberer M, Zechner R. Lipolysis - a highly regulated multi-enzyme complex mediates the catabolism of cellular fat stores. Progress in Lipid Research. 2011;**50**(1):14-27. DOI: 10.1016/j.plipres.2010.10.004

[52] Mittendorfer B. Origins of metabolic complications in obesity: Adipose tissue and free fatty acid trafficking. Current Opinion in Clinical Nutrition & Metabolic Care. 2011;**14**(6):535-541. DOI: 0.1097/MCO.0b013e32834ad8b6

[53] Namgaladze D, Lips S, Leiker TJ, et al. Inhibition of macrophage fatty acid β-oxidation exacerbates palmitate-induced inflammatory and endoplasmic reticulum stress responses. Diabetologia. 2014;**57**:1067. DOI: 10.1007/s00125-014-3173-4

[54] Gomillion CT, Burg KJL. Stem cells and adipose tissue engineering. Biomaterials. 2006;**27**:6052-6063. DOI: 10.1016/j.biomaterials.2006.07.033

[55] Wang L, Johnson JA, Zhang Q, Beahm EK. Combining decellularized human adipose tissue extracellular matrix and adipose-derived stem cells for adipose tissue engineering. Acta Biomaterialia. 2013;**9**(11):8921-8931. DOI 10.1016/j.actbio.2013.06.035

[56] Kim JS, Choi JS, Cho YW. Cell-Free hydrogel system based on a tissue-specific extracellular matrix for in situ adipose tissue regeneration. ACS Applied Materials & Interfaces. 2017;**9**(10):8581-8588. DOI: 10.1021/acsami.6b16783

Chapter 8

Peripheral Nerve Reconstruction Using Enriched Chitosan Conduits

Shimon Rochkind, Mira M. Mandelbaum-Livnat, Stefania Raimondo, Michela Morano, Giulia Ronchi, Nicoletta Viano, Moshe Nissan, Akiva Koren, Tali Biron, Yifat Bitan, Evgeniy Reider, Mara Almog, Ofra Ziv-Polat, Abraham Shahar and Stefano Geuna

Additional information is available at the end of the chapter

http://dx.doi.org/10.5772/intechopen.69882

Abstract

The repair of peripheral nerve traumatic lesions still represents a major cause of permanent motor and sensory impairment. In case of substance loss, a nerve guide should be used to bridge the proximal with the distal stump of the severed nerve. The effectiveness of hollow nerve guides is limited by the delay of axonal growth due to the absence of a regeneration substrate inside the conduit. To fasten up nerve regeneration, nerve guides should thus be enriched by a luminal filler. In this study, we investigated, in a 12-mm rat sciatic nerve defect experimental model, the effectiveness of chitosan-based conduits of different acetylation filled either with a hyaluronic acid gel (NVR gel) or with a magnetic fibrin hydrogel, in comparison with traditional autografts. Results showed that all types of artificial nerve conduits led to functional recovery not significantly different from autografts. By contrast, morphological and morphometrical analyses showed that the best results among nerve guides were found in medium degree of acetylation (DAII: ~5%) chitosan conduits enriched with the NVR gel.

Keywords: peripheral nerve injury, chitosan guidance conduit, conduit acetylation, hyaluronic acid gel, magnetic fibrin hydrogel

1. Introduction

Chitosan is a biopolymer derived from chitin that can be extracted in large quantities from the shells of the crustaceans. Among the different applications, the use of chitosan as a scaffold for neural repair is receiving growing interest from the scientific community and, in particular, many experimental studies have shown that chitosan-based conduits provide an effective scaffold for the reconstruction of peripheral nerve defects [1]. Recently, chitosan hollow nerve guides have been approved for clinical use (Reaxon® Nerve Guide) [2].

In spite of these positive results, the early growth of axons along hollow nerve guides is delayed since the conduit's lumen should first be colonized by connective tissue and Schwann cells migrating from the distal nerve stump [3, 4]. To prevent this occurrence, a luminal filler must be used for providing a substrate for early nerve fiber regeneration [5–7]. A number of different materials have been investigated as luminal fillers for nerve guides, including biological or artificial substrates [8–11].

The aim of the present study was to investigate rat sciatic nerve reconstruction with chitosan conduits filled up with two types of substrates which might potentially improve axon regeneration, namely (i) NVR gel and (ii) magnetic fibrin hydrogel.

NVR gel is a gel based on hyaluronic acid (HA), a non-immunogenic biopolymer that has been proposed for various tissue engineering applications [12, 13]. HA is a glycosaminoglycan composed of disaccharide D-glucuronic acid and N-acetyl-D-glucosamine produced by hyaluronan synthases, a membrane-bound enzyme. It is a component of ECM where it modulates cell adhesion, migration, and neuronal sprouting, and it controls tissue homeostasis and absorbs mechanical shock. HA action is related to its molecular weight; a long polymer (>500 kDA) is generally known as high molecular weight HA (HMW HA) and is found predominantly in the brain where it has a structural role and takes part in various biological process like silences inflammation, angiogenesis and neural differentiation [14]. In the peripheral nerves, HA is one of the prominent components of the Bunger band formed during Wallerian degeneration, and the ablation of HA receptors can reduce cell adhesion and migration [15]. Several studies showed that HA has a positive effects on peripheral nerve regeneration [16, 17].

Fibrin hydrogels are natural polymers made by mixing two blood coagulation components, fibrinogen and thrombin, which form a clot upon mixing.

In previous studies, it was demonstrated that appropriate conjugation of thrombin to iron oxide (γ-Fe2O3) magnetic nanoparticles preserved the thrombin-clotting activity, stabilized the thrombin against its major inhibitor, antithrombin III, and improved its storage stability [18, 19]. Moreover, these thrombin-conjugated γ-Fe_2O_3 nanoparticles were used to fabricate novel magnetic fibrin hydrogel scaffolds [20].

2. Materials and methods

2.1. Chitosan tube preparation

Chitosan conduits were provided by Medovent GmbH (Mainz, Germany). They were produced under ISO 13485 conditions from chitin tubes made by extrusion process. Distinctive washing and hydrolysis steps were then carried out to adjust the required low, medium, and high degree of acetylation (DA): (1) DAI tubes (low DA, ~2%); (2) DAII tubes (medium DA, ~5%); and (3) DAIII tubes (high DA, ~20%). Tubes were finally cut and sterilized by electron beam.

2.2. Preparation of NVR gel

NVR gel was prepared by mixing high molecular weight hyaluronic acid (3×106 Da, BTG Polymers, Kiryat Malachi, Israel) and laminin (Sigma, Rehovot, Israel). The NVR gel was then diluted with nutrient medium (composed of 86% Dulbecco's modified Eagle medium-nutrient mixture F-12), 10% heat-denatured fetal calf serum (FCS), 6 g/L D-glucose, 2 nM glutamine, 25 μg/mL gentamycin, and 50 ng/mL IGF-I, all purchased from Biological Industries, Israel, to a final concentration of 0.5%. NVR gel was then injected into the chitosan tubes immediately before surgical implantation.

2.3. Preparation of the magnetic fibrin hydrogel

Magnetic fibrin hydrogel was prepared by mixing 30 μL of bovine fibrinogen (from stock solution of 100 mg/mL, in PBS without Ca^{2+} and Mg^{2+}), 460 μL of a culture medium, and 10 μL of a $CaCl_2$ aqueous solution (from stock solution of 25 mM). To achieve coagulation, thrombin was conjugated to iron oxide nanoparticles [19–21] and added to the gel (50 μL from a stock solution of 100 μg/mL). The final liquid mixture was then injected into the chitosan conduits immediately before surgical implantation.

2.4. Experimental design and surgical technique

All animal experiments were approved by the Institutional Animal Care and Usage Committee (IACUC) and adhered strictly to the Animal Care Guidelines of University Tel Aviv Sourasky Medical Center (number of approval for animal testing 9-3-12). Female Wistar rats were brought to the vivarium 2 weeks prior to the surgery and housed two per cage with a 12-h light/dark cycle, with free access to food and water. The study was conducted on 45 female Wistar rats (weight 200–250 g), using an experimental model for producing a complete peripheral nerve injury with massive nerve defect that has been recently described [22].

General anesthesia was induced with intraperitoneal injection of xylazine (15 mg/kg) and ketamine (50 mg/kg). Surgical procedures were performed using a high magnification microscope. The left sciatic nerve was exposed and separated from biceps femoris and semimembranosus muscles beginning from the area of branches to the glutei and hamstring muscles and distally to the trifurcation into peroneal, tibial, and sural nerves. The sciatic nerve was completely transected at the third femur level using microsurgical scissors and a 6-mm nerve segment was removed. A 14-mm chitosan hollow conduit of different degrees of acetylation (DAI ~2%; DAII ~5%; DAIII ~20%) filled with 0.5% NVR gel was placed between the proximal and the distal parts of the transected nerve for reconstruction, enabling the nerve to enter the conduit 1 mm on each side, while providing a 12-mm gap between the proximal and distal ends. Two 9–0 non-absorbable sutures were used to anchor the conduit to the epineurium at the proximal and distal nerve stumps (**Figure 1**).

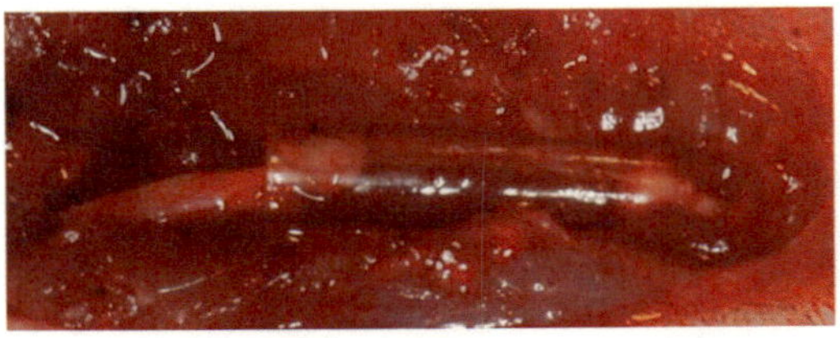

Figure 1. In vivo image of sciatic nerve reconstruction using chitosan-based scaffold immediately after its implantation.

Only one experimental group was used for testing magnetic fibrin hydrogel injected inside DAI tubes.

Autologous nerve graft reconstruction was adopted as control. The left sciatic nerve was exposed as described above and then sharply incised with microscissors at the femur level below the superior gluteal nerve and above the division of the sciatic nerve to the tibial nerve and the peroneal nerve. A nerve segment of 12 mm was harvested, turned upside down, and immediately replaced to bridge the nerve gap using 2–3 non-absorbable 10-0 sutures for each stump. Coaptation of nerve fascicles was carried out to preserve all the fascicles within the epineural sac. The muscular, subcutaneous and skin layers were closed.

Experimental groups can be then summarized as follows: DAI tubes filled with NVR gel (n = 10); DAII tubes filled with NVR gel (n = 10); DAIII tubes filled with NVR gel (n = 10); DAI tubes filled with magnetic fibrin hydrogel (n = 10); and autologous nerve grafts (n = 5).

2.5. Functional analysis and electrophysiological evaluation

Preoperative evaluation and postoperative follow-up were performed and consisted of motor assessment of the sciatic nerve utilizing sciatic functional index (SFI) (data not shown) and somatosensory evoked potentials (SSEP). SFI and SSEP were tested preoperatively and retested at 90 and 120 days postoperatively.

All assessments were carried out in a blinded manner without disclosure of rat's affiliation to the evaluating team.

SSEP were recorded in both the operated and intact hind limbs using a Dantec™ KEYPONT® PORTABLE. Conductivity of rat sciatic nerve and spinal cord was studied by stimulation of the sciatic nerve at the level of the tarsal joint with simultaneous recording from the skull over the somatosensory cortex in anesthetized rats. Two subcutaneous needle electrodes were inserted under the skin of the scalp with the active electrode over the somatosensory cortex along the midline and reference electrode between the eyes. The ground electrode was placed subcutaneously on the dorsal neck. The sciatic nerve was stimulated by a set of two polarized electrodes placed on the lateral aspect of the tarsal joint. Three hundred stimulation pulses of 0.2 ms in duration were generated at a rate of 3 Hz. The stimulus intensity was set on 2–5 mA, and a slight twitching of the limb was noted in all rats. The appearance of an evoked potential in at least two consecutive tests as a response to a stimulus was considered positive.

2.6. Morphological and stereological analysis

Nerve samples were collected 120 days after surgery. The complete conduits were harvested together with parts of proximal and distal nerves, and the whole specimens were fixed for 4–6 h at 4°C in 2.5% glutaraldehyde prepared in 0.1 M phosphate buffer (pH 7.4). A postfixation was then performed using 2% osmium tetroxide for 2 h, followed by dehydration in ethanol from 30 to 100%. After two washings with propylene oxide of 7 min each, samples were left for an hour in a mixture of propylene oxide and Glauerts' mixture of resins (50% Araldite HY964, 50% Araldite M, and 0.5% dibutylphthalate) mixed in equal parts. Then, samples were immersed in Glauerts' mixture of resins and left overnight. A second immersion with Glauerts' mixture of resin was then performed, leaving samples at 37°C for 1 h. Samples were then immersed twice for 30 min in resin with the addition of 2% of accelerator 964 and finally left in resin for three days at 60°C. Resin-embedded nerves were then processed for stereological analysis: transverse sciatic nerve sections were prepared starting from the distal part of the samples. For optical analysis, semi-thin sections (2.5 μm of thickness) were prepared using Ultracut UCT ultramicrotome (Leica Microsystems, Wetzlar, Germany) and stained with 1% toluidine blue.

The stereological analysis [23] was performed on semi-thin transverse nerve section stained with toluidine blue using DFC320 digital camera on DM4000B microscope. A random selected semi-thin section was analyzed for each sample with the help of specific image software (IM50 image manager system, Leica Microsystems). The total cross-sectional area of the nerve was measured. Then, 12–16 fields of the section were randomly selected following a systematic random protocol and proceeded as described previously [24] to measure the following parameters: total number of myelinated fibers, axon diameter and fiber diameter, myelin thickness, and g-ratio. Finally, the correlation between g-ratio and axon diameter of individual fibers was carried out as scatter plot graphs.

2.7. Statistical analysis

Functional and electrophysiological analyses were done using MATLAB software (v.2008b, The MathWorks, Inc.). Non-parametric statistics were used in this study. Hence, all figures are presented with Median ± MAD. Significance levels were calculated using a Mann-Whitney U test and a Wilcoxon signed-rank test. SSEP responses were analyzed as categorical parameters using χ^2 test.

For stereological analysis, one-way ANOVA followed by Bonferroni post hoc test was performed using SPSS Statistic Program. The GraphPad software was used to obtain the regression line in scatter plot and to test whether slopes are significantly different. Results are reported as a mean ± standard deviation.

3. Results

3.1. Functional analysis and electrophysiological evaluation

Due to autotomy which often occurs in the operated hind limb [25], the number of rats available for meaningful SFI testing was not sufficient for reliable statistics (data not shown). The group of low acetylation conduit (DAI) filled with NVR gel had the lowest number of autotomies (4 out of 10), while the group of low acetylation (DAI) conduit filled with magnetic fibrin hydrogel had the highest number of autotomies (8 out of 10 rats).

For the electrophysiological evaluation, the SSEP peak-to-peak (P2P) amplitude was calculated. Two consecutive electrophysiological recordings were performed on 38 and 39 rats during 90 and 120 days, respectively. In order to evaluate P2P amplitude in the operated and intact limbs, we calculated the results from each limb separately. No significant differences were found between limbs (data not shown). Rats with P2P measurement in only one of the limbs were ignored in calculations from limb comparison. Regarding the operated limb, no significant differences were observed between all groups after both 90 and 120 days (**Figure 2**).

The P2P values of the intact limb were subtracted from the values of the operated limb in order to evaluate the effectiveness of the various conduits and fillers. A value closer to "0" indicates similar activity in both intact and operated limbs, while a major shift is an indicator of neurological dysfunction. As can be seen in **Figure 3**, all groups evolved with a value close to "0," and no significant difference among the various acetylation conduits and fillers was detected.

3.2. Morphological and stereological analysis

During harvesting (120 days postoperation), a macroscopic observation was performed in order to evaluate the degradation conditions of the conduit and the presence of nerve filaments inside the conduit reconnecting the two nerve stumps. This qualitative analysis showed that 2 rats out of 9 of DAI-NVR gel group and 6 rats out of 10 of DAIII-NVR gel group showed conduits with sign of degradation. Moreover, 5 rats out of 7 of DAI-magnetic fibrin clot group, 7 rats out of 9 of DAI-NVR gel group, 8 rats out of 10 of DAII-NVR gel group, and 8 rats out of 10 of DAIII-NVR gel group showed macroscopically signs of regeneration. This macroscopic observation was then confirmed after histological analysis.

The morphological evaluation was carried out on regenerated nerves immediately downstream to the conduit, 120 days after nerve repair. Representative high-resolution light microscopy images of transverse semi-thin sections of regenerated sciatic nerves of all experimental groups are shown in **Figure 4**. A good regeneration was observed in most of

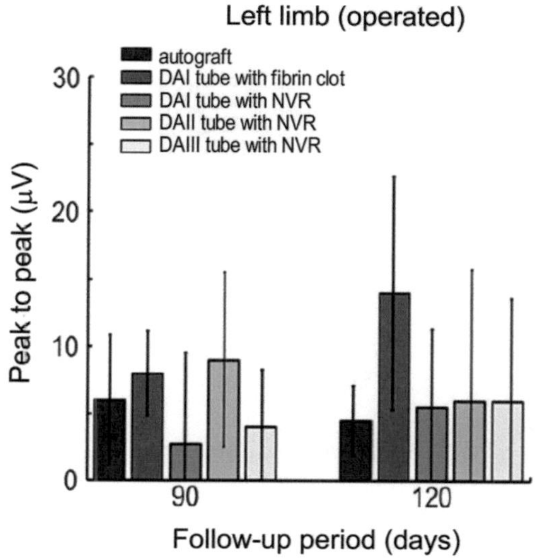

Figure 2. Somatosensory evoked potentials (SSEP) peak-to-peak (P2P) amplitude in operated limbs in various treatments during 90 and 120 days postoperatively. Values are presented with median ± MAD.

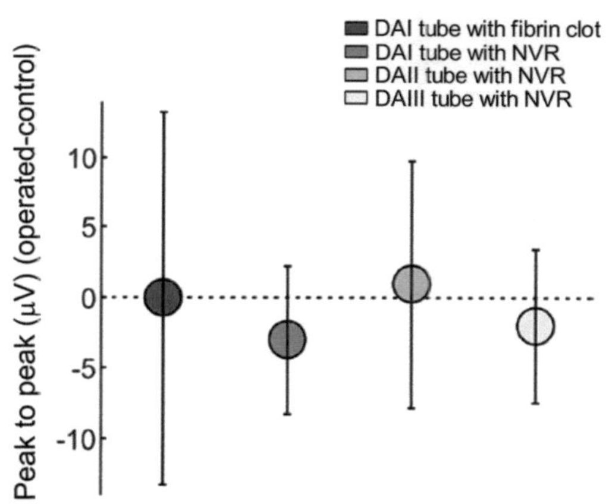

Figure 3. Evaluation of the effectiveness of various acetylation conduits and fillers. The P2P values of the intact limb were subtracted from the values of the operated limb.

the samples for each experimental group: autologous nerve graft (**Figure 4A**); DAI-magnetic fibrin clot (**Figure 4B**); DAI-NVR gel (**Figure 4C**); DAII-NVR gel (**Figure 4D**); DAIII-NVR gel (**Figure 4E**).

Semi-thin sections were also used to perform stereological analysis for the evaluation of the number of myelinated fibers, the axon and fiber size, and the myelin thickness.

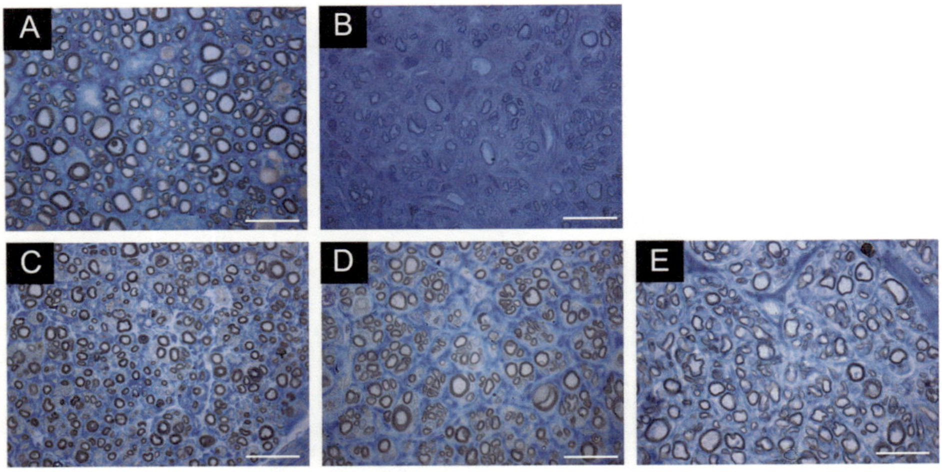

Figure 4. Representative semi-thin sections of the distal part of a sciatic nerve repaired with (A) autologous nerve graft; (B) low acetylation (DAI) conduit filled with magnetic fibrin clot; (C) low acetylation conduit (DAI) filled with NVR gel; (D) medium acetylation (DAII) conduit filled with NVR gel; (E) high acetylation (DAIII) conduit filled with NVR gel. A good regeneration after nerve repair is detectable in all groups. Scale bar: 20 μm.

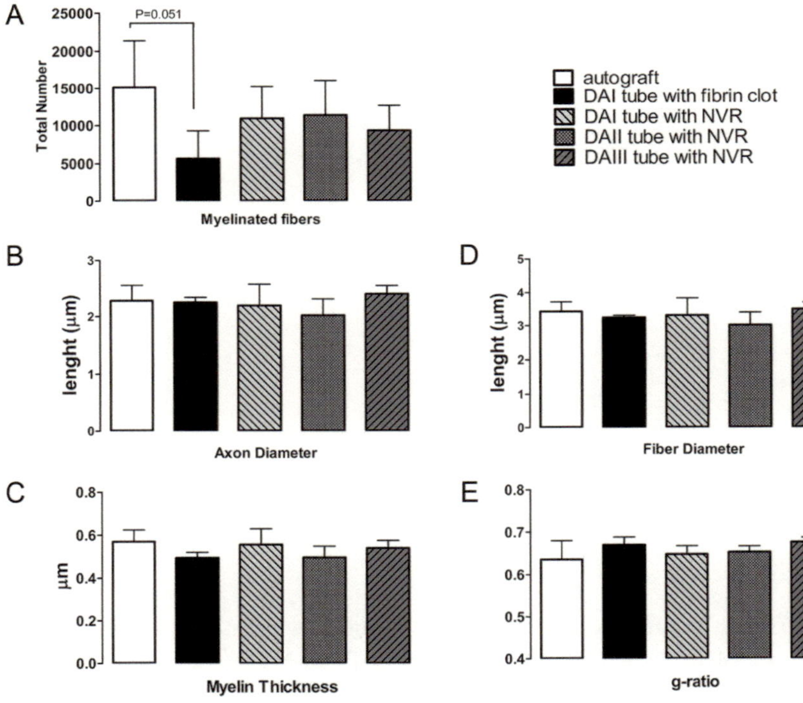

Figure 5. The graphs show the results of stereological analysis on semi-thin sections of the distal part of regenerated nerves. The number of myelinated fibers (A), axon diameter (B), the myelin thickness (C), the fiber diameter (D), and the g-ratio (E) were evaluated 120 days after surgery. All data are presented as means±SD; statistical analysis: one-way ANOVA with post hoc Bonferroni test.

http://dx.doi.org/10.5772/intechopen.69882

As we can observe from the graph (**Figure 5A**), the total number of myelinated fibers in DAI magnetic fibrin clot group is about two-thirds smaller than the autograft group ($p = 0.05$). Moreover, the number of myelinated fibers in all NVR gel groups is similar to each other and about one-third reduced compared to the autograft group. Although the differences are evident, they can only be interpreted as a statistical trend of $p < 0.1$ due to the high standard deviation (that is probably due to the high variability in the regeneration success observed for all-gel-filled conduits).

Axon and fiber diameter, myelin thickness, and g-ratio are not statistically different among all groups (**Figure 5B–E**).

Scatter plot graphs in **Figure 6** represent the correlation between g-ratio and axon diameter of individual fibers. All groups are compared to the autograft, considered as control. Data show that in DAI-magnetic fibrin clot, DAI-NVR gel, and DAIII-NVR gel group, the slope of the regression line is significantly different from the autograft (**Figure 6A, B, D**). Contrary, there is no difference between autograft and DAII-NVR gel group (**Figure 6C**) since the slope of the two regression lines is not significant different.

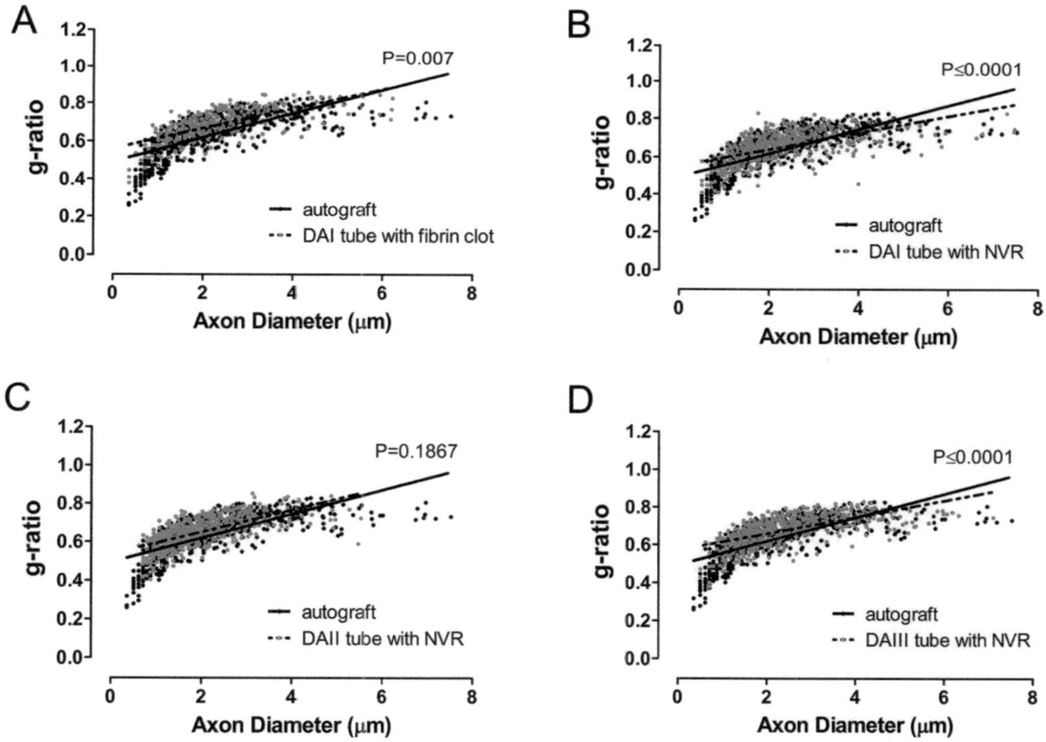

Figure 6. Scatter plot graphs display g-ratio (y-axis) in relation to axon diameter (x-axis) of individual fiber. More than 250 myelinated axons were considered for each group. All the experimental groups are compared to autograft. The slope of linear regression for (A) DAI conduit filled with magnetic fibrin clot, (B) DAI conduit filled with NVR gel, and (D) DAIII conduit filled with NVR gel is significantly different from the autograft. On the contrary, there are no significant differences between autograft and the group DAII conduit filled with NVR gel (C). The statistical analysis was performed using Prism Program, comparing the slope of the linear regression.

4. Discussion

Peripheral nerves are very often subjected to traumatic lesions both because of accidents (e.g., at work, on the road, and at home) and also because of iatrogenic damage (e.g., oncologic surgical excisions) [26, 27]. When the lesion causes substance loss, the two nerve stumps must be connected by a nerve guide in order to allow regenerating axons to bridge the gap [28, 29]. The nerve guide can be represented by an autologous nerve segment (traditional autograft), e.g., from the sural nerve or by a non-nervous conduit [5, 30]. Along the last years, an increasing number of papers describing innovative bio-artificial nerve guides has been published [7]. In general, nerve guides are composed of two main components: a tubular scaffold which can be sutured (or glued) to the nerve stumps and a luminal filler which provides the substrate for cell migration and axon regeneration inside the conduit.

In this study, we selected the chitosan as the biomaterial of choice for fashioning the nerve conduit based on the previous *in vitro* and *in vivo* experimental evidences. *In vitro,* it has been shown that chitosan membranes are a suitable substrate for survival and orientation of Schwann cell growth as well as survival and differentiation of neuronal cells [31, 32]. *In vivo,* studies showed that chitosan tubes can efficiently bridge peripheral nerve defects [33–36].

On the other hand, two types of luminal filler have been investigated in this study: (i) NVR gel and (ii) magnetic fibrin hydrogel.

The selection of NVR gel as a potentially bioactive luminal filler has been based on a number of studies that have proposed his employment for various tissue engineering applications [12, 13]. In peripheral nerves, HA has positive effects on peripheral nerve regeneration modulating glial cell adhesion and migration and neuronal sprouting [15–17]. Yet, topical application of HA is able to reduce the scar formation and create a more favorable environment for nerve regeneration [37, 38].

The results of this study, which compared HA gel enrichment of three different types of chitosan conduits with traditional autografts, showed no significant inter-group differences in the electrophysiological functional outcome and with respect to all histomorphometrical predictors of axon regeneration. By contrast, g-ratio and axon diameter correlation plots (a strong predictor of nerve fiber maturation) showed significant differences between autograft and all groups of conduits except for NVR gel-enriched medium degree of acetylation (DAII). The better performance of DAII (w5%) conduits is in line with a previous report which compared chitosan hollow conduits with different degrees of acetylation in the rat sciatic nerve model [35] and might be explained with the need of a medium-degree degradation time (several months). In fact, the degree of acetylation is directly related to the degradation velocity of the biomaterial, resulting in too slow degradation (more than 1 year) under low acetylation conditions and too fast degradation (several weeks) under high acetylation conditions [35].

As regards the choice of magnetic fibrin hydrogel, the second potentially bioactive luminal filler investigated in this study, this has been based on its previous successful use for tissue engineering applications in various tissues and organs [39–43]. Fibrin hydrogels combine some important advantages such as inherent flexibility, soft, high seeding efficiency, and they cast easily into three-dimensional shapes, can be injected directly into the site of an injury, and contain cell-binding sites which enhance cell adhesion [44, 45].

In this study, thrombin-conjugated γ-Fe_2O_3 nanoparticles were used to fabricate the magnetic fibrin hydrogel in order to enrich low acetylation (DAI) chitosan conduits. The selection of DAI conduits instead of DAII conduits, which showed a better performance in other experiments (35), was due to the fact that, when these experiments started, full data on the comparison among different acetylation chitosan scaffolds were not available yet.

Results of the electrophysiological assessment showed that no significant difference in functional recovery can be detected between magnetic fibrin hydrogel-enriched DAI conduits, NVR gel-enriched DAI conduits and traditional autografts. By contrast, as regards morphometrical analysis, magnetic fibrin hydrogel-enriched DAI conduits showed significantly less fibers than autograft.

5. Conclusion

Altogether, the results of our study showed that the enrichment of chitosan tubes with both NVR gel and magnetic fibrin hydrogel for the repair of 12 mm long rat sciatic nerve gaps leads to a degree of functional recovery (measured by electrophysiology) not significantly different from traditional autograft. The functional results were not completely matched by the histomorphometric investigation of nerve fibers that showed that best results were found in medium degree (w5%) of acetylation chitosan conduits enriched with the NVR gel. This occurrence, which is not surprising since several studies previously showed that morphological and functional predictors of nerve regeneration are often unrelated [46], must be taken into consideration when translating experimental results to the clinics.

Acknowledgements

This study was supported by the European Community's Seventh Framework Programme (FP7-HEALTH-2011) under grant agreement n° 278612 (BIOHYBRID) and Compagnia di San Paolo (InTheCure Project).

Medical grade chitosan for manufacturing the chitosan films and nerve guides was supplied by Altakitin SA (Lisbon, Portugal). The chitosan materials were supplied by Medovent GmbH (Mainz, Germany). We thank Alex Litvak DVM for technical assistance in animal treatment and care.

Conflict of interest

O. Ziv-Polat and A. Shahar were employees of NVR Research Ltd., while the experiments were conducted, that owns the patent of the NVR gel used in this study.

Author details

Shimon Rochkind[1], Mira M. Mandelbaum-Livnat[1], Stefania Raimondo[2*], Michela Morano[2], Giulia Ronchi[2], Nicoletta Viano[2], Moshe Nissan[1], Akiva Koren[1], Tali Biron[1], Yifat Bitan[1], Evgeniy Reider[1], Mara Almog[1], Ofra Ziv-Polat[3], Abraham Shahar[3] and Stefano Geuna[2]

*Address all correspondence to: stefania.raimondo@unito.it

1 Division of Peripheral Nerve Reconstruction, Department of Neurosurgery, Research Center for Nerve Reconstruction, Tel Aviv Sourasky Medical Center, Tel Aviv University, Israel

2 Department of Clinical and Biological Sciences, Neuroscience Institute Cavalieri Ottolenghi, University of Turin, Italy

3 NVR Research Ltd., Ness Ziona, Israel

References

[1] Gnavi S, Barwig C, Freier T, Haastert-Talini K, Grothe C, Geuna S. The use of chitosan-based scaffolds to enhance regeneration in the nervous system. International Review of Neurobiology. 2013;**109**:1-62

[2] Neubrech F, Heider S, Harhaus L, Bickert B, Kneser U, Kremer T. Chitosan nerve tube for primary repair of traumatic sensory nerve lesions of the hand without a gap: Study protocol for a randomized controlled trial. Trials. 2016;**17**:48

[3] Hall SM, Redford EJ, Smith KJ. Tumour necrosis factor-alpha has few morphological effects within the dorsal columns of the spinal cord, in contrast to its effects in the peripheral nervous system. Journal of Neuroimmunology. 2000;**106**(1-2):130-136

[4] Geuna S, Raimondo S, Ronchi G, Di Scipio F, Tos P, Czaja K, et al. Chapter 3: Histology of the peripheral nerve and changes occurring during nerve regeneration. International Review of Neurobiology. 2009;**87**:27-46

[5] Deumens R, Bozkurt A, Meek MF, Marcus MA, Joosten EA, Weis J, et al. Repairing injured peripheral nerves: Bridging the gap. Progress in Neurobiology. 2010;**92**(3):245-276

[6] Geuna S, Gnavi S, Perroteau I, Tos P, Battiston B. Tissue engineering and peripheral nerve reconstruction: An overview. International Review of Neurobiology. 2013;**108**:35-57

[7] Chiono V, Tonda-Turo C. Trends in the design of nerve guidance channels in peripheral nerve tissue engineering. Progress in Neurobiology. 2015;**131**:87-104

[8] Geuna S, Tos P, Battiston B, Guglielmone R, Giacobini-Robecchi MG. Morphological analysis of peripheral nerve regenerated by means of vein grafts filled with fresh skeletal muscle. Anatomy and Embryology (Berlin). 2000;**201**(6):475-482

[9] Geuna S, Tos P, Battiston B, Giacobini-Robecchi MG. Bridging peripheral nerve defects with muscle-vein combined guides. Neurological Research. 2004;**26**(2):139-144

[10] Tos P, Battiston B, Nicolino S, Raimondo S, Fornaro M, Lee JM, et al. Comparison of fresh and predegenerated muscle-vein-combined guides for the repair of rat median nerve. Microsurgery. 2007;**27**(1):48-55

[11] Sedaghati T, Seifalian AM. Nanotechnology and bio-functionalisation for peripheral nerve regeneration. Neural Regeneration Research. 2015;**10**(8):1191-1194

[12] Collins MN, Birkinshaw C. Hyaluronic acid based scaffolds for tissue engineering—A review. Carbohydrate Polymers. 2013;**92**(2):1262-1279

[13] Moshayedi P, Carmichael ST. Hyaluronan, neural stem cells and tissue reconstruction after acute ischemic stroke. Biomatter. 2013;**3**(1)pii: e23863

[14] Preston M, Sherman LS. Neural stem cell niches: Roles for the hyaluronan-based extracellular matrix. Frontiers in Bioscience (Scholar Edition). 2011;**3**:1165-1179

[15] Tona A, Perides G, Rahemtulla F, Dahl D. Extracellular matrix in regenerating rat sciatic nerve: A comparative study on the localization of laminin, hyaluronic acid, and chondroitin sulfate proteoglycans, including versican. Journal of Histochemistry and Cytochemistry. 1993;**41**(4):593-599

[16] Adanali G, Verdi M, Tuncel A, Erdogan B, Kargi E. Effects of hyaluronic acid-carboxymethylcellulose membrane on extraneural adhesion formation and peripheral nerve regeneration. Journal of Reconstructive Microsurgery. 2003;**19**(1):29-36

[17] Zor F, Deveci M, Kilic A, Ozdag MF, Kurt B, Sengezer M, et al. Effect of VEGF gene therapy and hyaluronic acid film sheath on peripheral nerve regeneration. Microsurgery. 2014;**34**(3):209-216

[18] Margel S, Sheichat L, Tennenbaum T, inventors. Biological Glues based on Thrombin Conjugated Nanoparticles. Israel: Bar-Ilan University [assignee]; 2009

[19] Ziv-Polat O, Lublin-Tennenbaum T, Margel S. Synthesis and characterization of thrombin conjugated γ-Fe_2O_3 magnetic nanoparticles for hemostasis. Advanced Engineering Materials. 2009;**11**:B251-B260

[20] Ziv-Polat O, Skaat H, Shahar A, Margel S. Novel magnetic fibrin hydrogel scaffolds containing thrombin and growth factors conjugated iron oxide nanoparticles for tissue engineering. International Journal of Nanomedicine. 2012;**7**:1259-1274

[21] Shahar A, Ziv-Polat O, Margel S, Freier T, editors. Developing a Customized Tissue-Engineered Implant for Peripheral Nerve Reconstruction: Role of Nanoparticles. USA: Studium Press LLC; 2015

[22] Shapira Y, Tolmasov M, Nissan M, Reider E, Koren A, Biron T, et al. Comparison of results between chitosan hollow tube and autologous nerve graft in reconstruction of peripheral nerve defect: An experimental study. Microsurgery. 2016;**36**(8):664-671

[23] Geuna S, Herrera-Rincon C. Update on stereology for light microscopy. Cell and Tissue Research. 2015;**360**(1):5-12

[24] Kaplan S, Geuna S, Ronchi G, Ulkay MB, von Bartheld CS. Calibration of the stereological estimation of the number of myelinated axons in the rat sciatic nerve: A multicenter study. Journal of Neuroscience Methods. 2010;**187**(1):90-99

[25] Geuna S. The sciatic nerve injury model in pre-clinical research. Journal of Neuroscience Methods. 2015;**243**:39-46

[26] Siemionow M, Brzezicki G. Chapter 8: Current techniques and concepts in peripheral nerve repair. International Review of Neurobiology. 2009;**87**:141-172

[27] Geuna S, Tos P, Battiston B. Emerging issues in peripheral nerve repair. Neural Regeneration Research. 2012;**7**(29):2267-2272

[28] Battiston B, Raimondo S, Tos P, Gaidano V, Audisio C, Scevola A, et al. Chapter 11: Tissue engineering of peripheral nerves. International Review of Neurobiology. 2009;**87**:227-249

[29] Tos P, Ronchi G, Geuna S, Battiston B. Future perspectives in nerve repair and regeneration. International Review of Neurobiology. 2013;**109**:165-192

[30] Raimondo S, Fornaro M, Tos P, Battiston B, Giacobini-Robecchi MG, Geuna S. Perspectives in regeneration and tissue engineering of peripheral nerves. Annals of Anatomy. 2011;**193**(4):334-340

[31] Freier T, Koh HS, Kazazian K, Shoichet MS. Controlling cell adhesion and degradation of chitosan films by N-acetylation. Biomaterials. 2005;**26**(29):5872-5878

[32] Simoes MJ, Gartner A, Shirosaki Y, Gil da Costa RM, Cortez PP, Gartner F, et al. In vitro and in vivo chitosan membranes testing for peripheral nerve reconstruction. Acta Médica Portuguesa. 2011;**24**(1):43-52

[33] Amado S, Simoes MJ, Armada da Silva PA, Luis AL, Shirosaki Y, Lopes MA, et al. Use of hybrid chitosan membranes and N1E-115 cells for promoting nerve regeneration in an axonotmesis rat model. Biomaterials. 2008;**29**(33):4409-4419

[34] Ao Q, Fung CK, Tsui AY, Cai S, Zuo HC, Chan YS, et al. The regeneration of transected sciatic nerves of adult rats using chitosan nerve conduits seeded with bone marrow stromal cell-derived Schwann cells. Biomaterials. 2011;**32**(3):787-796

[35] Haastert-Talini K, Geuna S, Dahlin LB, Meyer C, Stenberg L, Freier T, et al. Chitosan tubes of varying degrees of acetylation for bridging peripheral nerve defects. Biomaterials. 2013;**34**(38):9886-9904

[36] Meyer C, Stenberg L, Gonzalez-Perez F, Wrobel S, Ronchi G, Udina E, et al. Chitosan-film enhanced chitosan nerve guides for long-distance regeneration of peripheral nerves. Biomaterials. 2016;**76**:33-51

[37] Park JS, Lee JH, Han CS, Chung DW, Kim GY. Effect of hyaluronic acid-carboxymethylcellulose solution on perineural scar formation after sciatic nerve repair in rats. Clinics in Orthopedic Surgery. 2011;**3**(4):315-324

[38] Ozgenel GY. Effects of hyaluronic acid on peripheral nerve scarring and regeneration in rats. Microsurgery. 2003;**23**(6):575-581

[39] Ahmed TA, Dare EV, Hincke M. Fibrin: A versatile scaffold for tissue engineering applications. Tissue Engineering. Part B, Reviews. 2008;**14**(2):199-215

[40] Qian LM, Zhang ZJ, Gong AH, Qin RJ, Sun XL, Cao XD, et al. A novel biosynthetic hybrid scaffold seeded with olfactory ensheathing cells for treatment of spinal cord injuries. Chinese Medical Journal (England). 2009;**122**(17):2032-2040

[41] Uibo R, Laidmae I, Sawyer ES, Flanagan LA, Georges PC, Winer JP, et al. Soft materials to treat central nervous system injuries: Evaluation of the suitability of non-mammalian fibrin gels. Biochimica et Biophysica Acta. 2009;**1793**(5):924-930

[42] Straley KS, Foo CW, Heilshorn SC. Biomaterial design strategies for the treatment of spinal cord injuries. Journal of Neurotrauma. 2010;**27**(1):1-19

[43] Zuidema JM, Provenza C, Caliendo T, Dutz S, Gilbert RJ. Magnetic NGF-releasing PLLA/iron oxide nanoparticles direct extending neurites and preferentially guide neurites along aligned electrospun microfibers. ACS Chemical Neuroscience. 2015;**6**(11):1781-1788

[44] Mosesson MW, Siebenlist KR, Meh DA. The structure and biological features of fibrinogen and fibrin. Annals of the New York Academy of Sciences. 2001;**936**:11-30

[45] Laurens N, Koolwijk P, de Maat MP. Fibrin structure and wound healing. Journal of Thrombosis and Haemostasis. 2006;**4**(5):932-939

[46] Ikeda M, Oka Y. The relationship between nerve conduction velocity and fiber morphology during peripheral nerve regeneration. Brain and Behavior. 2012;**2**(4):382-390

Chapter 9

Scaffolds for Peripheral Nerve Regeneration, the Importance of *In Vitro* and *In Vivo* Studies for the Development of Cell-Based Therapies and Biomaterials: State of the Art

Sílvia Santos Pedrosa, Ana Rita Caseiro,
José Domingos Santos and Ana Colette Maurício

Additional information is available at the end of the chapter

http://dx.doi.org/10.5772/intechopen.69540

Abstract

Human adult peripheral nerve injuries are a high incidence clinical problem that greatly affects patients' quality of life. Although peripheral nervous system has intrinsic regenerative capacity, this occurs in an incomplete or poorly functional manner. When a nerve fiber loses its continuity with consequent damage of the basal lamina tubes, axon spontaneous regeneration is disorganized and mismatched. These phenomena translate in an inadequate nerve functional recovery and consequent musculoskeletal incapacity. Nerve grafts still remain the gold standard in peripheral injuries treatment. However, this approach contains its disadvantages such as the necessity of primary surgery to harvest the autografts, loss of a functional nerve, donor site morbidity and longer surgery procedures. Therefore, biomaterials and tissue engineering can provide efficient resources and alternatives to nerve injury repair not only by the development of biocompatible structures but also, introducing neurotrophic factors and cellular systems to stimulate optimum clinical outcome. In this chapter, a comprehensive state-of-the art picture of tissue-engineered nerve grafts scaffolds, their application in nerve regeneration along with latest advances in peripheral nerve repair and future perspectives will be discussed, including our own large experience in this field of knowledge.

Keywords: nerve regeneration, peripheral nerve, biomaterials, hydrogels, tube-guides, neurotrophic factors, cell-based therapies, functional assessment, mesenchymal stem cells, Schwann cells, tissue engineering, scaffolds

1. Introduction

Tissue engineering was originally defined by Skalak and Fox, in 1988 [1] as 'the application of the principles and methods of engineering and life sciences towards the fundamental understanding of structure-function relationships in normal and pathological mammalian tissues and the development of biological substitutes to restore, maintain, or improve functions'. Later, Langer and Vacanti [2], in a widespread review paper, defined three main pillars of tissue-engineering principles, the (a) isolated cells and substitutes—cellular systems; (b) tissue-inducing substances—bioactive molecules; and (c) scaffolds, biomaterials and/or matrices.

The first strategy concerns cell-based therapies in which cells in a small volume or in cell sheets are transplanted into the body. Cellular system includes a wide range of cells and most significantly stem cells [3–7]. Stem cells are responsive undifferentiated cells with varying degrees of self-proliferation and differentiation plasticity [8]. Although the number of stem cells is higher before birth, in adults there are still several 'niches' with significant number of stem cells [9]. The second pillar focuses on the [5, 7, 10] bioactive molecules that can be signalling molecules, proteins and oligonucleotides that can enhance cell migration, cell growth and/or differentiation. These bioactive molecules are roughly divided into mitogens, growth factors and morphogens. Finally, the third pillar is the three-dimensional structure that provides shelter and structure for the cellular system [5–7, 11]. Usually, the biomaterials or scaffold mimics the environment and natural extracellular matrix (ECM) of the place of implantation and should be biocompatible such as their metabolites. Also, scaffolds can be used as drug-delivery system in the controlled release of bioactive molecules [9, 12, 13].

Biomaterials according to the American National Institute of Health describes 'any substance or combination of substances, other than drugs, synthetic or natural in origin, which can be used for any period of time, which augments or replaces partially or totally any tissue, organ or function of the body, in order to maintain or improve the quality of life of the individual'. Earlier, the Williams Dictionary of Biomaterials [14] defined biomaterial as 'any substance intended to interact with the biological system in order to replace living matter which has lost its function. It can serve as a vehicle or not, matrix, support, or for stimulating new tissue growth'. The selection of the most adequate material for a given application should fulfil several requirements of physical, mechanical, chemical and biological properties. The most important features of biomaterials must be [7] (i) biocompatibility, that is, the biomaterial itself must not cause any harm in the living system; (ii) biofunctionality, since the biomaterial must feature mechanical and physico-chemical properties adequate to the function and intended application; and (iii) sterilizability, while materials must be able to undergo sterilization procedures, especially the polymeric materials. Biocompatibility is considered the main feature of a biomaterial and represents the response of the living system to the introduction of a foreigner material. It can be defined as the 'ability of a biomaterial to perform its desired function with respect to a medical therapy, without eliciting any undesirable local or systemic effects in

http://dx.doi.org/10.5772/intechopen.69540

the recipient or beneficiary of that therapy, but generating the most appropriate beneficial cellular or tissue response to that specific situation, and optimizing the clinically, relevant performance of that therapy' [14]. However, biocompatible biomaterials are not useful if they are not biofunctional. Similarly, bioactive devices cannot be used if they are not biocompatible. The term biofunctionality can be simply explained as the suitability to the function [7]. Scaffold porosity is an important desirable feature in the majority of scaffolds, as it promotes cell seeding and cell-matrix interaction and leads to increased neovascularization. However, the exact pore size depends on application, the average pore diameter of 20–125 μm is adequate of skin tissue and >300 μm, in bone tissue [11, 15]. The macro- and micro-topography and other physico-chemical properties of the scaffolds influence cell attachment, migration, proliferation and differentiation and promote protein and other factors adsorption and therefore the success of the system [11, 16]. The mechanical properties and specially the degradation kinetics also is a key feature, especially for musculoskeletal and neuromuscular repair due to the slower repair rates.

In this chapter, a comprehensive state of the art of tissue engineering focused on peripheral nerve repair and the advances on materials and nerve grafts will be discussed, as well as our own large experience in this field of knowledge and the future perspectives.

2. Peripheral nerve regeneration

Peripheral nerve injury remains a major public health problem with an estimated incidence of 13–23 cases per 100,000 persons [17, 18]. These injuries may have a traumatic or an iatrogenic cause and usually are associated with pain, decrease of function and sensory sensibility with devastating effects on patients and family lives [3, 17, 19, 20].

Aegineta et al. [21] for the first time performed nerve repair and wound closure in wounded soldiers, as a military surgeon. Then, in 1873, Hunter [22] first described the epineural nerve repair procedure, still in use today. Sunderland portrayed the principles of nerve repair resourcing to microsurgical techniques, and Kurze and Smith were able to apply those principles in 1964, thanks to the advance in microscopy [23–25].

Peripheral nerve injuries treatment understands the most challenging surgical procedures, and despite the major breakthroughs in this area, complete nerve recovery and nerve function in all clinical cases have not yet been achieved [17, 20, 26]. Despite the exquisite surgical techniques, poor recovery outcome results from nervous system intrinsic and extrinsic factors, such as the integrity of the surrounding tissues post lesion, type and level of the injury itself, the effect on the spinal cord and neurons, the compromising of end organs and with key importance the timing of the surgery [17, 27–29]. Also, although peripheral nervous system has spontaneous regeneration ability, there is a very limited prospective of spontaneous recovery, mostly concerning the complete functional neuromuscular recovery [20].

2.1. Peripheral nerve anatomy

The peripheral nerve system is composed by neurons, Schwann cells (SCs), fibroblasts, macrophages and interconnected blood system [20, 30]. Motor and sensory neurons are polarized cells whose bodies reside in the spinal cord and with long cytoplasm called axon. Their terminations, called dendrites, target a site of innervation. The signal conduction originates in the axon hillock, in the cell body and projects itself in synapses with target end organs. Axons plasma membrane is partially enclosed by the SCs that produce myelin that encapsulates the axon and helps with signal transmission. Myelin is therefore an insulator that enhances the signal transmission efficiency down the axon. Then, there is a connective tissue net that surrounds the individual axons called the endoneurium. An arrangement of axons, designed fascicles, is surrounded by the perineurium, and groups of fascicles are separated by the epineurium. External to this layer is the blood supply derived from major arteries and the latter involved by the mesoneurium (**Figure 1**). The conservation of fascicles patterns and connective tissue is vital for optimal nerve repair and regeneration. Therefore, more commonly, the epineurium is sutured in end-to-end suture (called epineural end-to-end suture), and all nerve surgical interventions are strictly directed at these connective tissue layers. The most important feature is the fact that these sutures must be tension free; otherwise, they will compromise the nerve blood supply and the process of regeneration itself [20, 30–32].

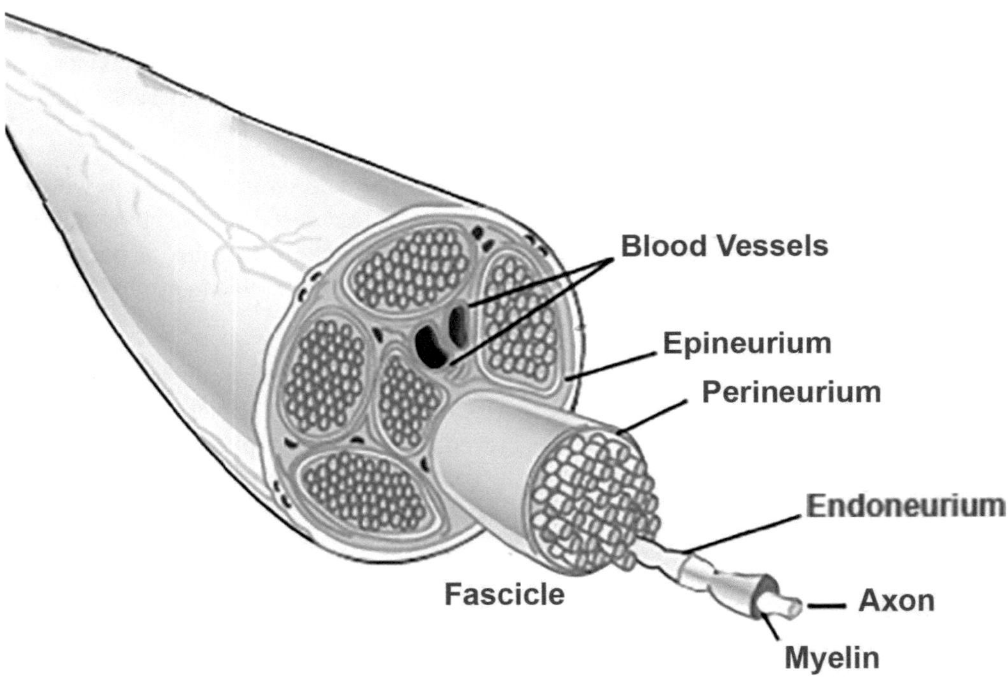

Figure 1. Peripheral nerve anatomy. The figure was adapted from Ref. [30].

http://dx.doi.org/10.5772/intechopen.69540

2.2. Nerve response to injury

Nerve injuries can be from a chronic or acute nature, and SCs are the major cause of chronic injuries [20, 30]. Acute injuries are mediated by axonal degeneration and occur in a sequence of events proximally and distally from the zone of trauma. The initial stages of degeneration occur proximally to the lesion as programmed cell death, called chromatolysis [33, 34], and distally to the lesion, through Wallerian degeneration of the distal axonal segment [35, 36]. Within 24–48 h of the injury, SCs degrade the myelin and phagocytes debris from distal axons [37, 38]. The proximal portion of the axon also degenerates up to the node of Ranvier, where the axonal regrowth occurs. Then macrophages recruitment occurs, originating growth factor release and fibroblast and SCs proliferation. The SCs form organized longitudinal structures inside the endoneurium—called bands of Büngner [30, 34]—that are critical to the axonal regeneration. At the distal site of lesion, the node of Ranvier, around 50–100 finger-like sprouts, starts to form a growth cone directed to the distal nerve stump [39]. Proteases are also released from the growth cone by the influence of several factors, clearing the way towards the target tissue. Nerve growth factor (NGF), brain-derived growth factor (BDNF) and other neurotrophic factors are upregulated by SCs, and their expression is increased which promotes the migration and proliferation of the SCs [20]. Axon existing actin allows axon elongation and which occurs in a 1–3 mm/day rate [36] until a receptor is reached. If no receptor or endoneurial tube is reached, axon continues to grow but in a disorganized manner, causing neuroma and clinically painful lesions [40]. In severe nerve injury, axon regeneration is further disorganized with additional scaring and pourer regeneration.

2.3. Nerve injury grading

Nerve injuries were firstly classified into neuropraxia, axonotmesis and neurotmesis by Seddon [41], after his World War II experience in treating nerve-injured soldiers. Neuropraxia is characterized by the segmentation of the myelin without disturbance of the axon, usually a consequence of a compression. Typically, it resolves itself, once the myelin is restored, within 12 weeks. Axonotmesis concerns axonal injury and occurs from a crush mechanism. In this case, connective tissue and nerve continuity are not affected but are followed by Wallerian degeneration. Axonal regeneration occurs at 1–3 mm/day rate [36], and depending on the distance of the lesion, an incomplete recovery may happen. Neurotmesis comprehends the anatomical and physiological section of the axons and connective tissues. Therefore, no spontaneous regeneration may occur and needs surgical reconstruction [30, 42]. Sunderland later expanded this classification by including five types of injury, based on histological knowledge that allowed further distinction between axonotmesis injuries [25, 43]. Sunderland grade Type I classification is equivalent to neurapraxia. The Type II, III and IV classifications differentiate axonotmesis injuries based on the commitment of the connective tissues. In Type II class injury, there is axonal damage without commitment of the endoneurium, and therefore, it is possible to achieve a full recovery. In Type III lesions, we find axon impairment affecting the endoneurium, and in Type IV besides endoneurium, there is perineurium damage. Sunderland grades III and IV may heal spontaneously, but there is attendant scaring and increasing axon and connective tissue damage that causes incomplete recovery. Type IV lesion usually implies surgical intervention and results in extensive scaring. Scars are associated with pain and nerve conduction impairment and may require reconstructive surgery. Type V Sunderland classification corresponds to neurotmesis [30, 42]. Finally, Mackinnon and Dellon [44] described a mixed type of lesion degree, a Type VI addition to the Sunderland classification. This classification represents probably the most common type of lesions, with several layers of injury and not necessarily traditional model as described by Sunderland. The recovery potential and also the treatment approach vary, according to the type of lesion, considering the three classifications mentioned earlier.

In **Table 1**, the major findings in nerve injury grading according to Seddon [41], Sunderland [25, 43] and Mackinnon and Dellon [44] addition are described.

2.4. Diagnosis

Nerve injury may involve variable lengths of nerve impairment, and the degree of the lesion is affected by the type of lesion [45]. Also, the prognosis is dependent on the age of the patient, location of the lesion—distal fare better than proximal lesions—and also demographics [42]. In terms of nerve electrical signal and electrophysiology, the absence of electrical conduction may not indicate severe nerve damage, since conduction may be recovered after just 1 week. Clinical examination (including the functional evaluation) and surgical inspection still are the most accurate means to obtain diagnosis. However, non-invasive procedures such as nerve conduction studies (NCSs) and electromyograms (EMGs) have a diagnostic role in the delayed setting, when muscle fibrillation occurs [30, 42]. NCS assesses both motor and sensory functions through a voltage stimulator applied to the skin at different points of

Sunderland	Seddon	Characteristics	Spontaneous recovery potential
Type I	Neuropraxia	Injury to myelin sheath only	Full
Type II	Axonotmesis	Injuries involve the axon only	Full
Type III	Axonotmesis	Injuries involve the axon and disrupt the endoneurium	Usually slow or incomplete
Type IV	Axonotmesis	Injuries involve the axon and disrupt the endoneurium and perineurium	Poor to none
Type V	Neurotmesis	Complete disruption of the nerve; with the epineurium	None
Type VI*	Mixed	Combination of Types II, III and IV	Variable, can be poor to none

*Mackinnon and Dellon [44] addition to the Sunderland [25, 43] classification.

Table 1. Nerve injury classification according to Sunderland and Seddon.

http://dx.doi.org/10.5772/intechopen.69540

the nerve. A sensor detects response at the muscle (motor function) or nerve (sensory function). This is the initial screening test for the presence or absence of conduction signal. The EMG assesses only the motor function, and in this test, a needle is inserted in the muscle to assess the resting electrical activity and voluntary motor unit analysis [30]. In **Figure 2**, the representation of an algorithm of best treating approach selection according to the type of lesion, length and also complementary diagnostic results is shown. Sciatic Function Index (SFI) is one of the most widely used forms of functional assessment. It compares parameters from footprints and mathematically infers about sensory-motor gait function mediated by the sciatic nerve, without requiring terminal assessment [46, 47]. Sciatic Static Index (SSI) was first introduced by Bervar [48, 49] and is another way of assessing recovery of function after sciatic injury in animal models. It also uses the footprints in a static position and minimizes bias related to gait's velocity. Also, the SSI improves the acquisition of footprints and is more repeatable and accurate than the SFI. Other motor performance index is measuring

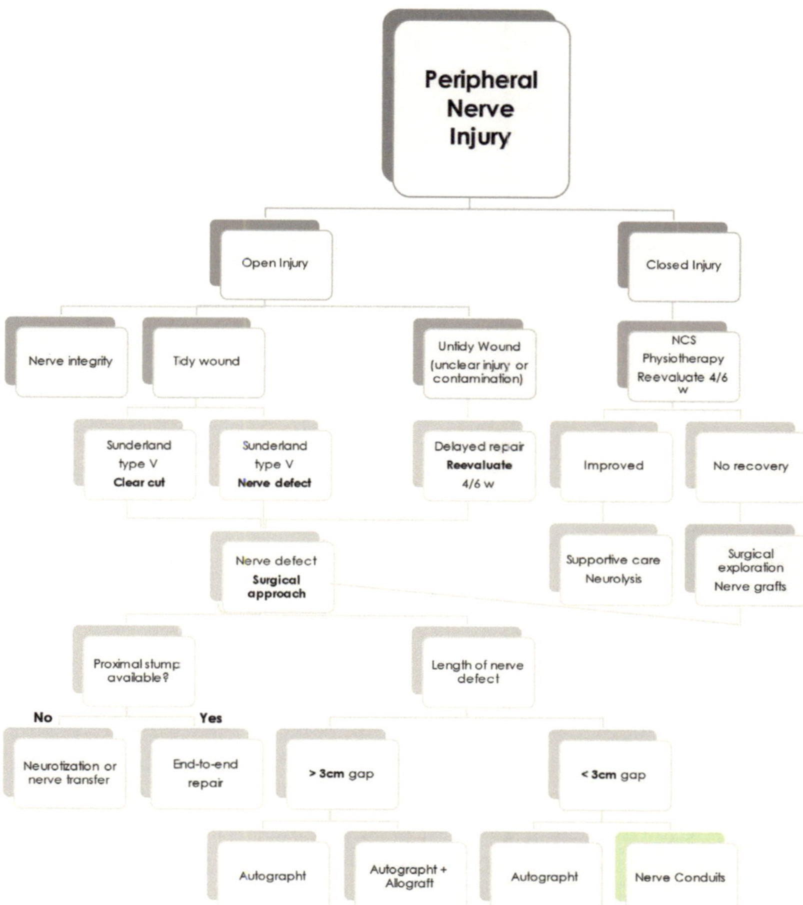

Figure 2. Peripheral nerve repair treatment diagram.

the extensor postural thrust (EPT) and nociceptive function using the withdrawal reflex latency (WRL). EPT is induced by lowering the affected hindlimb towards the platform of a digital balance supporting the animal by the thorax. During the test, the rat extends the hindlimb and the distal metatarsus and digits connect with digital platform balance [50–54]. Nociceptive function using the withdrawal reflex latency was described by Masters and colleagues [55] and is based on the fact that rats without sciatic nerve injury withdraw their paws from the hotplate within 4.3 s or less, when this period of time is increased, it is a symptom of impaired nerve conduction [51–53, 56–61]. Nuclear magnetic resonance (RMN) imaging could become a valuable tool in nerve injury diagnosis since it allows fine, detailed evaluation nerve anatomy and pathology due to excellent image resolution [42]. Also, functional diffusion tensor imaging (DTI), a new diagnosis tool, has been showing peripheral nerve sheath tumours; however, its application on differentiating various grades of injuries remains to be tested [42].

2.5. Timing of medical intervention

After peripheral nerve injury, repair events begin to take place. Primary repair events occur within the first couple of days. However, the rate of axon regeneration is very slow, as previously referred—it is 1–3 mm/day [36]—and no therapeutic methods have yet improved this regeneration rate [30]. However, it is consensual that early nerve repair results in improved functional outcomes, as described by Mackinnon and Dellon [44]. Furthermore, there is a set period of 12–18 months in which muscle re-enervation can occur before irreversible motor end-plate degeneration occurs, and the neurogenic atrophy takes place [30]. Slow axonal regeneration associated with muscle structural changes and increasingly degraded stromal environment contribute for an incomplete functional recovery. Muscle fibrosis and atrophy phenomena begin immediately after denervation and are called neurogenic atrophy. After 4 months, a plateau is reached, when 60–80% of muscle mass is lost, and although motor end plates increase, beyond a 12-month period, a functional muscle re-enervation is highly unlikely [30, 39, 62]. The time frame for sensory re-enervation is longer but not endless, and early repair also grants better results [30, 39]. In **Figure 2**, we present the peripheral nerve injuries repair algorithm which helps to understand the variables taken into account in the selection of best nerve repair strategy.

2.6. Nerve repair strategies

2.6.1. Direct nerve repair

Direct nerve repair with epineural end-to-end sutures using microsurgery techniques is still the gold-standard surgical treatment for severe neurotmesis injuries, but only in cases where well-vascularized tension-free coaptation can be achieved [30, 37, 63–65]. The procedure involves rough fascicular matching between proximal and distal nerve ends and the alignment of nerve fascicles and epineural blood vessels [30, 66]. Other types of direct repair consist in fascicular repair or grouped fascicular repair. This requires intranerval dissection and direct matching and suturing of fascicular groups. Despite better fascicle alignment, this procedure is no better than epineural repair in functional outcomes and in fact is associated with more traumas and scaring [30, 63, 66]. Complementary assistance techniques of histologic staining using acetylcholine esterase and carbonic anhydrase, electrical stimulation during the procedure in awaken patients and visual observation of surface vessels, are visual orientations to the surgeon that grant the success of the procedure [63, 67]. Another possible approach is the use of tissue adhesives such as fibrin glue to supplement or replace sutures, creating a gel-like clot at the nerve ends. The advantages of this procedure are the efficiency and practicality, the reduced trauma and scaring due to a barrier effect. The major disadvantage of this technique is the inferior holding strength more subject to stress [63, 68].

2.6.2. *Nerve grafts*

When peripheral nerve injury originates a significant gap (>3 cm) between the nerve ends with excessive tension for direct epineural repair and reversed interposition, nerve grafts are required [30, 63]. Such gaps may occur in severe neurotmesis lesions or in axonotmesis stretch injuries in which long regions of the nerve may be damaged in the setting of a lesion-in-continuity [63, 69]. Nerve grafts are single, cable, trunk, interfascicular or vascularized portions of the nerve with similar diameter to the affected [30, 70, 71]. Nerve grafting may be from autologous or allograph origin. Xenografts have been described as viable alternatives but require extensive immunosuppression and prionic diseases transmission if they are from ruminants [72, 73]. Nerve autografts are considered the gold standard since they provide appropriate neurotrophic factors and viable SCs, both essential for axonal regeneration without immune compromise [63, 74]. For the choice of the autologous grafts, many factors must be taken into account, such as the size of the nerve gap, the location of proposed nerve repair and associated donor-site morbidity [63, 74]. Grafts are either sutured to the epineurium of single nerves or more commonly to the perineurium of individual fascicles, depending on nerve calibre, type and location [30, 63, 74]. The interfascicular nerve graft was described by Millesi et al. [75]. Vascularized nerve graft was designed by Taylor and Ham, whereby the donor nerve is transposed with its arterial and venous supply into the graft site [76]. Terzis and Kostopoulos [71] clinically demonstrated that medium-sized trunk grafts, which would normally undergo central necrosis, could be transferred as vascularized nerve grafts and survive. However, autografts sacrifice a functioning nerve, usually a sensory nerve, to substitute a more important injured motor nerve. Therefore, sensory loss and scarring at the donor site, where neuroma and pain phenomena, are expected [30, 63, 77]. Autologous nerve graft undergoes Wallerian degeneration and therefore just provides support and guidance for the ingrowing axon. Also, fascicle mismatch, scarring and fibrosis of the repair site is unavoidable and is caused by the injury, tissue handling and suture itself. [30, 63, 77]. An alternative to autologous nerve grafting is the use of nerve allografts. The advantages of allografts are no donor supply limitations or donor-site morbidity, accessibility and unlimited supply of neuronal tissue. However, there are significant costs and complexity with their use, such as immunosuppression [30, 63, 72, 73, 77]. Several techniques have been used to reduce allograft antigenicity, such as cold preservation, irradiation and lyophilization and certainly patients' immunosuppressive therapy. However, it is proven that immune response is caused by SCs, and once their migration has occurred, approximately 24 months after nerve repair, systemic immunosuppression can be withdrawn [74, 78]. To avoid immunosuppression, nerve allografts are decellularized by a process of chemical detergent, enzyme degradation or irradiation, resulting in an acellular nerve scaffold [30, 63, 79]. Similarly, in tendon transfer, a distal function is treated at the expenses of a secondary function [30, 63, 79].

2.6.3. *Nerve transfers and free-functioning muscle transfer*

The definition of nerve transfer is the surgical coaptation of a healthy nerve donor to a denervated nerve [30, 63, 80]. The procedure was first described by Harris in 1921 [81], in the treatment of low median nerve injury suffered during World War I. The major disadvantage is finding an expendable donor nerve near the target muscle with a large enough motor fibre population [30, 63]. Free-functioning muscle transfer (FFMT) is another treatment approach, in severe injuries and especially in secondary reconstructions. The procedure entails the transfer of a healthy muscle and its neurovascular pedicle to a new location to assume a new function [71, 82]. Since it is a complex procedure, it is only considered as a secondary reconstructive surgery.

2.6.4. *Nerve conduits*

Nerve gap repair and nerve grafts have its complications. In the procedures described earlier, the nerve repair requires a second incision site for autograft harvest, donor-site morbidity, loss of a functional, usually sensory nerve, and long-surgery procedure [63, 65]. The described disadvantages triggered the development of nerve conduit or nerve guides to bring a new approach for nerve gap repair [40, 83]. Also, developments in tissue engineering and regenerative medicine, and research in artificial and natural biomaterials have enabled the development of the first nerve conduits [73]. Nerve tubulation approach to the repair of peripheral nerve gaps can be traced back to the nineteenth century. Gluck (Gluck, 1881) performed the first experiment of nerve tabulation, with a tube of decalcified bone to aid the approximation of transected nerve ends, in animals [78, 84, 85]. Later, Dahlin and Lundborg [86] developed the first synthetic tube, made of silicone. Their work was also pioneer in the characterization of the mechanism of regeneration within the lumen of the engineered tube [78, 86]. A nerve conduit is a tubular structure made of biological or synthetic materials designed to bridge the gap of a sectioned nerve. It is used when primary end-to-end direct repair is not possible, to protect the nerve from scar formation, to prevent fluid from leaking from the nerve stump and to guide the axon nerve cone into the distal nerve stump [63, 65, 78]. The fluid formed from the transected nerve ends is essentially made of fibrin, which forms a matrix or a hydrogel matrix between the nerve ends that is able to support cell migration. Cell migration within the fibrin matrix creates some linear bands—bands of Büngner—that steer the growth of the nerve cone [78, 86]. The mechanism through which neurite growth cone forms within the lumen of the conduit depends on the volumetric ratio [87, 88]. If the gap is too long or the diameter of the inner lumen is too large, the growth cables are too thin, and due to the fibrin matrix, the growth cone takes on an hourglass figure that affects axonal regeneration. Nectow et al. [89] also studied the effect of the defect size in the regenerative process through nerve conduits. Nerve conduits provide control environment to outgrowing axons, migration of SCs and neurotrophic stimulation by the distal stump crucial for optimal regeneration of nerve function [40, 90]. This approach is usually reserved for gap defects between 1.5 and 3 cm [91, 92]. As early as 1994, Brunelli et al. [91] defined four factors for an ideal nerve conduit material: (i) biocompatibility, (ii) easy preparation and tailoring, (iii) incorporation of neurotrophins and stimulating substances, and (iv) protection against scaring. Recently, Arslantunali et al. [93] defined the desirable features for a nerve conduit, as flexibility, biocompatibility, biodegradability, high porosity, neuroinductivity, neuroconductivity, easy handling and sufficient endurance. Nowadays, the second- and third-generation nerve conduits are becoming Food and Drug Administration (FDA) approved and reaching the marked, so, many more pre-clinical and clinical trials are demonstrating major breakthrough in this area. In **Figure 3**, we represent the major features of nerve conduits, and the modification researchers have made to introduce neurotrophic elements that enhance nerve regeneration and peripheral nerve repair.

2.6.5. Biological conduits

The use of non-neuronal tissue as conduits was first reported by Büngner (Büngner, 1891), when he successfully used a segment of human brachial artery to regenerate a gap in sciatic nerve [84]. The use of arterial grafts has been demonstrated [94, 95] but is associated with high morbidity. Also, lack of donor vessels makes this a less popular approach in nerve repair. However, it has been described that neurovascular injuries in the hand have benefited from the use of homolateral arteries in the repair [40, 84]. Frerichs et al. and Kim at al. [96, 97] have used acellular allogenic nerve grafts effectively in the regeneration nerve gaps in a rat sciatic model. A similar approach was approved by such Food and Drug Administration and is commercially available as Avance®, by AxoGen, Inc. (Alachua, FL, USA) (**Table 2**). Veins have also been a viable option for nerve repair. The risk of vein collapse led to their filling with

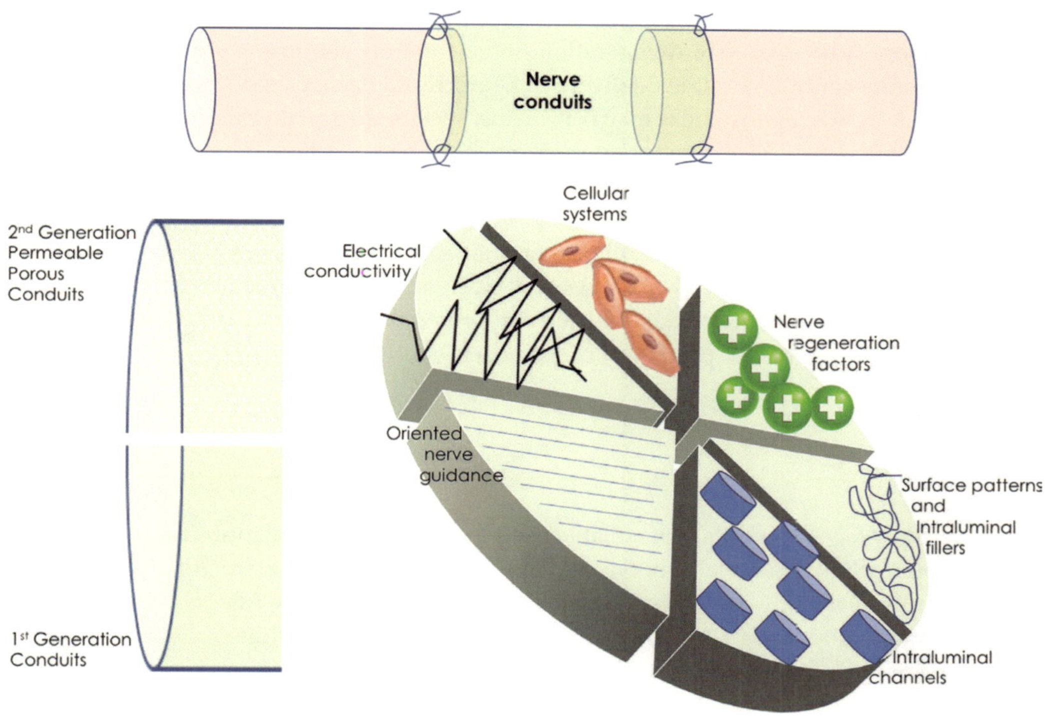

Figure 3. Different types of nerve conduits and their main features. Nerve conduits modification that functions as nerve regeneration enhancers.

nerve or muscle tissue. This supplementation has the added advantage of supplying neurotrophic factors and ECM (muscle fibres) and has been described to facilitate nerve regeneration across longer gaps by promoting SCs migration, cell proliferation and guidance of the axonal growth cone [84]. Another type of biological conduit is tendon autograft. Although with only theoretically and historic interest, ECM macrostructure and the presence of hyaluronic acid have been described to enhance regenerating in the nerve cone [84]. AxoGuard™ is the only FDA-approved device composed by small intestine submucosa (SIS) extracellular matrix (**Table 2**). Preliminary studies in rat sciatic model showed distally directed growth of the proximal nerve [98]. Later further in vivo studies revealed better EMG response for distal motor latency and amplitude [99].

2.6.6. Manufactured conduits

Manufactured nerve conduits can be divided as first-, second- and third-generation conduits. The first-generation conduits are non-resorbable, synthetic tubes made of silicone or polytetrafluoroethylene (ePTFE, Gore-Tex®) [65]. These conduits require a second surgery in order to remove the non-resorbable material. The original idea was to provide support, structure to guide axonal regrowth and form a stable barrier against connective tissue infiltration [73, 100, 101]. Synthetic nerve conduits made of ePTFE were successfully applied in a 4-cm nerve gap, in human [102]. Second-generation nerve conduits are resorbable, biocompatible tubes and are FDA approved and commercially available through different materials. The main advantage of these conduits is their permeability and resorbability that spares patients of a second surgery procedure. Third-generation nerve conduits contrary to second-generation may incorporate controlled release/delivery of neurotrophic factors, electroconductive material, stem cells or SCs, extracellular matrix proteins, surface micropatterning or luminal fillers [73, 103]. Already, two third-generation products have been approved by the FDA, namely NeuraGen® 3D from Integra LifeSciences Corporation and Nerbridge from Toyobo Co., Ltd. NeuraGen® 3D (K130557, approved in 2014) is a bovine Type I collagen conduit with a porous inner hydrogel matrix of collagen and glycosaminoglycan (chondroitin-6-sulfate). Nerbridge™ (K152967, approved in 2016) is a flexible, resorbable and semipermeable tubular membrane matrix filled with porous collagen that provides a non-constricting encasement for injured peripheral nerves for protection of the neural environment.

2.7. Materials

2.7.1. Collagen

Denaturated collagen conduits are available from a wide number of manufacturers and are in fact the most exploited material in nerve conduits [73, 78]. Collagen is a structural protein ubiquitous in the human body, particularly in the peripheral nerve system. Also, collagen supports cell proliferation and tissue regeneration [65, 73, 104]. As nerve conduits collagen allows the establishment of topographical cues that guide axons to regrow [105, 106] and has shown excellent cell adhesive properties that encourage cell attachment and proliferation [106, 107]. The degradation time of the collagen conduits is relatively prolonged and takes up to 48 months which can cause nerve compression and fibrosis [84]. The first commercially available collagen

http://dx.doi.org/10.5772/intechopen.69540

Name	Composition	Structure	Length (cm)	Degradation Time (months)	Manufacturer	FDA approval
NeuroTube	Polyglycolic acid	Absorbable woven mesh tube	2–4	3	Synovis Micro companies	1999
Saluleridge nerve cuff	Polyvinyl alcohol	Flexible tubular membrane	6.35	No degradation	Salumedica LLC	2000
NeuraGen	Collagen Type I	Fibrillar, semipermeable	2–3	3–4	Integra LifeSciences Co.	2001
NeuroFlex	Collagen Type I	Flexible, Semipermeable and tubular	2.5	4–8	Collagen Matrix, Inc.	2001
NeuroMatrix	Collagen Type I	Flexible, Semipermeable and tubular	2.5	4–8	Collagen Matrix, Inc.	2001
Surgesis nerve cuff	Type I, III, IV and VI collagen	Extracellular collagen matrix	5	Reabsorbable	Cook Biotech	2003
Neurolac	Poly-DL-lactide-caprolactone	Tubular	3	16	Polyganics BV	2003/2005
NeuraWrap	Collagen Type I	Flexible, Semipermeable longitudinal slit	2–4	36–48	Integra LifeSciences Co.	2004
NeuroMend	Collagen Type I	Semipermeable wrap that curls and enrols	2.5–5	4–8	Collagen Matrix, Inc.	2006
SaluTunnel	Polyvinyl alcohol	Tubular	6.35	No degradation	Salumedica LLC	2010
Avance	Processed human nerve allograft	Tubular	Variable	No data	AxoGen, Inc.	2010
Cova ORTHO-NERVE	Type I collagen	Rollable membrane	2.5–6	3–4	Biom'Up S.A.	2012
AxoGuard	Extracellular matrix derived from porcine intestine	Semipermeable, and absorbable tube	Variable	No data	AxoGen, Inc.	2013
Flexible Collagen Nerve Cuff	Collagen Type I	Flexible, Semipermeable and tubular	2.5	No data	Collagen Matrix, Inc.	2014

Name	Composition	Structure	Length (cm)	Degradation Time (months)	Manufacturer	FDA approval
Nerve cuff	Type I, III, IV and VI collagen	Extracellular collagen matrix	1–5	Reabsorbable	Cook Biotech	2014
Neuragen 3D	Type I collagen and glycosaminoglycan (chondroitin-6-sulphate)	Flexible, pliable tube with collagen-glycosaminoglycan inner matrix	6.35	9–12	Integra LifeSciences Corporation	2014
Reaxon Plus	Chitosan	Flexible, pliable tube	3	3	MEDOVENT GmbH	2015
Nerbridge	polyglycolic acid and Type I and III collagen	flexible, semipermeable tubular membrane filled with porous collagen	3–5	3–4	TOYOBO CO., LTD.	2016

Table 2. Commercially available and FDA-approved nerve conduits.

nerve conduit was NeuraGen®, from Integra Lifesciences, Princeton, NJ, FDA approved in 2001 (**Table 2**). Several studies regarding the efficacy of collagen conduits in peripheral nerve injuries have been stated. Bushnell and colleagues [108] performed in 2008 a retrospective study of the utilization of collagen conduits in digital sensory nerve gaps of up to 20 mm and demonstrated a significant recovery rate of 89%. In a retrospective review, Wangensteen and Kalliaine [109] reported on a large number of sensory nerve gaps of 2.5–20 mm repaired with collagen conduits in multiple body regions and concluded that clinically successful outcomes were only observed in 43% of the cases. A prospective cohort study was performed by Lohmeyer and colleagues in 2009 [110] in digital and palmar nerve gaps of 6–18 mm, and results showed meaningful recovery in 75% of patients. Later, in 2011, Taras et al. [111] reported a 73% meaningful recovery in 5- to 15-mm isolated digital nerve lacerations repaired with collagen conduits. In a niche approach, collagen conduits have been used in children suffering from plexus brachialis injury during birth [112]. Also, a significant number of in vivo studies on collagen conduits showed good functional outcomes in nerve reconstructions in rat, cat, dog and primate models [113–115]. Researchers and surgeons have, however, raised their concern about these conduits due to its high cost, conduit stiffness, lack of flexibility and poor enhancement of nerve regeneration [116, 117]. Also, collagen conduit application on major peripheral nerve injuries is limited to median and ulnar nerve repairs at the wrist and only observed as an alternative to the classic epineural suture repair [78]. In a recent study, Monaco and colleagues [118] investigated the effect of three different sterilization methods, dry heat, ethylene oxide and electron beam radiation, on the properties of cylindrical collagen scaffolds with longitudinally oriented pore channels, specifically designed for peripheral nerve regeneration. Ethylene oxide exposure demonstrated to be the most suitable method for the sterilization of the proposed scaffolds, since β-sterilization significantly augmented scaffold enzymatic degradation. Currently, there are 10 commercially available and/or FDA-approved collagen nerve conduits (**Table 2**).

2.7.2. Fibrin

Fibrin is a fibrous, non-globular protein involved in the clotting of blood. It is a commonly used biomaterial as fibrin glue in nerve repair. Nonetheless, it can also be engineered into a hydrogel with aligned matrix or shaped into tubular conduits [84]. As fibrin glue, it is widely used in sutures, thanks to its semisolid structure that enhances haemostasis and integrity of the repair and also due to its angiogenesis, chemotaxis, leucocytosis and macrophage proliferation stimulation [119, 120]. When used in a conduit, fibrin has shown to promote axon regeneration and functional recovery in small gaps [116, 121, 122]. Rafijah and colleagues [123] have reported the use of collagen conduits filled with fibrin glue in 10-mm rat sciatic nerve defect. After 12 weeks, no inhibitory effect was observed on function, axonal regeneration and compound motor action potential compared to hollow collagen conduit.

2.7.3. N-fibroin

N-fibroin is a soluble protein, derived from silk with great potential due to their superior biocompatibility and low immunogenicity and mechanical stability upon degradation [73, 124]. Several studies have demonstrated the potential of N-fibroin in the regeneration of peripheral nerve injuries [124–128]. Silk fibroin-based nerve guidance conduit with oriented filaments was produced by Yang and colleagues [129] and originated successful results in rat sciatic model. Researchers have also shown that silk-based conduits [130] and silk nanofibres [131] enhance cell-material interactions like cell adhesion, proliferation and differentiation. A new approach entails the development of conduits with multi-walled silk fibroin/silk sericin internal lumen, beneficial to nerve regeneration and outer sheath (the hollow poly(lactic-co-glycolic acid) conduits that provide strong mechanical protection. Engineered bionic conduit showed promising in vivo results [132]. Research has also reported the development of silk fibroin conduits loaded with nerve growth factors [133–138]. Despite the promising results, no silk conduit has yet been FDA approved.

2.7.4. Chitosan

Chitosan is a natural polymer currently under investigation as nerve conduits due to its favourable biocompatibility, biodegradability and bioactivity [139]. Our group research tested the nerve-regenerative potential of chitosan membrane with N1E-115 cellular system in rat sciatic nerve crush injury. Results showed that freeze-dried chitosan Type III without N1E-115 cell addition was the only type of membrane that significantly improved post-traumatic axonal regrowth and functional recovery [52]. Recent study showed meaningful motor and sensory recovery in 30-mm defect in the nerve of the distal right forearm [140].

2.7.5. Polyglycolic acid

Polyglycolic acid (PGA) was the first FDA approved and commercially available nerve conduit—NeuroTube®, from Synovis Micro Companies Alliance, Birmingham, Ala. PGA is a common suture material and is more flexible and porous than silicone. PGA is degraded into lactic acid in 6–12 months [73, 90]. Therefore, critics claim that PGA may degrade faster than the regeneration process and resulting lactic acid may a have toxic effects [84, 141–143]. Dellon and Mackinnon [144] were the first to report the use of PGA conduits as secondary reconstitution of digital nerve defects 3 cm or smaller in 15 clinical cases in monkeys. After 1 year, 86% meaningful recovery was reported. Later, in an attempt to compare PGA conduit results to conventional autograft repair, Weber et al. [145] conducted a randomized prospective multicentre trial in the reconstruction of long sensory nerve gaps up to 3 cm. After 1 year, 74% meaningful recovery was noted in the PGA group compared with 86% in the standard techniques group ($P > 0.05$). Further analysis showed that PGA conduits were equivalent or superior to traditional autografts in less than 4- and 9–30-mm gaps [145]. Other researchers also compared PGA conduits and vein grafts to repair digital nerve gaps up to 4 cm and equivalent or superior recovery was obtained [146, 147]. Further experiments have demonstrated the success of bioabsorbable PGA nerve conduits in the regeneration of nerve defects [148–151]. The most recent FDA-approved PGA conduit is Nerbridge®, from Toyobo Co., Ltd., which is a flexible, semipermeable tubular membrane filled with porous Type I and III collagens (**Table 2**).

2.7.6. Poly (D, L lactide-co-ε-caprolactone)

Poly-D, L lactide-co-epsilon-caprolactone (PCL) consists in lactic acid and caprolactone monomers. Their nerve conduits are resolvable polyester with the advantage of being transparent and with less acidic degradation product that therefore causes less toxic reaction [60, 84, 116, 152].

Also, PCL conduit is easy to produce, has a low-cost processing and has a long degradation time up to 16 months. However, resulting conduits have higher rigidity and more difficult to handle in clinical settings [84]. Also, Duda et al. [153] reported a strong foreign body response in PCL conduits. Currently, Neurolac® [154–156] is the only FDA-approved caprolactone conduit (**Table 2**). Several reports have been made in the application of PCL conduits in sciatic rat nerve repair [157–162]. Bertleff et al. [154] performed a randomized prospective multicentre study where PCL conduits were comparable to either primary end-to-end repair or nerve autograft. However, latter research showed no meaningful recovery in digital nerves repair [155, 156]. Secer and colleagues [163] studied the use of PCL in the recovery of 455 patients with ulnar nerve injuries. PCL conduits filled with muscle tissue showed superior results in comparison to single PCL conduits in 10- and 15-mm gap in sciatic nerve rat model [36, 156, 164–168].

2.7.7. Polyhydroxybutyrate

Polyhydroxybutyrate (PHB) is a polyester polymer, also used in sutures and wound dressings [116, 169]. PHB has a long degradation time, up to 24–30 months [116, 169]. It has been reported that PHB has a neuroprotective effect and can help axon regeneration [170, 171]. Recently, Axongen Pharmaceuticals has a pending approval order of a PHB conduit. A study in which PHB wrap implants were used in human patients showed promising results compared to epineural suture [172].

2.7.8. Polyvinyl alcohol

Polyvinyl alcohol (PVA) [36, 89, 93] is the only nondegradable synthetic nerve conduit approved by the FDA—SaluBridge and SaluTunnel, from SaluMedica LLC, Atlanta, GA, USA (**Table 2**).

2.7.9. Polyhedral oligomeric silsesquioxane

Polyhedral oligomeric silsesquioxane (POSS) can be described as the smallest particles of silica [116]. The combination of POSS and PCL has been employed in the fabrication of peripheral nerve conduits and in vivo clinical studies to show the potential translation into clinics [117].

2.8. Optimizing nerve regeneration

Although nerve conduits provide sufficient guidance for regeneration of nerve defects, the development of new generation of scaffolds is under way. Third-generation conduits include artificial conduits that may incorporate controlled release/delivery of neurotrophic factors, electroconductive material, stem cells, SCs, extracellular matrix proteins, surface micropatterning, luminal fillers and guidance structures [73, 103].

2.8.1. Conduits structure modulation

Surface micropatterning and the inclusion of extracellular matrix proteins are new approaches that can provide suitable nanostructure topography for adequate neural growth and simulate topographical dimensions that mimic native nerve extracellular matrix [103, 173]. Coating of peripheral nerve conduits can enhance nerve regeneration process and solve longer nerve gaps repair. Resource to extracellular matrix materials, such as fibronectin, laminin and collagen, give naturally hydrophobic scaffolds a hydrophilic surface that promotes cell adhesion [73, 116, 174]. Collagen Type I conduits coated with laminin and fibronectin have shown improved neural regeneration [175, 176]. With that intend, new peptides have been engineered to mimic the active binding domains of various extracellular matrix molecules [177, 178]. Coating with arginylglycylaspartic acid (RGD) sequences (the tripeptide Arg-Gly-Asp) also obtained very good results [117, 179–182]. Structure pore design is a great strategy to promote nerve regeneration. Controlling the architecture of the conduit wall is possible to develop a microporous inner layer and macroporous outer layer and obtain a bidirectional permeability [86, 103]. The diameter of the pores plays a critical role in the efficiency of the scaffolds since it influences cell attachment, axon regeneration and diffusion of nutrients [183, 184]. Electrospinning is a frequently used technique in bioengineering to produce imprinted micropatterns and can be used as a luminal guidance strategy. The advantages associate with electrospun conduits are (i) highly flexible and porous materials; (ii) high surface area-to-volume ration, thereby great availability for protein absorption, stem cells migration and regeneration of axons; (iii) fibers can be preferentially aligned, resulting in increased SC alignment, proliferation and growth, and the promotion of guided axonal growth [116, 185, 186]. Uneven fibre distribution and nerve growth inhibition caused by fibre overlapping are some of the disadvantages associated with this type of conduits. However, association with wall guides helps to avoid this problem [105, 187, 188].

2.8.2. Luminal fillers

Biomaterials and other strategies as conduits luminal fillers aim to change its microenvironment in a favourable way. The goal of this strategy is to promote axon regeneration inside the conduit and the restoration of motor and sensory nerve function.

2.8.2.1. Cellular systems

Stem cell therapies have received increased attention in regenerative medicine [189–201]. The use of cellular systems inside nerve conduits is intended to promote axon regeneration [202–206]. Several cells have been used with this intention, SCs, bone marrow stem cells (BMSCs) including more specifically the mesenchymal stem cells (MSCs). SCs are the natural glia of the peripheral nervous system and have been used successfully with beneficial effects in nerve reconstruction [204, 207–210]. However, there is limited availability, and it requires previous surgery. Stem cells have the opportunity to show their potential since they are able to secrete neurotrophic factors and provide a favourable microenvironment for neurogenesis and the proliferation of SCs in peripheral nerve repair [211] and also are able to differentiate into Schwann cell-like phenotype [200, 212, 213]. MSCs therefore seem the most attractive approach as cellular system in peripheral nerve regeneration. MSCs all share mesenchymal markers that differentiate them from other cells that are positive staining for CD10, CD13, CD29, CD44, CD90 and CD105, and negative expression of haematopoietic markers [214]. Their main feature is their ability to proliferate and self-renew in a sustained manner, high plasticity and low immunogenicity, and differentiate into multiple mesodermal cells, including neuron-like cells [215, 216]. Also, they are major histocompatibility complex (MHC) class II negative and also their MHC class I expression levels can be manipulated; therefore they do not require the use of immunosuppressive drugs [217, 218]. MSCs can be isolated from several origins, including skin, hair follicle, periosteum, amniotic fluid, umbilical cord blood adipose tissue and dental pulp [53, 56, 200, 215, 219–221]. Bone marrow represents the most commonly used tissue source of adult MSCs. BMSCs have limited availability especially in adult life and there is donor-site morbidity associated with its harvesting. Therefore, alternative sources have been reached. MSCs exist in the connective tissue (Wharton's jelly (WJ)) of human umbilical cord and can be harvested easily [214, 218, 222, 223]. Several researchers have demonstrated that umbilical cord–derived stem cells can be differentiated into neuronal phenotype [214, 218, 222, 223] and have demonstrated potential utility in neurodegenerative diseases [224, 225]. Our group research has vast experience in the application of mesenchymal stem cells in peripheral nerve regeneration [50, 52, 53, 56–61]. In our most recent work [56], we report the therapeutic value of MSCs isolated from the Wharton jelly in nerve repair associated to different tube guides made of biodegradable and biocompatible biomaterials. Biomaterials like PVA, PVA loaded with functionalized carbon nanotubes (PVA-CNTs), PVA loaded with polypyrrole (PVA-PPy) and PLC associated to MSCs were tested in terms of cytocompatibility and *in vivo* in the rat sciatic nerve neurotmesis injury model. The functional recovery was assessed serially for gait biomechanical analysis, by EPT, SFI and SSI, and by WRL. Results showed that MSCs enhanced the recovery of sensory and motor function in neurotmesis injuries showing a thicker myelin sheath. Other authors have reported the application of bone marrow stem cells and Schwann-like cells in the regeneration of facial nerves in rats [219].

Other sources of stem cells have been used in peripheral nerve injuries, such as adipose-derived stem cells [203, 206, 212, 226–231] and dental pulp stem cells (DPSCs) [232–236]. Dental pulp stem cells have also demonstrated differentiation capacity towards multiple other mesodermal and endodermal lineages, under appropriate conditions: adipogenic, osteo/dentinogenic, chondrogenic, neurogenic, endothelial, myofibroblastic [237] and hepatocytes [238]. DPSCs can be isolated from both the perivascular and the sub-odontoblastic compartments (the inner surface of the tooth and the outer part of the pulp tissue), by separate digestion of the tooth and extracted pulp tissue. Both populations presented identical cell size, doubling times and karyotype stability, differing only in morphology with rounder cells in the sub-odontoblastic compartment versus spindle-shaped cells with long processes in the perivascular one [239]. Pre-clinical experiments have demonstrated that stem cells show promising results in differentiation into neuronal-like cells [226, 228, 232, 233, 240] and their secretion of growth factors [229, 241, 242]. Shi and colleagues [243] stated the potential of glia-derived neurotrophic factor expressing neural stem cells in the regeneration of facial nerve gap in rats.

2.8.2.2. Neurotrophic factors

Growth factors and cytokines have a complex enhancing effect in tissue regeneration which can be exploited with great potential in nerve regeneration [73]. To date, several neurotrophic factors have been identified: transforming growth factor beta superfamily (TGF-β), nerve growth factor, neurotrophins 3, 4 and 5 (NT-3/4/5), ciliary neurotrophic factor (CNTF), neuregulin-1 (NRG1), brain-derived neurotrophic factor, glial cell line–derived neurotrophic factor (GDNF) and vascular endothelial growth factor (VEGF). The controlled release of the neurotrophic factors in nerve conduits may take several approaches, namely topical application, subcutaneous injection, microosmotic pump and diffusion or affinity-based polymer microspheres [244–251]. In another approach, Cui and colleagues [252] loaded the neurotrophic factors into the wall of the nerve conduit with controlled diffusion into the lumen. NGF has been reported to promote neurite outgrowth in vitro in co-culture of neurons and astrocytes [253] and in vivo–enhanced axonal regeneration in sciatic nerve in rat model [248, 254]. Several researchers have demonstrated BDNF efficacy in the regeneration of rat motor nerves [253, 255, 256]. The dosage of BDNS has also been studied [257]. A high dosage, set as 12–20 mg/day, has been stated to interact with p75 receptors and therefore inhibits axonal regeneration. In fact, single exogenous dosage of BDNF has shown better results than continuous long-term applications [258]. It is reported that muscle regeneration causes the overexpression of GDNF which increases the number of motor axons in the neuromuscular junctions in vivo [259–266]. Also, GDNF is a potent protective factor against axotomy-induced motor and sensory neuron death [267–270]. CNTF has the ability to improve and regenerate muscle function after nerve injury in vivo [271, 272]. It has also demonstrated to enable peripheral nerve regeneration by promoting axon elongation and sprouting from axon distal stump [273]. Six main isoforms of NRG-1 are described. Due to their importance in nerve regeneration isoforms I, II and III are the most studied in the last few years [274]. Nerve-regenerative effect of NRG-1 is highly dependent on the isoform and its dosage [275, 276]. Several studies indicate that NRG-1 isoforms are capable of stimulating SCs proliferation a remyelinization [275–278]. Gambarrota and colleagues [279] recommended that initial high dosages of NRGA-1 stimulate SCs differentiation.

2.8.2.3. Pharmacological agents

Until now, there is no pharmacological method that can effectively enhance nerve regeneration. However, as mentioned earlier, neurotrophic and growth factors have demonstrated potential in enhancing nerve repair and regeneration by reducing neuronal death and promoting axonal outgrowth. Recent advances in molecular biology have indicated that targeting specific steps in molecular pathways may allow for purposeful pharmacologic intervention, potentially leading to a better functional recovery after nerve injury [63, 280]. Major molecular pathways implicated in neuron survival and neurite outgrowth include PI3K (phosphatidylinositol-3 kinase)/Akt (protein kinase B)-signalling cascade, Ras-ERK (rat sarcoma-extracellular signal-regulated kinase) pathway, the cyclic adenosine monophosphate (cAMP)/protein kinase A (PKA) and Rho-ROK signalling [280]. PI3K/Akt cascade seems to provide trophic support for neurons, block apoptosis, facilitate signal transmission and mediate cell growth and differentiation in neurons. Also, it is reported that NGF is mediated by PI3K/Akt/mTOR pathway [281–283]. Ras-ERK pathway is a key promoter of neurite outgrowth and also has been found to enhance axonal survival [284]. Rho–ROK pathway participates in neural growth and modulates neurite outgrowth [285]. Nectins and nectin-like molecules are cell-cell adhesion molecules that participate in cell communication. Nectin-like molecule 1 (NECL1) is restricted to the nervous system and is responsible for the synapses formation, axon bundles, myelinated axons and cerebellar morphogenesis [286, 287]. In a recent study, Xu and colleagues [288] developed a PLGA scaffold coated with NECL1 to enhance adhesion of rat SCs and applied in the treatment of transected sciatic nerve in rat. Results also revealed that the final outcome of both motor and sensory regeneration and reinnervation. Other molecules with different clinical applications have, however, demonstrated beneficial effect in nerve regeneration process: erythropoietin (EPO) [289], tacrolimus (FK506) [290], acetyl-L-carnitine (ALCAR) [291], N-acetylcysteine (NAC) [292], ibuprofen [293], melatonin [294] and transthyretin [295].

2.8.2.4. Channels

The use of longitudinal channels inside nerve conduits is the usual strategy to promote axon guidance towards distal stump. The artificial micro-tubular structure mimics the endoneural tubes and fascicles of a peripheral nerve anatomy and therefore enhances neuroregeneration [296–298].

2.8.2.5. Hydrogels

As described earlier, conduits lumen can be filled with ECM components such as collagen and laminin which are involved in the process of regeneration by forming a substrate for neuron cell migration. Laminin-filled silicone conduits have demonstrated enhanced nerve regeneration [299, 300]. Collagen has also demonstrated to increase nerve regeneration [301]. BD Matrigel® and other *lamini* and collagen gels have been widely used as conduit fillings to incorporate or support cells and neurotrophic factors [302–304].

2.8.2.6. Conductive conduits

Electrical stimulation as a therapeutic in nerve injuries has been widely discussed in the academic community [305–313]. External electrical stimulation as peripheral nerve regeneration strategy has been demonstrated. Several conductive polymers such as polyaniline (PANI) and polypyrrole have been described with great potential for nerve regeneration due to their biocompatibility, tuneable conductivity, environmental stability and facility to produce. The great advantage of this material is the ability to continue to transmit the electrical signal in impaired nerves and physically stimulate cell growth and regeneration [314–316].

2.9. Future perspectives

Gene therapy involves the introduction of a foreign gene into living cells with the intention to overcome a disease [317]. The most efficient way to deliver transgenes into cells is through a vector, such as herpes simplex, adenovirus, lentivirus and adeno-associated viral vectors [318, 319]. Gene can reprogramme cells to produce neurotrophic factors, cell adhesion or extracellular matrix molecules, and transcription factors. Therefore, in peripheral nerve injury, Schwann cells, fibroblasts and denervated muscle are potential targets for this breakthrough approach [63].

Bioactive glasses have been widely used as bone-filling materials and dental implants and have demonstrated its potential in soft-tissue regeneration. A comprehensive study on the application of bioactive glass in peripheral nerve regeneration has been conducted by Novajra and colleagues [320, 321]. Polymer-glass composite devices are made with bioactive glass powder and successfully applied in peripheral nerve repair. In other approach, this material can be used as fibres to produce nerve wraps, topographic patterns at conduit lumen or guidance channel to guide axonal growth. Their major advantage is the ions release from bioactive glasses, which have demonstrated angiogenic effects and the possibility of manipulation of glass composition; I order to include antibacterial ions, such as silver, gallium, zinc and copper, that can be useful in the prevention and treatment of infections resulting from the clinical intervention.

In conclusion, the vast field for improvement in peripheral nerve regeneration strategies has been well recognized. The ideal therapeutic alternative should be readily available, able to confine and direct axonal growth and be biofunctional, by the supplementation with bioactive molecules and/or cellular systems. Therefore, there is scope for improvement in the development of new and better alternatives of bioactive peripheral nerve repair complexes.

Acknowledgements

This research was supported by Programa Operacional Regional do Norte (ON.2—O Novo Norte), QREN, FEDER with the project 'iBone Therapies: Terapias inovadoras para a regeneração óssea', ref. NORTE-01-0247-FEDER-003262, and by the programme COMPETE—Programa Operacional Factores de Competitividade, Projects PEst-OE/AGR/UI0211/2011 and PEst-C/EME/UI0285/2013 funding from FCT. This research was also supported by Programa Operacional Competitividade e Internacionalização (P2020), Fundos Europeus Estruturais e de Investimento (FEEI) and FCT with the project 'BioMate—A novel bio-manufacturing system to produce bioactive scaffolds for tissue engineering' with reference PTDC/EMS-SIS/7032/2014 and by COMPETE 2020, from ANI—Projectos ID&T Empresas em Copromoção, Programas Operacionais POCI, by the project 'insitu.Biomas—Reinvent biomanufacturing systems by using an usability approach for in situ clinic temporary implants fabrication' with the reference POCI-01-0247-FEDER-017771. Ana Rita Caseiro (SFRH/BD/101174/2014) acknowledges FCT for financial support.

Conflicts of interest

The authors confirm that they have no conflicts of interest to declare for this publication.

http://dx.doi.org/10.5772/intechopen.69540

Abbreviations

Akt	Protein kinase B
BDNF	Brain-derived neurotrophic factor
BMSCs	Bone marrow mesenchymal stem cells
cAMP	Cyclic adenosine monophosphate
CNTF	Ciliary neurotrophic factor
DTI	Diffusion tensor imaging
DPSCs	Dental pulp stem cells
ECM	Extracellular matrix
EMG	Electromyograms
ePTFE	Polytetrafluoroethylene
FDA	Food and Drug Administration
FFMT	Free-functioning muscle transfer
GDNF	Glial cell line-derived neurotrophic factor
MHC	Major histocompatibility complex
MSCs	Mesenchymal stem cells
NAC	N-acetylcysteine
NCS	Nerve conduction studies
NGF	Nerve growth factor
NRG1	Neuregulin-1
NT-3/4/5	Neurotrophins 3, 4 and 5
PANI	Polyaniline
PCL	Poly D, L lactide-co-epsilon-caprolactone
PGA	Polyglycolic acid
PHB	Polyhydroxybutyrate
PI3K	Phosphatidylinositol-3 kinase
PKA	Protein kinase A
POSS	Polyhedral oligomeric silsesquioxane

PPy	Polypyrrole
PVA	Polyvinyl alcohol
PVA-CNTs	PVA-functionalized carbon nanotubes
Ras-ERK	Rat sarcoma-extracellular signal-regulated kinase
RGD	Arginylglycylaspartic acid
SC	Schwann cells
SFI	Sciatic Functional Index
SIS	Small intestine submucosa
SSI	Static Sciatic Index
TGF-β	Transforming growth factor beta superfamily
VEGF	Vascular endothelial growth factor
WJ	Wharton jelly
WRL	Withdrawal reflex latency

Author details

Sílvia Santos Pedrosa[1,2], Ana Rita Caseiro[1,2,3], José Domingos Santos[3] and Ana Colette Maurício[1,2*]

*Address all correspondence to: ana.colette@hotmail.com

1 Centro de Estudos de Ciência Animal (CECA), Instituto de Ciências e Tecnologias Agrárias e Agroalimentares (ICETA) of University of Porto (ICETA), Praça Gomes Teixeira, Apartado, Porto, Portugal

2 Department of Veterinary Clinics, Instituto de Ciências Biomédicas de Abel Salazar (ICBAS), University of Porto (UP), Rua de Jorge Viterbo Ferreira, Porto, Portugal

3 CEMMPRE - Centre for Mechanical Engineering, Materials and Processes, University of Coimbra (UC), Pinhal de Marrocos, Coimbra, Portugal

References

[1] Skalak R, Fox CF. Tissue Engineering: Proceedings of a Workshop, Granlibakken, Lake Tahoe, CA; February 26-29, 1988; Liss; 1988, ISBN: 0845147064, 9780845147061

[2] Langer R, Vacanti J. Tissue engineering. Science. 1993;**260**(5110):920-926

[3] Bianco P, Robey PG. Stem cells in tissue engineering. Nature. 2001;**414**(6859):118-121

[4] Tuan RS, Boland G, Tuli R. Adult mesenchymal stem cells and cell-based tissue engineering. Arthritis Research & Therapy. 2003;**5**(1):32-45

[5] Nerem RM, Sambanis A. Tissue engineering: From biology to biological substitutes. Tissue Engineering. 1995;**1**(1):3-13

[6] Muschler GF, Nakamoto C, Griffith LG. Engineering principles of clinical cell-based tissue engineering. The Journal of Bone and Joint Surgery America. 2004;**86-A**(7):1541-1558

[7] Zavaglia CAC, Prado da Silva MH. Feature Article: Biomaterials. Reference Module in Materials Science and Materials Engineering. Elsevier; 2016, DOI: 10.1016/B978-0-12-803581-8.04109-6

[8] Graziano A, d'Aquino R, Laino G, Papaccio G. Dental pulp stem cells: A promising tool for bone regeneration. Stem Cell Reviews. 2008;**4**(1):21-26

[9] Demarco FF, Conde MC, Cavalcanti BN, Casagrande L, Sakai VT, Nor JE. Dental pulp tissue engineering. Brazilian Dental Journal. 2011;**22**(1):3-13

[10] Lee K, Silva EA, Mooney DJ. Growth factor delivery-based tissue engineering: General approaches and a review of recent developments. Journal of the Royal Society, Interface. 2011;**8**(55):153-170

[11] Henkel J, Hutmacher Dietmar W. Design and fabrication of scaffold-based tissue engineering. BioNanoMaterials. 2013;**14**(3-4):171-193, DOI:10.1515/bnm-2013-0021

[12] Potdar PD, Jethmalani YD. Human dental pulp stem cells: Applications in future regenerative medicine. World Journal of Stem Cells. 2015;**7**(5):839-851

[13] Barnes CP, Sell SA, Boland ED, Simpson DG, Bowlin GL. Nanofiber technology: Designing the next generation of tissue engineering scaffolds. Advanced Drug Delivery Reviews. 2007;**59**(14):1413-1433

[14] Williams DF. The Williams Dictionary of Biomaterials: Liverpool University Press; 1999

[15] Yannas IV. Emerging rules for inducing organ regeneration. Biomaterials. 2013;**34**(2):321-330

[16] Ponche A, Bigerelle M, Anselme K. Relative influence of surface topography and surface chemistry on cell response to bone implant materials. Part 1: Physico-chemical effects. Proceedings of the Institution of Mechanical Engineers, Part H. 2010;**224**(12):1471-1486

[17] Muheremu A, Ao Q. Past, present, and future of nerve conduits in the treatment of peripheral nerve injury. Biomedical Research International. 2015;**2015**:237507

[18] Katirji B, Kaminski HJ, Ruff RL. Neuromuscular Disorders in Clinical Practice. Springer Science & Business Media. 2013, ISBN: 978-1-4614-6567-6

[19] Isaacs J. Major peripheral nerve injuries. Hand Clinics. 2013;**29**(3):371-382

[20] Palispis WA, Gupta R. Surgical repair in humans after traumatic nerve injury provides limited functional neural regeneration in adults. Experimental Neurology. 2017;**290**:106-114

[21] Aegineta P. De re medica. Paulou Aiginetou iatrou aristoubiblia hepta. En arche hekastou ton biblion deiknytai ta en ekeino pereichomena. Pavli Aeginetae medici optimi. Libri septem. 1528

[22] Millesi H. Microsurgery of peripheral nerves. The Hand. 1973;**5**(2):157-160

[23] Kurze T. Microtechniques in neurological surgery. Clinical Neurosurgery. 1964;**11**:128-137

[24] Smith JW. Microsurgery of peripheral nerves. Plastic and Reconstructive Surgery. 1964;**33**:317-329

[25] Sunderland SS, Williams HB. Nerve injuries and their repair: A critical appraisal. Plastic and Reconstructive Surgery. 1992;**89**(6):1170

[26] Rochkind S, Nevo Z. Recovery of peripheral nerve with massive loss defect by tissue engineered guiding regenerative gel. Biomedical Research International. 2014;**2014**:327578

[27] He B, Zhu Z, Zhu Q, Zhou X, Zheng C, Li P, et al. Factors predicting sensory and motor recovery after the repair of upper limb peripheral nerve injuries. Neural Regeneration Research. 2014;**9**(6):661-672

[28] He C-q, Zhang L-h, Liu X-f, Tang P-f. A 2-year follow-up survey of 523 cases with peripheral nerve injuries caused by the earthquake in Wenchuan, China. Neural Regeneration Research. 2015;**10**(2):252-259

[29] Ma CH, Omura T, Cobos EJ, Latr, xE, moli, et al. Accelerating axonal growth promotes motor recovery after peripheral nerve injury in mice. The Journal of Clinical Investigation. 2011;**121**(11):4332-4347, DOI: 10.1172/JCI58675

[30] Grinsell D, Keating CP. Peripheral nerve reconstruction after injury: A review of clinical and experimental therapies. Biomedical Research International. 2014;**2014**:698256

[31] Yegiyants S, Dayicioglu D, Kardashian G, Panthaki ZJ. Traumatic peripheral nerve injury: A wartime review. Journal of Craniofacial Surgery. 2010;**21**(4):998-1001

[32] Flores AJ, Lavernia CJ, Owens PW. Anatomy and physiology of peripheral nerve injury and repair. The American Journal of Orthopedics (Belle Mead, NJ). 2000;**29**(3):167-173

[33] Menorca RM, Fussell TS, Elfar JC. Nerve physiology: Mechanisms of injury and recovery. Hand Clinics. 2013;**29**(3):317-330

[34] Pfister BJ, Gordon T, Loverde JR, Kochar AS, Mackinnon SE, Cullen DK. Biomedical engineering strategies for peripheral nerve repair: Surgical applications, state of the art, and future challenges. Critical Reviews in Biomedical Engineering. 2011;**39**(2):81-124

[35] Schwartz LM. Atrophy and programmed cell death of skeletal muscle. Cell Death and Differentiation. 2008;**15**(7):1163-1169

[36] Deumens R, Bozkurt A, Meek MF, Marcus MA, Joosten EA, Weis J, et al. Repairing injured peripheral nerves: Bridging the gap. Progress in Neurobiology. 2010;**92**(3):245-276

[37] Griffin JW, Hogan MV, Chhabra AB, Deal DN. Peripheral nerve repair and reconstruction. The Journal of Bone and Joint Surgery America. 2013;**95**(23):2144-2151

http://dx.doi.org/10.5772/intechopen.69540

[38] Fu SY, Gordon T. The cellular and molecular basis of peripheral nerve regeneration. Molecular Neurobiology. 1997;**14**(1-2):67-116

[39] Lee SK, Wolfe SW. Peripheral nerve injury and repair. Journal of the American Academy of Orthopaedic Surgeons. 2000;**8**(4):243-252

[40] Siemionow M, Brzezicki G. Chapter 8: Current techniques and concepts in peripheral nerve repair. International Review of Neurobiology. 2009;**87**:141-172

[41] Seddon HJ. Three types of nerve injury. Brain. 1943;**66**(4):237-288

[42] Chhabra A, Ahlawat S, Belzberg A, Andreseik G. Peripheral nerve injury grading simplified on MR neurography: As referenced to Seddon and Sunderland classifications. The Indian Journal of Radiology & Imaging. 2014;**24**(3):217-224

[43] Sunderland S. A classification of peripheral nerve injuries producing loss of function. Brain. 1951;**74**(4):491-516

[44] Mackinnon SE, Dellon AL. Surgery of the Peripheral Nerve. Thieme Medical Publishers; 1988, ISBN:1588905136

[45] Glaus SW, Johnson PJ, Mackinnon SE. Clinical strategies to enhance nerve regeneration in composite tissue allotransplantation. Hand Clinics. 2011;**27**(4):495-509, ix

[46] Leoni AS, Mazzer N, Guirro RR, Jatte FG, Chereguini PA, Monte-Raso VV. High voltage pulsed current stimulation of the sciatic nerve in rats: Analysis by the SFI. Acta Ortopedica Brasiliera. 2012;**20**(2):93-97

[47] Merle M. SFI = sciatic functional index? Or some feebler imitation? Microsurgery. 1995;**16**(3):171-172

[48] Bervar M. Video analysis of standing—an alternative footprint analysis to assess functional loss following injury to the rat sciatic nerve. Journal of Neuroscience Methods. 2000;**102**(2):109-116

[49] Bervar M. An alternative video footprint analysis to assess functional loss following injury to the rat sciatic nerve. Acta Chirurgiae Plasticae. 2002;**44**(3):86-89

[50] Amado S, Armada-da-Silva PAS, João F, Maurício AC, Luís AL, Simões MJ, et al. The sensitivity of two-dimensional hindlimb joint kinematics analysis in assessing functional recovery in rats after sciatic nerve crush. Behavioural Brain Research. 2011;**225**(2):562-573

[51] Amado S, Rodrigues JM, Luis AL, Armada-da-Silva PA, Vieira M, Gartner A, et al. Effects of collagen membranes enriched with in vitro-differentiated N1E-115 cells on rat sciatic nerve regeneration after end-to-end repair. Journal of Neuroengineering and Rehabilitation. 2010;**7**:7

[52] Amado S, Simões MJ, Armada da Silva PAS, Luís AL, Shirosaki Y, Lopes MA, et al. Use of hybrid chitosan membranes and N1E-115 cells for promoting nerve regeneration in an axonotmesis rat model. Biomaterials. 2008;**29**(33):4409-4419

[53] Gärtner A, Pereira T, Armada-da-Silva PAS, Amorim I, Gomes R, Ribeiro J, et al. Use of poly(DL-lactide-ε-caprolactone) membranes and mesenchymal stem cells from the

Wharton's jelly of the umbilical cord for promoting nerve regeneration in axonotmesis: In vitro and in vivo analysis. Differentiation. 2012;**84**(5):355-365

[54] Thalhammer JG, Vladimirova M, Bershadsky B, Strichartz GR. Neurologic evaluation of the rat during sciatic nerve block with lidocaine. Anesthesiology. 1995;**82**(4):1013-1025

[55] Masters DB, Berde CB, Dutta SK, Griggs CT, Hu D, Kupsky W, et al. Prolonged regional nerve blockade by controlled release of local anesthetic from a biodegradable polymer matrix. Anesthesiology. 1993;**79**(2):340-346

[56] Caseiro AR, Pereira T, Ribeiro J, Amorim I, Faria F, Bártolo PJ, et al. Neuro-muscular regeneration using scaffolds with mesenchymal stem cells (MSCs) isolated from human umbilical cord Wharton's jelly: Functional and morphological analysis using rat sciatic nerve neurotmesis injury model. Procedia Engineering. 2015;**110**:106-113

[57] Costa LM, Simões MJ, Maurício AC, Varejão ASP. Chapter 7 Methods and Protocols in Peripheral Nerve Regeneration Experimental Research: Part IV—Kinematic Gait Analysis to Quantify Peripheral Nerve Regeneration in the Rat. International Review of Neurobiology. Academic Press; 2009;**87**:127-139. DOI: 10.1016/S0074-7742(09)87007-4.

[58] Gärtner A, Pereira T, Alves MG, Armada-da-Silva PAS, Amorim I, Gomes R, et al. Corrigendum to "use of poly(DL-lactide-ε-caprolactone) membranes and mesenchymal stem cells from the Wharton's jelly of the umbilical cord for promoting nerve regeneration in axonotmesis: In vitro and in vivo analysis" [Differentiation 84 (2012) 355-365]. Differentiation. 2013;**85**(3):119

[59] Luís AL, Amado S, Geuna S, Rodrigues JM, Simões MJ, Santos JD, et al. Long-term functional and morphological assessment of a standardized rat sciatic nerve crush injury with a non-serrated clamp. Journal of Neuroscience Methods. 2007;**163**(1):92-104

[60] Luis AL, Rodrigues JM, Amado S, Veloso AP, Armada-Da-Silva PA, Raimondo S, et al. PLGA 90/10 and caprolactone biodegradable nerve guides for the reconstruction of the rat sciatic nerve. Microsurgery. 2007;**27**(2):125-137

[61] Ribeiro J, Gartner A, Pereira T, Gomes R, Lopes MA, Gonçalves C, et al. Chapter four—Perspectives of employing mesenchymal stem cells from the Wharton's jelly of the umbilical cord for peripheral nerve repair. In: Stefano Geuna IPPT, Bruno B, editors. International Review of Neurobiology. Academic Press; 2013;**108**:79-120. DOI: 10.1016/B978-0-12-410499-0.00004-6

[62] Trehan SK, Model Z, Lee SK. Nerve repair and nerve grafting. Hand Clinics. 2016;**32**(2):119-125

[63] Panagopoulos GN, Megaloikonomos PD, Mavrogenis AF. The present and future for peripheral nerve regeneration. Orthopedics. 2016;**40**(1):e141-e156, DOI: 10.3928/01477447-20161019-01

[64] Gerth DJ, Tashiro J, Thaller SR. Clinical outcomes for conduits and scaffolds in peripheral nerve repair. World Journal of Clinical Cases. 2015;**3**(2):141-147

[65] Konofaos P, Ver Halen JP. Nerve repair by means of tubulization: Past, present, future. Journal of Reconstructive Microsurgery. 2013;**29**(3):149-164

[66] Lundborg G. A 25-year perspective of peripheral nerve surgery: Evolving neuroscientific concepts and clinical significance. Journal of Hand Surgery America. 2000;**25**(3):391-414

[67] Isaacs J. Treatment of acute peripheral nerve injuries: Current concepts. Journal of Hand Surgery America. 2010;**35**(3):491-497; quiz 8

[68] Isaacs JE, McDaniel CO, Owen JR, Wayne JS. Comparative analysis of biomechanical performance of available "nerve glues". Journal of Hand Surgery America. 2008;**33**(6):893-899

[69] Daroff RB, Jankovic J, Mazziotta JC, Pomeroy SL. Bradley's Neurology in Clinical Practice. Elsevier Health Sciences; 2015, ISBN: 9780323287838

[70] Colen KL, Choi M, Chiu DT. Nerve grafts and conduits. Plastic and Reconstructive Surgery. 2009;**124**(6 Suppl):e386–e394

[71] Terzis JK, Kostopoulos VK. Vascularized nerve grafts and vascularized fascia for upper extremity nerve reconstruction. Hand (New York). 2010;**5**(1):19-30

[72] Choi D, Raisman G. Immune rejection of a facial nerve xenograft does not prevent regeneration and the return of function: An experimental study. Neuroscience. 2003;**121**(2): 501-507

[73] Gaudin R, Knipfer C, Henningsen A, Smeets R, Heiland M, Hadlock T. Approaches to peripheral nerve repair: Generations of biomaterial conduits yielding to replacing autologous nerve grafts in craniomaxillofacial surgery. Biomedical Research International. 2016;**2016**:3856262

[74] Ray WZ, Mackinnon SE. Management of nerve gaps: Autografts, allografts, nerve transfers, and end-to-side neurorrhaphy. Experimental Neurology. 2010;**223**(1):77-85

[75] Millesi H, Meissl G, Berger A. The interfascicular nerve-grafting of the median and ulnar nerves. The Journal of Bone and Joint Surgery America. 1972;**54**(4):727-750

[76] Taylor GI, Ham FJ. The free vascularized nerve graft. A further experimental and clinical application of microvascular techniques. Plastic and Reconstructive Surgery. 1976;**57**(4):413-426

[77] Moore AM, Ray WZ, Chenard KE, Tung T, Mackinnon SE. Nerve allotransplantation as it pertains to composite tissue transplantation. Hand (New York). 2009;**4**(3):239-244

[78] Safa B, Buncke G. Autograft substitutes: Conduits and processed nerve allografts. Hand Clinics. 2016;**32**(2):127-140

[79] Karabekmez FE, Duymaz A, Moran SL. Early clinical outcomes with the use of decellularized nerve allograft for repair of sensory defects within the hand. Hand (New York). 2009;**4**(3):245-249

[80] Lee SK, Wolfe SW. Nerve transfers for the upper extremity: New horizons in nerve reconstruction. Journal of the American Academy of Orthopedic Surgeons. 2012;**20**(8):506-517

[81] Harris RI. The treatment of irreparable nerve injuries. Canadian Medical Association Journal. 1921;**11**(11):833-841

[82] Carlsen BT, Bishop AT, Shin AY. Late reconstruction for brachial plexus injury. Neurosurgery Clinics of North America. 2009;**20**(1):51-64, vi

[83] Kehoe S, Zhang XF, Boyd D. FDA approved guidance conduits and wraps for peripheral nerve injury: A review of materials and efficacy. Injury. 2012;**43**(5):553-572

[84] Lin MY, Manzano G, Gupta R. Nerve allografts and conduits in peripheral nerve repair. Hand Clinics. 2013;**29**(3):331-348

[85] FF IJpma, Van De Graaf RC, Meek MF. The early history of tubulation in nerve repair. Journal of Hand Surgery European Volume. 2008;**33**(5):581-586, DOI: 10.1177/1753193408091349

[86] Dahlin LB, Lundborg G. Use of tubes in peripheral nerve repair. Neurosurgery Clinics of North America. 2001;**12**(2):341-352

[87] Isaacs J, Mallu S, Yan W, Little B. Consequences of oversizing: Nerve-to-nerve tube diameter mismatch. Journal of Bone and Joint Surgery America. 2014;**96**(17):1461-1467

[88] Moore AM, Kasukurthi R, Magill CK, Farhadi HF, Borschel GH, Mackinnon SE. Limitations of conduits in peripheral nerve repairs. Hand (New York). 2009;**4**(2):180-186

[89] Nectow AR, Marra KG, Kaplan DL. Biomaterials for the development of peripheral nerve guidance conduits. Tissue Engineering Part B Reviews. 2012;**18**(1):40-50

[90] Meek MF, Coert JH. US Food and Drug Administration/Conformit Europe-approved absorbable nerve conduits for clinical repair of peripheral and cranial nerves. Annals in Plastic Surgery. 2008;**60**(1):110-116

[91] Brunelli GA, Vigasio A, Brunelli GR. Different conduits in peripheral nerve surgery. Microsurgery. 1994;**15**(3):176-178

[92] Ichihara S, Inada Y, Nakamura T. Artificial nerve tubes and their application for repair of peripheral nerve injury: An update of current concepts. Injury. 2008;**39**(Suppl 4):29-39

[93] Arslantunali D, Dursun T, Yucel D, Hasirci N, Hasirci V. Peripheral nerve conduits: Technology update. Medical Devices (Auckland). 2014;**7**:405-424

[94] Anderson PN, Turmaine M. Axonal regeneration through arterial grafts. Journal of Anatomy. 1986;**147**:73-82

[95] Kosutic D, Krajnc I, Pejkovic B, Solman L. Autogenous digital artery graft for repair of digital nerve defects in emergency hand reconstruction: Two-year follow-up. Journal of Plastic, Reconstructive & Aesthetic Surgery. 2009;**62**(4):553

[96] Frerichs O, Fansa H, Schicht C, Wolf G, Schneider W, Keilhoff G. Reconstruction of peripheral nerves using acellular nerve grafts with implanted cultured Schwann cells. Microsurgery. 2002;**22**(7):311-315

[97] Kim BS, Yoo JJ, Atala A. Peripheral nerve regeneration using acellular nerve grafts. Journal of Biomedical Material Research A. 2004;**68**(2):201-209

[98] Smith RM, Wiedl C, Chubb P, Greene CH. Role of small intestine submucosa (SIS) as a nerve conduit: Preliminary report. Journal of Investigative Surgery. 2004;**17**(6):339-344

[99] Yi JS, Lee HJ, Lee HJ, Lee IW, Yang JH. Rat peripheral nerve regeneration using nerve guidance channel by porcine small intestinal submucosa. Journal of Korean Neurosurgery Society. 2013;**53**(2):65-71

[100] Carriel V, Alaminos M, Garzon I, Campos A, Cornelissen M. Tissue engineering of the peripheral nervous system. Expert Review of Neurotherapeutics. 2014;**14**(3):301-318

[101] Quigley AF, Bulluss KJ, Kyratzis IL, Gilmore K, Mysore T, Schirmer KS, et al. Engineering a multimodal nerve conduit for repair of injured peripheral nerve. Journal of Neural Engineering. 2013;**10**(1):016008

[102] Stanec S, Stanec Z. Reconstruction of upper-extremity peripheral-nerve injuries with ePTFE conduits. Journal of Reconstructive Microsurgery. 1998;**14**(4):227-232

[103] Zhang BG, Quigley AF, Myers DE, Wallace GG, Kapsa RM, Choong PF. Recent advances in nerve tissue engineering. International Journal of Artificial Organs. 2014;**37**(4):277-291

[104] Kadler KE, Holmes DF, Trotter JA, Chapman JA. Collagen fibril formation. Biochemistry Journal. 1996;**316** (Pt 1):1-11

[105] Stang F, Fansa H, Wolf G, Reppin M, Keilhoff G. Structural parameters of collagen nerve grafts influence peripheral nerve regeneration. Biomaterials. 2005;**26**(16):3083-3091

[106] Stang F, Fansa H, Wolf G, Keilhoff G. Collagen nerve conduits--assessment of biocompatibility and axonal regeneration. Bio-Medical Materials and Engineering. 2005;**15**(1-2):3-12

[107] Sulong AF, Hassan NH, Hwei NM, Lokanathan Y, Naicker AS, Abdullah S, et al. Collagen-coated polylactic-glycolic acid (PLGA) seeded with neural-differentiated human mesenchymal stem cells as a potential nerve conduit. Advances in Clinical and Experimental Medicine. 2014;**23**(3):353-362

[108] Bushnell BD, McWilliams AD, Whitener GB, Messer TM. Early clinical experience with collagen nerve tubes in digital nerve repair. Journal of Hand Surgery America. 2008;**33**(7):1081-1087

[109] Wangensteen KJ, Kalliainen LK. Collagen tube conduits in peripheral nerve repair: A retrospective analysis. Hand (New York). 2010;**5**(3):273-277

[110] Lohmeyer JA, Sommer B, Siemers F, Mailander P. Nerve injuries of the upper extremity-expected outcome and clinical examination. Plastic Surgical Nursing. 2009;**29**(2):88-93; quiz 4-5

[111] Taras JS, Jacoby SM, Lincoski CJ. Reconstruction of digital nerves with collagen conduits. Journal of Hand Surgery America. 2011;**36**(9):1441-1446

[112] Ashley WW, Jr., Weatherly T, Park TS. Collagen nerve guides for surgical repair of brachial plexus birth injury. Journal of Neurosurgery. 2006;**105**(6 Suppl):452-456

[113] Archibald SJ, Shefner J, Krarup C, Madison RD. Monkey median nerve repaired by nerve graft or collagen nerve guide tube. Journal of Neuroscience. 1995;**15**(5 Pt 2):4109-4123

[114] Bekler HI, Rosenwasser MP, Akilina Y, Bulut G. The use of an absorbable collagen cover (NeuraWrap) improves patency of interpositional vein grafts. Acta Orthopaedica et Traumatologica Turcica. 2010;**44**(2):157-161

[115] Kitahara AK, Suzuki Y, Qi P, Nishimura Y, Suzuki K, Kiyotani T, et al. Facial nerve repair using a collagen conduit in cats. Scandinavian Journal of Plastic and Reconstructive Surgery, Hand Surgery. 1999;**33**(2):187-193

[116] Dalamagkas K, Tsintou M, Seifalian A. Advances in peripheral nervous system regenerative therapeutic strategies: A biomaterials approach. Material Science and Engineering C: Materials for Biological Applications. 2016;**65**:425-432

[117] Sedaghati T, Jell G, Seifalian A. Investigation of Schwann cell behaviour on RGD-functionalised bioabsorbable nanocomposite for peripheral nerve regeneration. Nature Biotechnology. 2014;**31**(3):203-213

[118] Monaco G, Cholas R, Salvatore L, Madaghiele M, Sannino A. Sterilization of collagen scaffolds designed for peripheral nerve regeneration: Effect on microstructure, degradation and cellular colonization. Materials Science and Engineering: C. 2017;**71**:335-344

[119] Hall H. Modified fibrin hydrogel matrices: Both, 3D-scaffolds and local and controlled release systems to stimulate angiogenesis. Current Pharmaceutical Design. 2007;**13**(35):3597-3607

[120] Sameem M, Wood TJ, Bain JR. A systematic review on the use of fibrin glue for peripheral nerve repair. Plastic and Reconstructive Surgery. 2011;**127**(6):2381-2390

[121] Kalbermatten DF, Pettersson J, Kingham PJ, Pierer G, Wiberg M, Terenghi G. New fibrin conduit for peripheral nerve repair. Journal of Reconstructive Microsurgery. 2009;**25**(1):27-33

[122] Pettersson J, Kalbermatten D, McGrath A, Novikova LN. Biodegradable fibrin conduit promotes long-term regeneration after peripheral nerve injury in adult rats. Journal of Plastic, Reconstructive & Aesthetic Surgery. 2010;**63**(11):1893-1899

[123] Rafijah G, Bowen AJ, Dolores C, Vitali R, Mozaffar T, Gupta R. The effects of adjuvant fibrin sealant on the surgical repair of segmental nerve defects in an animal model. Journal of Hand Surgery America. 2013;**38**(5):847-855

[124] Huang W, Begum R, Barber T, Ibba V, Tee NC, Hussain M, et al. Regenerative potential of silk conduits in repair of peripheral nerve injury in adult rats. Biomaterials. 2012;**33**(1):59-71

[125] Lu Y, Chi F, Zhao X, Shao Z, Cao Z. Experimental study on facial nerve regeneration by porous silk fibroin conduit. Zhonghua er bi yan hou tou jing wai ke za zhi= Chinese Journal of Otorhinolaryngology Head and Neck Surgery. 2006;**41**(8):603-606

[126] Park SY, Ki CS, Park YH, Lee KG, Kang SW, Kweon HY, et al. Functional recovery guided by an electrospun silk fibroin conduit after sciatic nerve injury in rats. Journal of Tissue Engineering and Regenerative Medicine. 2015;**9**(1):66-76

[127] Xie H, Yang W, Chen J, Zhang J, Lu X, Zhao X, et al. A silk sericin/silicone nerve guidance conduit promotes regeneration of a transected sciatic nerve. Advanced Healthcare Materials. 2015;**4**(15):2195-2205

[128] Mottaghitalab F, Farokhi M, Zaminy A, Kokabi M, Soleimani M, Mirahmadi F, et al. A biosynthetic nerve guide conduit based on silk/SWNT/fibronectin nanocomposite for peripheral nerve regeneration. PLoS One. 2013;**8**(9):e74417

[129] Yang Y, Ding F, Wu J, Hu W, Liu W, Liu J, et al. Development and evaluation of silk fibroin-based nerve grafts used for peripheral nerve regeneration. Biomaterials. 2007;**28**(36):5526-5535

[130] Das S, Sharma M, Saharia D, Sarma KK, Sarma MG, Borthakur BB, et al. In vivo studies of silk based gold nano-composite conduits for functional peripheral nerve regeneration. Biomaterials. 2015;**62**:66-75

[131] Cohen-Karni T, Jeong KJ, Tsui JH, Reznor G, Mustata M, Wanunu M, et al. Nanocomposite gold-silk nanofibers. Nano Letters. 2012;**12**(10):5403-5406

[132] Rao J, Cheng Y, Liu Y, Ye Z, Zhan B, Quan D, et al. A multi-walled silk fibroin/silk sericin nerve conduit coated with poly(lactic-co-glycolic acid) sheath for peripheral nerve regeneration. Materials Science and Engineering: C. 2017;**73**:319-332

[133] Dinis T, Elia R, Vidal G, Dermigny Q, Denoeud C, Kaplan D, et al. 3D multi-channel bi-functionalized silk electrospun conduits for peripheral nerve regeneration. Journal of the Mechanical Behavior of Biomedical Materials. 2015;**41**:43-55

[134] Dinis TM, Vidal G, Jose RR, Vigneron P, Bresson D, Fitzpatrick V, et al. Complementary effects of two growth factors in multifunctionalized silk nanofibers for nerve reconstruction. PloS One. 2014;**9**(10):e109770

[135] Lin Q, Cai Y, Li H. Experimental study on gradient of nerve growth factor immobilized conduits promoting peripheral nerve regeneration in rats. Zhongguo xiu fu chong jian wai ke za zhi= Zhongguo xiufu chongjian waike zazhi= Chinese Journal of Reparative and Reconstructive Surgery. 2014;**28**(2):167-172

[136] Lin Y-C, Ramadan M, Hronik-Tupaj M, Kaplan DL, Philips BJ, Sivak W, et al. spatially controlled delivery of neurotrophic factors in silk fibroin–based nerve conduits for peripheral nerve repair. Annals of Plastic Surgery. 2011;**67**(2):147-155

[137] Sivak WN, White JD, Bliley JM, Tien LW, Liao HT, Kaplan DL, et al. Delivery of chondroitinase ABC and glial cell line-derived neurotrophic factor from silk fibroin conduits enhances peripheral nerve regeneration. Journal of Tissue Engineering and Regenerative Medicine. 2014;**11**(3):733-742, DOI: 10.1002/term.1970

[138] Zhang H, Wang K, Xing Y, Yu Q. Lysine-doped polypyrrole/spider silk protein/poly(L-lactic) acid containing nerve growth factor composite fibers for neural application. Materials Science and Engineering: C. 2015;**56**:564-573

[139] Zheng L, Cui HF. Use of chitosan conduit combined with bone marrow mesenchymal stem cells for promoting peripheral nerve regeneration. The Journal of Material Science: Materials in Medicine. 2010;**21**(5):1713-1720

[140] Gu J, Hu W, Deng A, Zhao Q, Lu S, Gu X. Surgical repair of a 30 mm long human median nerve defect in the distal forearm by implantation of a chitosan-PGA nerve guidance conduit. Journal of Tissue Engineering and Regenerative Medicine. 2012;**6**(2):163-168

[141] Clavijo-Alvarez JA, Nguyen VT, Santiago LY, Doctor JS, Lee WP, Marra KG. Comparison of biodegradable conduits within aged rat sciatic nerve defects. Plastic and Reconstructive Surgery. 2007;**119**(6):1839-1851

[142] Rampichova M, Kostakova E, Filova E, Prosecka E, Plencner M, Ocheretna L, et al. Non-woven PGA/PVA fibrous mesh as an appropriate scaffold for chondrocyte proliferation. Physiology Research. 2010;**59**(5):773-781

[143] Schlosshauer B, Dreesmann L, Schaller HE, Sinis N. Synthetic nerve guide implants in humans: A comprehensive survey. Neurosurgery. 2006;**59**(4):740-747; discussion 7-8

[144] Dellon AL, Mackinnon SE. An alternative to the classical nerve graft for the management of the short nerve gap. Plastic and Reconstructive Surgery. 1988;**82**(5):849-856

[145] Weber RA, Breidenbach WC, Brown RE, Jabaley ME, Mass DP. A randomized prospective study of polyglycolic acid conduits for digital nerve reconstruction in humans. Plastic and Reconstructive Surgery. 2000;**106**(5):1036-1045; discussion 46-48

[146] Battiston B, Geuna S, Ferrero M, Tos P. Nerve repair by means of tubulization: Literature review and personal clinical experience comparing biological and synthetic conduits for sensory nerve repair. Microsurgery. 2005;**25**(4):258-267

[147] Rinker B, Liau JY. A Prospective randomized study comparing woven polyglycolic acid and autogenous vein conduits for reconstruction of digital nerve gaps. Journal of Hand Surgery. 2011;**36**(5):775-781

[148] Crawley WA, Dellon AL. Inferior alveolar nerve reconstruction with a polyglycolic acid bioabsorbable nerve conduit. Plastic and Reconstructive Surgery. 1992;**90**(2):300-302

[149] Dellon AL, Maloney CT, Jr. Salvage of sensation in a hallux-to-thumb transfer by nerve tube reconstruction. Journal of Hand Surgery America. 2006;**31**(9):1495-1498

[150] Donoghoe N, Rosson GD, Dellon AL. Reconstruction of the human median nerve in the forearm with the Neurotube. Microsurgery. 2007;**27**(7):595-600

[151] Rosson GD, Williams EH, Dellon AL. Motor nerve regeneration across a conduit. Microsurgery. 2009;**29**(2):107-114

[152] Gu X, Ding F, Williams DF. Neural tissue engineering options for peripheral nerve regeneration. Biomaterials. 2014;**35**(24):6143-6156

[153] Duda S, Dreyer L, Behrens P, Wienecke S, Chakradeo T, Glasmacher B, et al. Outer electrospun polycaprolactone shell induces massive foreign body reaction and impairs axonal regeneration through 3D multichannel chitosan nerve guides. Biomedical Research International. 2014;**2014**:835269

[154] Bertleff MJ, Meek MF, Nicolai JP. A prospective clinical evaluation of biodegradable Neurolac nerve guides for sensory nerve repair in the hand. Journal of Hand Surgery America. 2005;**30**(3):513-518

[155] Hernandez-Cortes P, Garrido J, Camara M, Ravassa FO. Failed digital nerve reconstruction by foreign body reaction to Neurolac nerve conduit. Microsurgery. 2010;**30**(5):414-416

[156] Meek MF, Den Dunnen WF. Porosity of the wall of a Neurolac nerve conduit hampers nerve regeneration. Microsurgery. 2009;**29**(6):473-478

[157] Den Dunnen W, Meek M. Sensory nerve function and auto-mutilation after reconstruction of various gap lengths with nerve guides and autologous nerve grafts. Biomaterials. 2001;**22**(10):1171-1176

[158] Den Dunnen W, Meek M, Robinson P, Schakernraad J. Peripheral nerve regeneration through P (DLLA-ε-CL) nerve guides. Journal of Materials Science: Materials in Medicine. 1998;**9**(12):811-814

[159] Den Dunnen W, Stokroos I, Blaauw E, Holwerda A, Pennings A, Robinson P, et al. Light-microscopic and electron-microscopic evaluation of short-term nerve regeneration using a biodegradable poly (DL-lactide-ϵ-caprolacton) nerve guide. Journal of Biomedical Materials Research Part A. 1996;**31**(1):105-115

[160] Meek M, Den Dunnen W, Schakenraad J, Robinson P. Long-term evaluation of functional nerve recovery after reconstruction with a thin-walled biodegradable poly(DL-lactide-ε-caprolactone) nerve guide, using walking track analysis and electrostimulation tests. Microsurgery. 1999;**19**(5):247-253

[161] Meek MF. More than just sunshine with implantation of resorbable (p(DLLA-epsilon-CL)) biomaterials. Biomedical Materials and Engineering. 2007;**17**(5):329-334

[162] Meek MF, Van Der Werff JF, Nicolai JPA, Gramsbergen A. Biodegradable p (DLLA-ϵ-CL) nerve guides versus autologous nerve grafts: Electromyographic and video analysis. Muscle & Nerve. 2001;**24**(6):753-759

[163] Secer HI, Solmaz I, Anik I, Izci Y, Duz B, Daneyemez MK, et al. Surgical outcomes of the brachial plexus lesions caused by gunshot wounds in adults. Journal of Brachial Plexus and Peripheral Nerve Injury. 2009;**4**:11

[164] Meek MF, den Dunnen WF, Robinson PH, Pennings AJ, Schakenraad JM. Evaluation of functional nerve recovery after reconstruction with a new biodegradable poly(DL-lactide-epsilon-caprolactone) nerve guide. International Journal of Artificial Organs. 1997;**20**(8):463-468

[165] Meek MF, Dijkstra JR, Den Dunnen WF, Ijkema-Paassen J, Schakenraad JM, Gramsbergen A, et al. Functional assessment of sciatic nerve reconstruction: Biodegradable poly (DLLA-epsilon-CL) nerve guides versus autologous nerve grafts. Microsurgery. 1999;**19**(8):381-388

[166] Meek MF, Robinson PH, Stokroos I, Blaauw EH, Kors G, den Dunnen WF. Electron microscopical evaluation of short-term nerve regeneration through a thin-walled biodegradable poly(DLLA-epsilon-CL) nerve guide filled with modified denatured muscle tissue. Biomaterials. 2001;**22**(10):1177-1185

[167] Varejao AS, Cabrita AM, Geuna S, Patricio JA, Azevedo HR, Ferreira AJ, et al. Functional assessment of sciatic nerve recovery: Biodegradable poly (DLLA-epsilon-CL) nerve guide filled with fresh skeletal muscle. Microsurgery. 2003;**23**(4):346-353

[168] Varejao AS, Cabrita AM, Meek MF, Fornaro M, Geuna S, Giacobini-Robecchi MG. Morphology of nerve fiber regeneration along a biodegradable poly (DLLA-epsilon-CL) nerve guide filled with fresh skeletal muscle. Microsurgery. 2003;**23**(4):338-345

[169] Chen GQ, Wu Q. The application of polyhydroxyalkanoates as tissue engineering materials. Biomaterials. 2005;**26**(33):6565-6578

[170] Xiao XQ, Zhao Y, Chen GQ. The effect of 3-hydroxybutyrate and its derivatives on the growth of glial cells. Biomaterials. 2007;**28**(25):3608-3616

[171] Young RC, Wiberg M, Terenghi G. Poly-3-hydroxybutyrate (PHB): A resorbable conduit for long-gap repair in peripheral nerves. British Journal of Plastic Surgery. 2002;**55**(3):235-240

[172] Aberg M, Ljungberg C, Edin E, Millqvist H, Nordh E, Theorin A, et al. Clinical evaluation of a resorbable wrap-around implant as an alternative to nerve repair: A prospective, assessor-blinded, randomised clinical study of sensory, motor and functional recovery after peripheral nerve repair. Journal of Plastic, Reconstructive & Aesthetic Surgery. 2009;**62**(11):1503-1509

[173] Subramanian A, Krishnan UM, Sethuraman S. Development of biomaterial scaffold for nerve tissue engineering: Biomaterial mediated neural regeneration. Journal of Biomedical Science. 2009;**16**:108

[174] Neal RA, Tholpady SS, Foley PL, Swami N, Ogle RC, Botchwey EA. Alignment and composition of laminin-polycaprolactone nanofiber blends enhance peripheral nerve regeneration. Journal of Biomedical Materials Research Part A. 2012;**100**(2):406-423

[175] Armstrong SJ, Wiberg M, Terenghi G, Kingham PJ. ECM molecules mediate both Schwann cell proliferation and activation to enhance neurite outgrowth. Tissue Engineering. 2007;**13**(12):2863-2870

[176] Patel S, Kurpinski K, Quigley R, Gao H, Hsiao BS, Poo MM, et al. Bioactive nanofibers: Synergistic effects of nanotopography and chemical signaling on cell guidance. Nano Letters. 2007;**7**(7):2122-2128

[177] Ahmed MR, Jayakumar R. Peripheral nerve regeneration in RGD peptide incorporated collagen tubes. Brain Research. 2003;**993**(1):208-216

[178] Hersel U, Dahmen C, Kessler H. RGD modified polymers: Biomaterials for stimulated cell adhesion and beyond. Biomaterials. 2003;**24**(24):4385-4415

[179] Ni HC, Lin ZY, Hsu SH, Chiu IM. The use of air plasma in surface modification of peripheral nerve conduits. Acta Biomaterialia. 2010;**6**(6):2066-2076

[180] Ahmed MR, Basha SH, Gopinath D, Muthusamy R, Jayakumar R. Initial upregulation of growth factors and inflammatory mediators during nerve regeneration in the presence of cell adhesive peptide-incorporated collagen tubes. Journal of the Peripheral Nervous System. 2005;**10**(1):17-30

[181] Pierschbacher MD, Ruoslahti E. Cell attachment activity of fibronectin can be duplicated by small synthetic fragments of the molecule. Nature. 1984;**309**(5963):30

[182] Pierschbacher MD, Ruoslahti E. Variants of the cell recognition site of fibronectin that retain attachment-promoting activity. Proceedings of the National Academy of Sciences. 1984;**81**(19):5985-5988

[183] Vleggeert-Lankamp CL, de Ruiter GC, Wolfs JF, Pego AP, van den Berg RJ, Feirabend HK, et al. Pores in synthetic nerve conduits are beneficial to regeneration. Journal of Biomedical Materials Research Part A. 2007;**80**(4):965-982

[184] O'Brien FJ, Harley BA, Yannas IV, Gibson LJ. The effect of pore size on cell adhesion in collagen-GAG scaffolds. Biomaterials. 2005;**26**(4):433-441

[185] Chew SY, Mi R, Hoke A, Leong KW. Aligned protein-polymer composite fibers enhance nerve regeneration: A potential tissue-engineering platform. Advanced Functional Materials. 2007;**17**(8):1288-1296

[186] Panseri S, Cunha C, Lowery J, Del Carro U, Taraballi F, Amadio S, et al. Electrospun micro- and nanofiber tubes for functional nervous regeneration in sciatic nerve transections. BMC Biotechnology. 2008;**8**:39

[187] Clements IP, Kim YT, English AW, Lu X, Chung A, Bellamkonda RV. Thin-film enhanced nerve guidance channels for peripheral nerve repair. Biomaterials. 2009;**30**(23-24):3834-3846

[188] Ngo TT, Waggoner PJ, Romero AA, Nelson KD, Eberhart RC, Smith GM. Poly(L-Lactide) microfilaments enhance peripheral nerve regeneration across extended nerve lesions. Journal of Neuroscience Research. 2003;**72**(2):227-238

[189] Araki T, Nagarajan R, Milbrandt J. Identification of genes induced in peripheral nerve after injury expression profiling and novel gene discovery. Journal of Biological Chemistry. 2001;**276**(36):34131-34141

[190] Chang L-W, Viader A, Varghese N, Payton JE, Milbrandt J, Nagarajan R. An integrated approach to characterize transcription factor and microRNA regulatory networks

involved in Schwann cell response to peripheral nerve injury. BMC Genomics. 2013;**14**(1):84

[191] Evans GR, Brandt K, Katz S, Chauvin P, Otto L, Bogle M, et al. Bioactive poly (L-lactic acid) conduits seeded with Schwann cells for peripheral nerve regeneration. Biomaterials. 2002;**23**(3):841-848

[192] Gottlieb DI. Large-scale sources of neural stem cells. Annual Review of Neuroscience. 2002;**25**(1):381-407

[193] Han SS, Fischer I. Neural stem cells and gene therapy: Prospects for repairing the injured spinal cord. Journal of American Medical Association. 2000;**283**(17):2300-2301

[194] Lobsiger CS, Taylor V, Suter U. The early life of a Schwann cell. Biological Chemistry. 2002;**383**(2):245-253

[195] Morrison SJ, White PM, Zock C, Anderson DJ. Prospective identification, isolation by flow cytometry, and in vivo self-renewal of multipotent mammalian neural crest stem cells. Cell. 1999;**96**(5):737-749

[196] Morrissey TK, Kleitman N, Bunge RP. Isolation and functional characterization of Schwann cells derived from adult peripheral nerve. Journal of Neuroscience. 1991;**11**(8):2433-2442

[197] Park KI, Liu S, Flax JD, Nissim S, Stieg PE, Snyder EY. Transplantation of neural progenitor and stem cells: Developmental insights may suggest new therapies for spinal cord and other CNS dysfunction. Journal of Neurotrauma. 1999;**16**(8):675-687

[198] Rao MS, Anderson DJ. Immortalization and controlled in vitro differentiation of murine multipotent neural crest stem cells. Journal of Neurobiology. 1997;**32**(7):722-746

[199] Rappa G, Kunke D, Holter J, Diep D, Meyer J, Baum C, et al. Efficient expansion and gene transduction of mouse neural stem/progenitor cells on recombinant fibronectin. Neuroscience. 2004;**124**(4):823-830

[200] Tohill M, Mantovani C, Wiberg M, Terenghi G. Rat bone marrow mesenchymal stem cells express glial markers and stimulate nerve regeneration. Neuroscience Letters. 2004;**362**(3):200-203

[201] Viader A, Chang L-W, Fahrner T, Nagarajan R, Milbrandt J. MicroRNAs modulate Schwann cell response to nerve injury by reinforcing transcriptional silencing of dedifferentiation-related genes. Journal of Neuroscience. 2011;**31**(48):17358-17369

[202] Chang CJ, Hsu SH. The effects of low-intensity ultrasound on peripheral nerve regeneration in poly(DL-lactic acid-co-glycolic acid) conduits seeded with Schwann cells. Ultrasound in Medicine & Biology. 2004;**30**(8):1079-1084

[203] di Summa PG, Kingham PJ, Raffoul W, Wiberg M, Terenghi G, Kalbermatten DF. Adipose-derived stem cells enhance peripheral nerve regeneration. Journal of Plastic, Reconstructive & Aesthetic Surgery. 2010;**63**(9):1544-1552

[204] Hadlock T, Sundback C, Hunter D, Cheney M, Vacanti JP. A polymer foam conduit seeded with Schwann cells promotes guided peripheral nerve regeneration. Tissue Engineering. 2000;**6**(2):119-127

[205] Kalbermatten DF, Kingham PJ, Mahay D, Mantovani C, Pettersson J, Raffoul W, et al. Fibrin matrix for suspension of regenerative cells in an artificial nerve conduit. Journal of Plastic, Reconstructive & Aesthetic Surgery. 2008;**61**(6):669-675

[206] Santiago LY, Clavijo-Alvarez J, Brayfield C, Rubin JP, Marra KG. Delivery of adipose-derived precursor cells for peripheral nerve repair. Cell Transplantation. 2009;**18**(2):145-158

[207] Madduri S, Gander B. Schwann cell delivery of neurotrophic factors for peripheral nerve regeneration. Journal of the Peripheral Nervous System. 2010;**15**(2):93-103

[208] Schlosshauer B, Muller E, Schroder B, Planck H, Muller HW. Rat Schwann cells in bioresorbable nerve guides to promote and accelerate axonal regeneration. Brain Research. 2003;**963**(1-2):321-326

[209] Euler de Souza Lucena E, Guzen FP, Lopes de Paiva Cavalcanti JR, Galvao Barboza CA, Silva do Nascimento Junior E, Cavalcante Jde S. Experimental considerations concerning the use of stem cells and tissue engineering for facial nerve regeneration: A systematic review. Journal of Oral and Maxillofacial Surgery. 2014;**72**(5):1001-1012

[210] Hadlock T, Elisseeff J, Langer R, Vacanti J, Cheney M. A tissue-engineered conduit for peripheral nerve repair. Archives of Otolaryngology—Head and Neck Surgery. 1998;**124**(10):1081-1086

[211] Wang J, Ding F, Gu Y, Liu J, Gu X. Bone marrow mesenchymal stem cells promote cell proliferation and neurotrophic function of Schwann cells in vitro and in vivo. Brain Research. 2009;**1262**:7-15

[212] Kingham PJ, Kalbermatten DF, Mahay D, Armstrong SJ, Wiberg M, Terenghi G. Adipose-derived stem cells differentiate into a Schwann cell phenotype and promote neurite outgrowth in vitro. Experimental Neurology. 2007;**207**(2):267-274

[213] Dezawa M, Takahashi I, Esaki M, Takano M, Sawada H. Sciatic nerve regeneration in rats induced by transplantation of in vitro differentiated bone-marrow stromal cells. European Journal of Neuroscience. 2001;**14**(11):1771-1776

[214] Wang HS, Hung SC, Peng ST, Huang CC, Wei HM, Guo YJ, et al. Mesenchymal stem cells in the Wharton's jelly of the human umbilical cord. Stem Cells. 2004;**22**(7):1330-1337

[215] Park HW, Lim MJ, Jung H, Lee SP, Paik KS, Chang MS. Human mesenchymal stem cell-derived Schwann cell-like cells exhibit neurotrophic effects, via distinct growth factor production, in a model of spinal cord injury. Glia. 2010;**58**(9):1118-1132

[216] Joshi CV, Enver T. Plasticity revisited. Current Opinion in Cell Biology. 2002;**14**(6):749-755

[217] Thuret S, Moon LD, Gage FH. Therapeutic interventions after spinal cord injury. Nature Review Neuroscience. 2006;**7**(8):628-643

[218] Sarugaser R, Lickorish D, Baksh D, Hosseini MM, Davies JE. Human umbilical cord perivascular (HUCPV) cells: A source of mesenchymal progenitors. Stem Cells. 2005;**23**(2):220-229

[219] Costa HJZR, Bento RF, Salomone R, Azzi-Nogueira D, Zanatta DB, Costa MP, et al. Mesenchymal bone marrow stem cells within polyglycolic acid tube observed in vivo after six weeks enhance facial nerve regeneration. Brain Research. 2013;**1510**:10-21

[220] Colter DC, Class R, DiGirolamo CM, Prockop DJ. Rapid expansion of recycling stem cells in cultures of plastic-adherent cells from human bone marrow. Proceedings of the National Academy of Sciences United States of America. 2000;**97**(7):3213-3218

[221] Gnecchi M, Melo LG. Bone marrow-derived mesenchymal stem cells: Isolation, expansion, characterization, viral transduction, and production of conditioned medium. Methods in Molecular Biology. 2009;**482**:281-294

[222] Fu YS, Cheng YC, Lin MY, Cheng H, Chu PM, Chou SC, et al. Conversion of human umbilical cord mesenchymal stem cells in Wharton's jelly to dopaminergic neurons in vitro: Potential therapeutic application for Parkinsonism. Stem Cells. 2006;**24**(1):115-124

[223] Mitchell KE, Weiss ML, Mitchell BM, Martin P, Davis D, Morales L, et al. Matrix cells from Wharton's jelly form neurons and glia. Stem Cells. 2003;**21**(1):50-60

[224] Weiss ML, Medicetty S, Bledsoe AR, Rachakatla RS, Choi M, Merchav S, et al. Human umbilical cord matrix stem cells: Preliminary characterization and effect of transplantation in a rodent model of Parkinson's disease. Stem Cells. 2006;**24**(3):781-792

[225] Weiss ML, Mitchell KE, Hix JE, Medicetty S, El-Zarkouny SZ, Grieger D, et al. Transplantation of porcine umbilical cord matrix cells into the rat brain. Experimental Neurology. 2003;**182**(2):288-299

[226] Fujimura J, Ogawa R, Mizuno H, Fukunaga Y, Suzuki H. Neural differentiation of adipose-derived stem cells isolated from GFP transgenic mice. Biochemical and Biophysical Research Communications. 2005;**333**(1):116-121

[227] Liu BS, Yang YC, Shen CC. Regenerative effect of adipose tissue-derived stem cells transplantation using nerve conduit therapy on sciatic nerve injury in rats. Journal of Tissue Engineering and Regenerative Medicine. 2014;**8**(5):337-350

[228] Xu Y, Liu Z, Liu L, Zhao C, Xiong F, Zhou C, et al. Neurospheres from rat adipose-derived stem cells could be induced into functional Schwann cell-like cells in vitro. BMC Neuroscience. 2008;**9**:21

[229] Zavan B, Vindigni V, Gardin C, D'Avella D, Della Puppa A, Abatangelo G, et al. Neural potential of adipose stem cells. Discovery Medicine. 2010;**10**(50):37-43

[230] di Summa PG, Kingham PJ, Campisi CC, Raffoul W, Kalbermatten DF. Collagen (NeuraGen®) nerve conduits and stem cells for peripheral nerve gap repair. Neuroscience Letters. 2014;**572**:26-31

[231] Georgiou M, Golding JP, Loughlin AJ, Kingham PJ, Phillips JB. Engineered neural tissue with aligned, differentiated adipose-derived stem cells promotes peripheral nerve regeneration across a critical sized defect in rat sciatic nerve. Biomaterials. 2015;**37**:242-251

[232] Arthur A, Rychkov G, Shi S, Koblar SA, Gronthos S. Adult human dental pulp stem cells differentiate toward functionally active neurons under appropriate environmental cues. Stem Cells. 2008;**26**(7):1787-1795

[233] Huang AH, Snyder BR, Cheng PH, Chan AW. Putative dental pulp-derived stem/stromal cells promote proliferation and differentiation of endogenous neural cells in the hippocampus of mice. Stem Cells. 2008;**26**(10):2654-2663

[234] Kim BC, Jun SM, Kim SY, Kwon YD, Choe SC, Kim EC, et al. Engineering three dimensional micro nerve tissue using postnatal stem cells from human dental apical papilla. Biotechnology and Bioengineering. 2017;**114**(4):903-914

[235] Yamamoto T, Osako Y, Ito M, Murakami M, Hayashi Y, Horibe H, et al. Trophic effects of dental pulp stem cells on schwann cells in peripheral nerve regeneration. Cell Transplantation. 2016;**25**(1):183-193

[236] Sasaki R, Aoki S, Yamato M, Uchiyama H, Wada K, Ogiuchi H, et al. PLGA artificial nerve conduits with dental pulp cells promote facial nerve regeneration. Journal of Tissue Engineering and Regenerative Medicine. 2011;**5**(10):823-830

[237] Karbanová J, Soukup T, Suchánek J, Pytlík R, Corbeil D, Mokrý J. Characterization of dental pulp stem cells from impacted third molars cultured in low serum-containing medium. Cells Tissues Organs. 2011;**193**(6):344

[238] Atari M, Barajas M, Hernández-Alfaro F, Gil C, Fabregat M, Ferrés Padró E, et al. Isolation of pluripotent stem cells from human third molar dental pulp. Histology and Histopathology. 2011;**26**(7):1057

[239] Suchánek J, Soukup T, Ivancaková R, Karbanová J, Hubková V, Pytlík R, et al. Human dental pulp stem cells—isolation and long term cultivation. Acta Medica (Hradec Králové)/Universitas Carolina, Facultas Medica Hradec Králové. 2007;**50**(3):195-201

[240] McLeod M, Hong M, Mukhida K, Sadi D, Ulalia R, Mendez I. Erythropoietin and GDNF enhance ventral mesencephalic fiber outgrowth and capillary proliferation following neural transplantation in a rodent model of Parkinson's disease. European Journal of Neuroscience. 2006;**24**(2):361-370

[241] Nosrat IV, Smith CA, Mullally P, Olson L, Nosrat CA. Dental pulp cells provide neurotrophic support for dopaminergic neurons and differentiate into neurons in vitro;

implications for tissue engineering and repair in the nervous system. European Journal of Neuroscience. 2004;**19**(9):2388-2398

[242] Rehman J, Traktuev D, Li J, Merfeld-Clauss S, Temm-Grove CJ, Bovenkerk JE, et al. Secretion of angiogenic and antiapoptotic factors by human adipose stromal cells. Circulation. 2004;**109**(10):1292-1298

[243] Shi Y, Zhou L, Tian J, Wang Y. Transplantation of neural stem cells overexpressing glia-derived neurotrophic factor promotes facial nerve regeneration. Acta Oto-Laryngologica. 2009;**129**(8):906-914

[244] Allen C, Eisenberg A, Mrsic J, Maysinger D. PCL-b-PEO micelles as a delivery vehicle for FK506: Assessment of a functional recovery of crushed peripheral nerve. Drug Delivery. 2000;**7**(3):139-145

[245] Fine EG, Decosterd I, Papaloïzos M, Zurn AD, Aebischer P. GDNF and NGF released by synthetic guidance channels support sciatic nerve regeneration across a long gap. European Journal of Neuroscience. 2002;**15**(4):589-601

[246] Kanje M, Lundborg G, Edström A. A new method for studies of the effects of locally applied drugs on peripheral nerve regeneration in vivo. Brain Research. 1988;**439**(1):116-121

[247] Laird J, Mason G, Thomas K, Hargreaves R, Hill R. Acidic fibroblast growth factor stimulates motor and sensory axon regeneration after sciatic nerve crush in the rat. Neuroscience. 1995;**65**(1):209-216

[248] Lee AC, Vivian MY, Lowe JB, Brenner MJ, Hunter DA, Mackinnon SE, et al. Controlled release of nerve growth factor enhances sciatic nerve regeneration. Experimental Neurology. 2003;**184**(1):295-303

[249] Marquardt LM, Sakiyama-Elbert SE. Engineering peripheral nerve repair. Current Opinion in Biotechnology. 2013;**24**(5):887-892

[250] Newman JP, Verity AN, Hawatmeh S, Fee WE, Terris DJ. Ciliary neurotrophic factor enhances peripheral nerve regeneration. Archives of Otolaryngology–Head & Neck Surgery. 1996;**122**(4):399-403

[251] Santos X, Rodrigo J, Hontanilla B, Bilbao G. Evaluation of peripheral nerve regeneration by nerve growth factor locally administered with a novel system. Journal of Neuroscience Methods. 1998;**85**(1):119-127

[252] Cui Q. Actions of neurotrophic factors and their signaling pathways in neuronal survival and axonal regeneration. Molecular Neurobiology. 2006;**33**(2):155-179

[253] Sendtner M, Holtmann B, Hughes RA. The response of motoneurons to neurotrophins. Neurochemical Research. 1996;**21**(7):831-841

[254] Jones MG, Munson JB, Thompson SW. A role for nerve growth factor in sympathetic sprouting in rat dorsal root ganglia. Pain. 1999;**79**(1):21-29

[255] Koliatsos VE, Clatterbuck RE, Winslow JW, Cayouette MH. Evidence that brain-derived neurotrophic factor is a trophic factor for motor neurons in vivo. Neuron. 1993;**10**(3):359-367

[256] Novikov L, Novikova L, Kellerth J-O. Brain-derived neurotrophic factor promotes axonal regeneration and long-term survival of adult rat spinal motoneurons in vivo. Neuroscience. 1997;**79**(3):765-774

[257] Boyd J, Gordon T. A dose-dependent facilitation and inhibition of peripheral nerve regeneration by brain-derived neurotrophic factor. European Journal of Neuroscience. 2002;**15**(4):613-626

[258] Chai H, Wu W, So K-F, Prevette D, Oppenheim R. Long-term effects of a single dose of brain-derived neurotrophic factor on motoneuron survival following spinal root avulsion in the adult rat. Neuroscience Letters. 1999;**274**(3):147-150

[259] Bozkurt A, Lassner F, O'Dey D, Deumens R, Bocker A, Schwendt T, et al. The role of microstructured and interconnected pore channels in a collagen-based nerve guide on axonal regeneration in peripheral nerves. Biomaterials. 2012;**33**(5):1363-1375

[260] Cobianchi S, Casals-Diaz L, Jaramillo J, Navarro X. Differential effects of activity dependent treatments on axonal regeneration and neuropathic pain after peripheral nerve injury. Experimental Neurology. 2013;**240**:157-167

[261] Cunha C, Panseri S, Antonini S. Emerging nanotechnology approaches in tissue engineering for peripheral nerve regeneration. Nanomedicine: Nanotechnology, Biology and Medicine. 2011;**7**(1):50-59

[262] Dubový P. Wallerian degeneration and peripheral nerve conditions for both axonal regeneration and neuropathic pain induction. Annals of Anatomy-Anatomischer Anzeiger. 2011;**193**(4):267-275

[263] Ramburrun P, Kumar P, Choonara YE, Bijukumar D, du Toit LC, Pillay V. A review of bioactive release from nerve conduits as a neurotherapeutic strategy for neuronal growth in peripheral nerve injury. BioMed Research International. 2014;**2014**:132350. DOI:10.1155/2014/132350

[264] Strauch RJ, Strauch B. Nerve conduits: An update on tubular nerve repair and reconstruction. Journal of Hand Surgery. 2013;**38**(6):1252-1255

[265] Wu P, Spinner RJ, Gu Y, Yaszemski MJ, Windebank AJ, Wang H. Delayed repair of the peripheral nerve: A novel model in the rat sciatic nerve. Journal of Neuroscience Methods. 2013;**214**(1):37-44

[266] Cho MS, Rinker BD, Weber RV, Chao JD, Ingari JV, Brooks D, et al. Functional outcome following nerve repair in the upper extremity using processed nerve allograft. Journal of Hand Surgery America. 2012;**37**(11):2340-2349

[267] Henderson CE, Phillips HS, Pollock RA, Davies AM, Lemeulle C, Armanini M, et al. GDNF: A potent survival factor for motoneurons present in peripheral nerve and muscle. Science. 1994;**266**(5187):1062-1064

[268] Johnson EO, Charchanti A, Soucacos PN. Nerve repair: Experimental and clinical evaluation of neurotrophic factors in peripheral nerve regeneration. Injury. 2008;**39**(Suppl 3):S37–S42

[269] Oppenheim RW, Houenou LJ, Johnson JE, Lin LF, Li L, Lo AC, et al. Developing motor neurons rescued from programmed and axotomy-induced cell death by GDNF. Nature. 1995;**373**(6512):344-346

[270] Wu W, Li L, Yick LW, Chai H, Xie Y, Yang Y, et al. GDNF and BDNF alter the expression of neuronal NOS, c-Jun, and p75 and prevent motoneuron death following spinal root avulsion in adult rats. Journal of Neurotrauma. 2003;**20**(6):603-612

[271] Sendtner M, Dittrich F, Hughes RA, Thoenen H. Actions of CNTF and neurotrophins on degenerating motoneurons: Preclinical studies and clinical implications. Journal of Neurology Science. 1994;**124**(Suppl):77-83

[272] Helgren ME, Squinto SP, Davis HL, Parry DJ, Boulton TG, Heck CS, et al. Trophic effect of ciliary neurotrophic factor on denervated skeletal muscle. Cell. 1994;**76**(3):493-504

[273] Siegel SG, Patton B, English AW. Ciliary neurotrophic factor is required for motoneuron sprouting. Experimental Neurology. 2000;**166**(2):205-212

[274] Fricker FR, Lago N, Balarajah S, Tsantoulas C, Tanna S, Zhu N, et al. Axonally derived neuregulin-1 is required for remyelination and regeneration after nerve injury in adulthood. Journal of Neuroscience. 2011;**31**(9):3225-3233

[275] Syed N, Reddy K, Yang DP, Taveggia C, Salzer JL, Maurel P, et al. Soluble neuregulin-1 has bifunctional, concentration-dependent effects on Schwann cell myelination. Journal of Neuroscience. 2010;**30**(17):6122-6131

[276] Stassart RM, Fledrich R, Velanac V, Brinkmann BG, Schwab MH, Meijer D, et al. A role for Schwann cell-derived neuregulin-1 in remyelination. Nature Neuroscience. 2013;**16**(1):48-54

[277] Michailov GV, Sereda MW, Brinkmann BG, Fischer TM, Haug B, Birchmeier C, et al. Axonal neuregulin-1 regulates myelin sheath thickness. Science. 2004;**304**(5671):700-703

[278] Taveggia C, Zanazzi G, Petrylak A, Yano H, Rosenbluth J, Einheber S, et al. Neuregulin-1 type III determines the ensheathment fate of axons. Neuron. 2005;**47**(5):681-694

[279] Gambarotta G, Fregnan F, Gnavi S, Perroteau I. Neuregulin 1 role in Schwann cell regulation and potential applications to promote peripheral nerve regeneration. International Review of Neurobiology. 2013;**108**:223-256

[280] Chan KM, Gordon T, Zochodne DW, Power HA. Improving peripheral nerve regeneration: From molecular mechanisms to potential therapeutic targets. Experimental Neurology. 2014;**261**:826-835

[281] Liu L, Sun T, Xin F, Cui W, Guo J, Hu J. Nerve growth factor protects against alcohol-induced neurotoxicity in PC12 cells via PI3K/Akt/mTOR pathway. Alcohol and Alcoholism. 2017;**52**(1):12-18

[282] Wu W, Liu Q, Liu Y, Yu Z, Wang Y. Dixdc1 targets CyclinD1 and p21 via PI3K pathway activation to promote Schwann cell proliferation after sciatic nerve crush. Biochemical and Biophysical Research Community. 2016;**478**(2):956-963

[283] Wu W, Liu Y, Wang Y. Sam68 promotes Schwann cell proliferation by enhancing the PI3K/Akt pathway and acts on regeneration after sciatic nerve crush. Biochemical and Biophysical Research Community. 2016;**473**(4):1045-1051

[284] Liu A, Prenger MS, Norton DD, Mei L, Kusiak JW, Bai G. Nerve growth factor uses Ras/ERK and phosphatidylinositol 3-kinase cascades to up-regulate the N-methyl-D-aspartate receptor 1 promoter. Journal of Biological Chemistry. 2001;**276**(48):45372-45379

[285] Li J, O'Connor KL, Greeley GH, Jr., Blackshear PJ, Townsend CM, Jr., Evers BM. Myristoylated alanine-rich C kinase substrate-mediated neurotensin release via protein kinase C-delta downstream of the Rho/ROK pathway. Journal of Biological Chemistry. 2005;**280**(9):8351-8357

[286] Ogita H, Takai Y. Nectins and nectin-like molecules: Roles in cell adhesion, polarization, movement, and proliferation. IUBMB Life. 2006;**58**(5-6):334-343

[287] Takai Y, Irie K, Shimizu K, Sakisaka T, Ikeda W. Nectins and nectin-like molecules: Roles in cell adhesion, migration, and polarization. Cancer Science. 2003;**94**(8):655-667

[288] Xu F, Zhang K, Lv P, Lu R, Zheng L, Zhao J. NECL1 coated PLGA as favorable conduits for repair of injured peripheral nerve. Materials Science and Engineering: C. 2017;**70**, Part 2:1132-1140

[289] Elfar JC, Jacobson JA, Puzas JE, Rosier RN, Zuscik MJ. Erythropoietin accelerates functional recovery after peripheral nerve injury. The Journal of Bone and Joint Surgery. 2008;**90**(8):1644-1653

[290] Yan Y, Sun HH, Hunter DA, Mackinnon SE, Johnson PJ. Efficacy of short-term FK506 administration on accelerating nerve regeneration. Neurorehabilitation and Neural Repair. 2012;**26**(6):570-580

[291] Kostopoulos VK, Davis CL, Terzis JK. Effects of acetylo-L-carnitine in end-to-side neurorrhaphy: A pilot study. Microsurgery. 2009;**29**(6):456-463

[292] Reid AJ, Shawcross SG, Hamilton AE, Wiberg M, Terenghi G. N-Acetylcysteine alters apoptotic gene expression in axotomised primary sensory afferent subpopulations. Neuroscience Research. 2009;**65**(2):148-155

[293] Mohammadi R, Hirsaee M-A, Amini K. Improvement of functional recovery of transected peripheral nerve by means of artery grafts filled with diclofenac. International Journal of Surgery. 2013;**11**(3):259-264

[294] Odaci E, Kaplan S. Chapter 16 Melatonin and Nerve Regeneration. International Review of Neurobiology. Vol. 87. Academic Press; 2009. pp. 317-335, ISBN: 978-0-12-375084-6

[295] Fleming CE, Saraiva MJ, Sousa MM. Transthyretin enhances nerve regeneration. Journal of Neurochemistry. 2007;**103**(2):831-839

[296] Sabri F, Gerth D, Tamula GR, Phung TC, Lynch KJ, Boughter JD, Jr. Novel technique for repair of severed peripheral nerves in rats using polyurea crosslinked silica aerogel scaffold. Journal of Investigative Surgery. 2014;**27**(5):294-303

[297] Farjah GH, Heshmatian B, Karimipour M, Saberi A. Using eggshell membrane as nerve guide channels in peripheral nerve regeneration. Iranian Journal of Basic Medical Sciences. 2013;**16**(8):901-905

[298] Bellamkonda RV. Peripheral nerve regeneration: An opinion on channels, scaffolds and anisotropy. Biomaterials. 2006;**27**(19):3515-3518

[299] Madison RD, Da Silva CF, Dikkes P. Entubulation repair with protein additives increases the maximum nerve gap distance successfully bridged with tubular prostheses. Brain Research. 1988;**447**(2):325-334

[300] Madison RD, da Silva C, Dikkes P, Sidman RL, Chiu TH. Peripheral nerve regeneration with entubulation repair: Comparison of biodegradable nerve guides versus polyethylene tubes and the effects of a laminin-containing gel. Experimental Neurology. 1987;**95**(2):378-390

[301] Verdu E, Labrador RO, Rodriguez FJ, Ceballos D, Fores J, Navarro X. Alignment of collagen and laminin-containing gels improve nerve regeneration within silicone tubes. Restorative Neurology and Neuroscience. 2002;**20**(5):169-179

[302] Rodriguez FJ, Verdu E, Ceballos D, Navarro X. Nerve guides seeded with autologous Schwann cells improve nerve regeneration. Experimental Neurology. 2000;**161**(2):571-584

[303] Sinis N, Schaller HE, Becker ST, Schlosshauer B, Doser M, Roesner H, et al. Long nerve gaps limit the regenerative potential of bioartificial nerve conduits filled with Schwann cells. Restorative Neurology and Neuroscience. 2007;**25**(2):131-141

[304] Sinis N, Schaller HE, Schulte-Eversum C, Schlosshauer B, Doser M, Dietz K, et al. Nerve regeneration across a 2-cm gap in the rat median nerve using a resorbable nerve conduit filled with Schwann cells. Journal of Neurosurgery. 2005;**103**(6):1067-1076

[305] Al-Majed AA, Neumann CM, Brushart TM, Gordon T. Brief electrical stimulation promotes the speed and accuracy of motor axonal regeneration. Journal of Neuroscience. 2000;**20**(7):2602-2608

[306] Brushart TM, Hoffman PN, Royall RM, Murinson BB, Witzel C, Gordon T. Electrical stimulation promotes motoneuron regeneration without increasing its speed or conditioning the neuron. Journal of Neuroscience. 2002;**22**(15):6631-6638

[307] Gordon T, Brushart TM, Chan KM. Augmenting nerve regeneration with electrical stimulation. Neurology Research. 2008;**30**(10):1012-1022

[308] Huang J, Hu X, Lu L, Ye Z, Wang Y, Luo Z. Electrical stimulation accelerates motor functional recovery in autograft-repaired 10 mm femoral nerve gap in rats. Journal of Neurotrauma. 2009;**26**(10):1805-1813

[309] Lal D, Hetzler LT, Sharma N, Wurster RD, Marzo SJ, Jones KJ, et al. Electrical stimulation facilitates rat facial nerve recovery from a crush injury. Otolaryngology—Head and Neck Surgery. 2008;**139**(1):68-73

[310] Sharma N, Coughlin L, Porter RG, Tanzer L, Wurster RD, Marzo SJ, et al. Effects of electrical stimulation and gonadal steroids on rat facial nerve regenerative properties. Restorative Neurology and Neuroscience. 2009;**27**(6):633-644

[311] Sinis N, Horn F, Genchev B, Skouras E, Merkel D, Angelova SK, et al. Electrical stimulation of paralyzed vibrissal muscles reduces endplate reinnervation and does not promote motor recovery after facial nerve repair in rats. Annals of Anatomy. 2009;**191**(4):356-370

[312] Sisken BF, Walker J, Orgel M. Prospects on clinical applications of electrical stimulation for nerve regeneration. Journal of Cellular Biochemistry. 1993;**51**(4):404-409

[313] Skouras E, Merkel D, Grosheva M, Angelova SK, Schiffer G, Thelen U, et al. Manual stimulation, but not acute electrical stimulation prior to reconstructive surgery, improves functional recovery after facial nerve injury in rats. Restorative Neurology and Neuroscience. 2009;**27**(3):237-251

[314] Hardy JG, Lee JY, Schmidt CE. Biomimetic conducting polymer-based tissue scaffolds. Current Opinion in Biotechnology. 2013;**24**(5):847-854

[315] Lee JY, Bashur CA, Milroy CA, Forciniti L, Goldstein AS, Schmidt CE. Nerve growth factor-immobilized electrically conducting fibrous scaffolds for potential use in neural engineering applications. IEEE Transactions on Nanobioscience. 2012;**11**(1):15-21

[316] Wang H, Yi SI, Pu X, Yu C. Simultaneously improving electrical conductivity and thermopower of polyaniline composites by utilizing carbon nanotubes as high mobility conduits. ACS Applied Materials & Interfaces. 2015;**7**(18):9589-9597

[317] de Winter F, Hoyng S, Tannemaat M, Eggers R, Mason M, Malessy M, et al. Gene therapy approaches to enhance regeneration of the injured peripheral nerve. European Journal of Pharmacology. 2013;**719**(1-3):145-152

[318] Mason MR, Tannemaat MR, Malessy MJ, Verhaagen J. Gene therapy for the peripheral nervous system: A strategy to repair the injured nerve? Current Gene Therapy. 2011;**11(**2):75-89

[319] Hastie E, Samulski RJ. Adeno-associated virus at 50: A golden anniversary of discovery, research, and gene therapy success—A personal perspective. Human Gene Therapy. 2015;**26**(5):257-265

[320] Boccaccini AR, Brauer DS, Hupa L. Bioactive Glasses: Fundamentals, Technology and Applications. Royal Society of Chemistry. 2016, ISVN: 978-1-78262-976-4

[321] Baino F, Novajra G, Miguez-Pacheco V, Boccaccini AR, Vitale-Brovarone C. Bioactive glasses: Special applications outside the skeletal system. Journal of Non-Crystalline Solids. 2016;**432**, Part A:15-30

Chapter 10

Three-Dimensional and Biomimetic Technology in Cardiac Injury After Myocardial Infarction: Effect of Acellular Devices on Ventricular Function and Cardiac Remodelling

Marco V. Chaud, Thais F. R. Alves, Márcia A. Rebelo, Juliana F. de Souza, Venâncio A. Amaral, Cecilia T. Barros, Katiusca S. Pontes, Carolina Santos, Patricia Severino and Lindemberg M. Silveira Filho

Additional information is available at the end of the chapter

http://dx.doi.org/10.5772/intechopen.69952

Abstract

Dilated cardiomyopathy (DMC) of ischemic or non-ischemic aetiology remains a lethal condition nowadays. Despite early percutaneous or medical revascularization after an acute myocardial infarct (AMI), many patients still develop DMC and severe heart failure due to cardiac remodelling. Possibility of regenerating myocardium already damaged or at least inducing a more positive cardiac remodelling with use of biodegradable scaffolds has been attempted in many experimental studies, which can be cellular or acellular. In the cellular scaffolds, the cells are incorporated in the structure prior to implantation of the same into the injured tissue. Acellular scaffolds, in turn, are composites that use one or more biomaterials present in the extracellular matrix (ECM), such as proteoglycans non-proteoglycan polysaccharide, proteins and glycoproteins to stimulate the chemotaxis of cellular/molecular complexes as growth factors to initiate specific regeneration. For the development of scaffold, the choice of biomaterials to be used must meet specific biological, chemical and architectural requirements like ECM of the tissue of interest. In acute myocardial infarction, treating the root of the problem by repairing injured tissue is more beneficial to the patient. Inducing more constructive forms of endogenous repair. Thus, patches of acellular scaffolds capable of mimicking the epicardium and ECM should be able to attenuate both cardiac remodelling and adverse cardiac dysfunction.

Keywords: tissue regeneration, myocardium regeneration, acellular scaffold, biomaterial

1. Introduction

Cardiomyopathy was originally defined as heart muscle disease of unknown cause, and was distinguished from other disease caused by a specific aetiology, such as ischemic heart disease. It is well known that cardiovascular disease is a main cause of morbidity and mortality worldwide, and many lives are lost due to heart attack. The understanding of cardiomyopathies in both the public and medical communities has been impaired by persistent confusion surrounding definitions and nomenclature. Generally, the cardiomyopathies are associated with mechanical or electrical dysfunction that exhibit inappropriate ventricular hypertrophy or dilatation [1, 2].

Many definitions of heart failure (HF) have been put forward over last 60 years or more. These highlight one or several features of this complex syndrome such as hemodynamic, oxygen consummation and exercise capacity [3]. HF is characterized by breathlessness, ankle swelling and fatigue that may be accompanied by elevated jugular venous pressure, pulmonary crackles and peripheral oedema caused by a structural and/or functional cardiac abnormality, resulting in a reduced cardiac output and/or elevated intra-cardiac pressures at rest or during stress [4]. The cell loss in the myocardium leads to dilation of ventricular wall and remodelling of the heart and, eventually, to congestive HF [5]. New-onset HF may also present acutely, for example, as a consequence of acute myocardial infarction (AMI), or sub-acute in a way, in patients with a dilated cardiomyopathy (DCM). The history is key in making the diagnosis of HF, grading symptom severity and not only establishing the underlying cause but also identifying factors that may have precipitated decompensation [6, 7].

DCM is the most common cardiomyopathy and has many causes. In the absence of abnormal loading condition and severe coronary artery disease, the DCM is characterized by abnormal findings of chamber size and wall thickness, left ventricular dilation and impaired contraction of the left or both ventricles [8]. This reorganization results in abnormal levels of resting tension with activation of the cell death pathway and a further reduction in myocardial performance thus establishing a vicious feedback loop [9, 10]. The DCM ischemic or non-ischemic aetiology remains a lethal condition and the most common indication for heart transplantation. This disorder occurs mostly in adults and for 60% of childhood cardiomyopathies, with infants younger than 12 months having the highest incidence [11]. Although genetic causes are important at all ages of primary cardiomyopathies, the DCM is predominantly of acquired cause. However, familial disease with a genetic origin has been reported in a minority of cases [1]. The DCM phenotype with sporadic occurrence may derive from a broad range of primary and secondary causes, including infectious agents and parasitic. Other causes include toxins, chronic excessive consumption of alcohol, chemotherapeutics agents, autoimmune disorders, collagen vascular disorders, neuromuscular disorders caused by mutations in the structural protein dystrophin [12, 13], muscular dystrophy caused by mutations in the nuclear membrane proteins, lamin A/lamin C rate, metabolic, endocrine and nutritional disorders [14].

DCM is a mixture of primary cardiomyopathies acquired (no genetic) and genetic causes, and it is related to left ventricle (LV) dysfunction. Following myocardial infarction (MI), the LV undergoes a series of cardiac wound healing responses that involve stimulation of robust inflammation to clear necrotic myocytes and tissue debris and induction of extracellular matrix (ECM)

http://dx.doi.org/10.5772/intechopen.69952

protein synthesis to generate a scar. The LV is a complex mixture of cell types, including cardiomyocytes, fibroblasts, immune cells, endothelial and vascular smooth muscle cells, as well as ECM that surrounds these cell types. Cell death was less studied, and whether myocyte cell loss participated in the initiation and evolution of heart failure remains to be established [10].

The cardiac fibrosis is the main event after myocardial infarction, which is characterized by deposition in the infarcted area, with a consequent increases of stiffness, the contraction as relaxation behaviour of the heart are affects resulting in a decrease in cardiac function. The continuous increase of cardiac fibrosis leads to a progressive decrease in heart tissue contractility and heart failure [15–17]. No efficient therapies that can inhibit cardiac fibrosis from progressing in the infarcted hearts to preserve cardiac function and prevent heart failure are available.

Myofibroblasts are widely accepted to be responsible for cardiac fibrosis. Thus, rationally designed anti-fibrotic therapies are likely to be invaluable in curbing this health problem. However, there is no therapy for fibrotic disease in general largely because the underlying basis of fibrosis is unclear, but may result from growth factor-mediated differentiation of resident mesenchymal cells or recruitment of microvascular pericytes-like progenitor cells [18]. The myofibroblasts also express highly contractile protein smooth muscle actin (α-SMA) that remodels the surrounding ECM because they are connected to ECM trough specialized cell surface structures called focal adhesion. In the context of the heart, excessive scarring can cause increases in tissue stiffness, cardiomyocyte atrophy, arrhythmia and hypoxia. Abundant data suggest that a complex interaction involving TGF-β (transforming growth factor β), ET-1 (endothelin-1), Ang-II (angiotensin II) and PDGF (platelet-derived growth factor) causes fibrogenesis [19–21].

2. Cardiac repair

The adult mammalian heart has negligible regenerative capacity, and thus, normal cardiac repair, for example, post infarction, is dependent on the clearance of dead cells and on the formation of a scar tissue to help preserve heart integrity. Myofibroblast is of mesenchymal origin, it has a key role in cardiac fibrogenesis, and after heart attack, it promotes connective tissue remodelling. Myofibroblasts contract fibroblasts that express α-smooth muscle actin (αSMA) to facilitate wound closure [22]. Different approaches have been explored to treat cardiac fibrosis, such as systemic delivery of anti-fibrotic drugs, localized transplantation of biomaterials, localized delivery of antibiotic drugs using biomaterials and localized delivery of cells. Localized transplantation of biomaterials controls cardiac fibrosis by decreasing left ventricular wall stress to decrease the elevated wall stress-induced inflammation. Selection of biomaterials with suitable mechanical properties is critical to decrease wall stress. The ideal biomaterials with suitable physical, chemistry and physio-mechanical properties are critical to decrease wall stress. It should have elasticity and stiffness matching those of the heart tissue. The biodegradation of this material should occur slowly from 6 to 8 weeks and simultaneously the cells from surrounding tissue may penetrate into the infarcted area for attenuation of cardiac fibrosis.

Many people survive myocardial infarction. If help happens quickly, treatment can limit damage to your heart muscle, less heart damage improves your chances for a better quality of life after a myocardial infarction, but a delay of 2 hours or more after symptoms start can result in lasting heart damage or death. However, medical follow up as lifestyle changes, medicines to control chest pain or discomfort and anticlotting medicines will help prevent another heart attack, but this will not rehabilitate cardiac function. Additionally, the biomaterial may provide adequate mechanical support to prevent tissue rupture, and high efficacy can be achieved to prevent new myofibroblast formation and ECM synthesis especially collagen.

3. Tissue regeneration

Therapeutic potential of heart transplant is limited by very small numbers of donor hearts available relative to the need and is complicated by long-term allograft vasculopathy. Interventional cardiology for acute MI has yielded significant advances over the past two decades, while there is still a considerable number of patients who either arrive too late to the clinics or are resistant to angioplasty. Localized drug delivery, yet the widespread clinical application of current approaches, is obstructed by the low therapeutic efficacy. Development of new and translational delivery approaches to improve therapeutic efficacy is essential to push the anti-cardiac fibrosis therapy towards clinical applications.

The human body is a complex system and has very large surface area in which multitude of cell-based interactions contribute to the viability and function of its parts. Complexity is obviously creating an environment that favours the interactions between a cell for sharing and propagation of important information exchange, often leading to complex and every detail very carefully planned from a target issue or cell [23]. Classic examples of such cell-based information exchange in the body include as just a few examples [24] as satellite cells (myoblasts), cardiomyocytes and cardiac pacemaker cells.

The heart possesses regenerative capacity attributed to endogenous and exogenous progenitor cell populations [25]. Thus, there is a growing interest in developing new approaches to treat MI, cardiovascular tissue engineering is currently considered as a promising alternative therapy to restore the structure and function of infarcted adult myocardium via application of a biological device, onto the ischemic tissue [26].

Cellular cardiomyoplasty is considered a novel therapy, in which stem cells are used for cardiac repair. Stem cells are potential therapeutic approach that could be the ultimate solution for salvaging damaged cardiomyocyte. However, more evidence is needed to widely advance the use of this modality. The concentration of stem cells, dose-effect relationship and safety of therapy need to be further investigated. One particular topic in regard to stem cell safety is the tumorigenicities of embryonic stem cells [27].

Unlike heart valves or blood vessels, heart muscle has no replacement alternatives. Evidence suggest that stem-cell-based cardiac therapy has centred on the premise that functional myocardium may be restored by transplanting cardiovascular cells derived from exogenous stem cells into injured hearts [28, 29]. Recent advances in methods of stem cell isolation and culture in bioreactors, the synthesis of bioactive materials and the use of proteomics to better understand matrix metalloproteinase role in post myocardium infarction LV remodelling show promise to contribute to creation of engineered cardiac tissue *in vitro* [30–32].

New discoveries in stem cell biology suggest that stem cells are a potential source of heart muscle cells and blood vessels and can be used by clinicians to rebuild or replace damaged heart tissue [18]. The ideal clinical intervention would either avoid such scar formation, or simply replace formed scar tissue with functioning cardiac muscle tissue [33]. In a first approach to such therapy, investigators have used injections of new cells into damaged areas of cardiac tissue [34]. These studies have met with limited success due to cell death, exit of cells from the heart, and poor cellular integration with the receiving heart tissue [29, 35]. However, multiple attempts have been made at injecting stem cells into myocardial tissue and injection of bone marrow cells after acute MI has even been tested in various clinical studies [27].

The possibility that adult stem cells can repair infarcted myocardium was indicated in experiments using animal models [36, 37]. Initial results from trials in human patients using autologous bone marrow cells were encouraging improvements in cardiac function that were only temporary [38–40]. In light of this enormous clinical burden, strategies for prevention, molecular genetics and cell therapy for cardiac repair and regeneration have attracted the attention. Regenerative strategies have moved rapidly for clinical application to patients. With aims repair or replacement of dysfunctional substrate, results from various animal models of MI and cardiomyopathy suggest that therapy with adult bone marrow cells (BMCs) improves LV function and attenuate LV remodelling [41]. Moreover, because the basis for improved recovery is unlikely mediated by (re)muscularization of damaged myocardium, the need to evaluate cells capable of differentiating into contractile tissue has been emphasized.

Contractile restoration of myocardial scars remains a challenge with important clinical implications. A unique feature of the cardiac muscle is the presence of transverse lines responsible for the contraction force and velocity of propagation of the cardiac impulse. Soler-Botija et al. developed a bioactive implant of biodegradable elastomeric membranes that act as a scaffold. These membranes filled with a self-assembling peptide hydrogel and cell with cardiac potential were fully vascularized with functional vessel and observed via echocardiography positive effects on global cardiac motility [42]. The results suggested, in this case, that the benefits of biodegradable scaffolds were not only due to local reduction in scar size, but to a more general event [43, 44]. Recent experimental studies have failed to answer important aspects of cell therapy. Cell therapy has more safety, improved cardiac function, increased healing, vascular density and increased regional circulation [45]. However, the efficiency of delivery and retention is lower than expected, and the retention and survival of cells at sites of delivery has been limited [5, 46, 47].

4. Tissue regeneration: a biomimetic design

In native tissue, cell growth and structural development are supported by an ECM that consistently assists in coordinating the contractility and maintenance of cardiac shape and size, as well as the function of cardiomyocytes. The common interactions and orchestrated information exchange between cells associated into precise temporal and spatial context are the systematic presentation of biomolecules from biomaterials able to send biomolecular signals. Further, even the size, shape and mechanical properties of a cell may be essential to the proper presentation of these signals to elicit the appropriate response and build the devices to cells orchestral arrangement [24]. Tissues organics repair is an exciting therapeutic conceit in the field of tissue engineering and biomaterial engineering. In the cardiac repair area, the most challenging aim is the creation of an engineered heart muscle. Tissue engineering is a hybrid technological science, in which overall approach combines biology with materials science and has concentrated its efforts on the development of biomedical devices compatible with the ECM of target tissue. Similar to the matrix, the scaffolds must be able to anchor the cells of the native tissue.

Biomimetic design has started innovation in design as well as pointed to ways of improving existing biological devices. A method for finding and using these ideas would make biomimetic innovation more accessible by use of biomaterials that mimics the cardiac ECM, or other tissues. Biomimetic design is that, fully or partially, imitates or to make remember any biological natural phenomenon include all levels of organization belonging to the biological phenomenon that wishes to mimic the health tissue [48].

The major challenge of biomimetic design and develop of 3D devices biologically compatible, highly porous with interconnected porous that favour the transport of both nutrients and metabolic products while providing the analogous function, and mechanical stability. These devices are so-called scaffolds [49, 50]. The scaffold manufacturing design should take into account its purpose and represent important components for tissue engineering; in any case, it must achieve structures exhibiting the aforementioned characteristics. In the architecture of scaffold used for tissue regeneration, the choice of biomaterial is of critical importance. The variety of processes used in the manufacture of scaffolds modifying the surface and bulk properties influences both the architectural and the similarity of the scaffold with the native organic tissues, and may actively provide bioactive cues to the residing cells for regulations of the activities [50, 51].

The ECM is made up collagen type I (80%) and collagen type III (10%). The collagen serves to maintain normal cardiac architecture by surrounding and bridging myocytes, which consistently assist in coordinating the contractility and maintenance of cardiac shape and size as well as the function of the cardiomyocytes [52]. In order of importance, the aim of regenerative medicine is the use of synthetic, natural or composites biomaterial scaffolds to replace or repair damaged tissues, which have been investigated as candidates for cardiac regeneration. These natural and composite scaffolds are modelled on the natural extracellular matrix, which is a porous hydrogel consisting of collagen, fibroin, chitosan, gelatine, fibrin glue, alginate, glycosaminoglycans associated with biocompatible synthetic co-polymers as poly(lactic-co-glycolic acid), polycaprolactone, ePTFE, PET, PUs, titanium, stainless steel and gold silver. These scaffolds provide both biomechanical support and biochemical signals to cells, and provide a biodegradable physical environment so as to allow neo-vascularization and remodelling in response to developmental, physiological and pathological challenges during tissue dynamic process morphogenesis, homeostasis and the built of a new tissue, inhibit apoptosis and attenuate LV dilatation and disease progression. Implantation of a novel biodegradable polyester urethane urea (PEUU) patch onto a sub-acute myocardial infarction promoted contractile phenotype smooth muscle tissue formation and improved cardiac remodelling and contractile function at the chronic stage [53].

Synthetic or natural biomaterials application in cardiovascular tissue repair and regeneration has shown increasing potential as a tool for such procedures, due its properties favourable for implantation while eliciting minimal side effect. Somewhat ambiguous, the biomaterial needs biocompatible and must interact organically and synergistically with the healthy part of the organ in order to reconstruct a new and neo-vascularized tissue over the area damaged by ischemia [54–57].

Separating the synthetic biomaterials is a shadow in comparison with the functional capabilities of natural biomaterials, especially the biopolymers. Synthetic biomaterials are constructed of polycaprolactone, polylactic acid, polyglicolid, metals (titanium, stainless steel, gold silver) or a composite, for example poly(L-lactic acid)/poly(ε-caprolactone)/collagen to mimic the native microenvironment of the myocardium [5]. A nanoscale PLA-co-poly(ε-caprolactone)/collagen biocomposite scaffold was used to culture and support isolated rabbit cardiomyocytes. The results showed that adult rabbit cardiomyocytes attached to the scaffold exhibited growth and cell organization comparable to that found in native myocardium [5]. The main benefits of synthetic materials are their strength and durability, although their biocompatibility tissues and surface properties, which are generally poor to achieve a favourable environment for cell attachment [58]. Toxicity is the outmost concern with synthetic materials, especially in the case of biodegradable materials, which can release potentially harmful by-products of degradation into the body. The linear aliphatic polyesters as PGA, PLA, poly(lactide-coglycolide) (PLGA) degrade through hydrolysis and are the most used for tissue engineering [59].

Each of tissue in the body is uniquely optimized to its specific organ system and offers an innate biocompatibility. Autologous tissue, or tissue harvested from and used for the same patient, is the current best solution for its superior functionality and no immunogenicity, these tissue are called homograft's a utopia in the cardiac regeneration. Main ECM materials such as collagens, hyaluronan, fibronectin, elastin, fibrillin and proteoglycans among others are natural biomaterials for damaged tissue to repair in and around the area. In cardiovascular applications, bovine porcine and equine tissue sources are playing a prominent role in establishing and maintaining an ideal microenvironment for tissue regeneration [54, 60].

Although the self-regeneration capacity of adult myocardium is insufficient to prevent the progression towards heart failure after various insults. To deal with limitation, the strategy can transfer progenitor cells from healthy area to disease heart area through scaffold implantation. The utility of this approach is called into question by human and murine studies showing that progenitor cell quantities are in fact normal or increased in diseased versus healthy hearts. Evidence supports the existence of endogenous cardiac renewal and repair mechanisms in adult mammalian hearts that could contribute to normal homeostasis and the responses to pathological insults [60–62]. Studies demonstrate there is some amount of cardiac turnover, both in healthy and injured tissues, and this subject is important on cardiac regeneration after IM [63], as it also occurs cardiac pressure overload and idiopathic dilated cardiomyopathy [49, 64]. Naqvi et al. reported that cardiac cell proliferation is highest in youth, but in an average healthy adult, the cardiomyocyte turnover rate is controversial [65]. In an average healthy adult, the cardiomyocyte turnover rate is controversial with reports predicting annual turnover rates to more than 50% [66, 67]. Even although, it has been proposed that limited cardiac regeneration mechanism is pathways of to protect the heart from developing cancer, in fact, primary cardiac tumours are extremely rare [68]. Methods have been questioned in studies where turnover is on the higher end, in the meantime, regenerative biology will bring together basic scientists and clinicians, developmental biologists and engineers, compelling us to expand our understanding of cell biology in order to grow new tissues [69, 70].

Biomaterials that mimic the ECM are used as a feasible alternative to cellular and molecular therapy in the field of tissue engineering. Biomaterials can be delivered alone, or serve as a scaffold or carrier for cells or growth factors. Scaffolds carefully design to support cellular growing has been development with finally of make a propitious environment for cardiomyocyte renewal. The biomaterials having individual, purpose-specific cues that can stimulate cells to behave in a predictable manner and in a pre-determined time course would have tremendous benefit to tissue engineering. When cells adhere and grow on the polymers substrates, cells sense, interpret and integrate extracellular signals through electrical connections between cardiomyocyte and surface of protein biomaterial and respond to them [27].

Cardiac scaffold is a therapeutic intervention with low cost and efficacy and has significantly improved patients' quality of life and prolonged their longevity [71–74]. The scaffold must reduce local micro-environment hostility persist for a sufficient time period over injured area to facilitate native cell migration and integration with native tissue to be feasible for cardiac repair/remodelling.

Delivery of cells using biomaterials has been shown to be an effective approach to control cardiac fibrosis [33]. Fibrosis is a dynamic process, at the molecular and cellular levels, changes are characterized by generation of ROS, and data suggest fibrogenesis is a result of the complex interaction involving TGF-β, ET-1, Ang-II and PDGF. Notably, TGF-β plays important roles in other biological processes, including homeostasis and normal repair. Pirfenidone (multiple target) and Nintedanib (BIBF-1120) inhibit multiple tyrosine kinases, a broad anti-inflammatory and anti-fibrotic effect and have an effect blocking TGF-β and Ang-II–induced fibrosis [21, 75, 76]. While a new tissue is growing, it may release anti-inflammatory factors to control inflammation thus indirectly controlling cardiac fibrosis, released angiogenic factors promote tissue vascularization and regeneration [16]. To control of cardiac fibrosis for long-term, high rate of cell survival in the infarcted hearts, high nutrition, high oxygen environment and catabolites' elimination it is a necessary condition.

4.1. Effect of acellular scaffold on ventricular function and cardiac remodelling

A human cardiac organoid injury model reveals innate regenerative potential [77]. Under normal conditions, the ECM provides structural support for the heart, acts as a reservoir for cytokines and growth factors and provides a connection with surrounding cells that is important for transmission of extracellular cues [27, 78]. Following pathologic stimulation, the ECM undergoes remodelling of its structural components and matricellular protein levels, for example; fibroblasts are influenced by autocrine and paracrine signal, and they are responsible for secretion and regulation of the ECM.

Myofibroblasts secret important ECM proteins including collagens, fibronectin, periostin, metalloproteinases and tissue inhibitors of metalloproteinases, which collectively regulate ECM components in the process of cardiac remodelling. The ECM plays a critical role in the maintenance of the functional myocardium as well as the regulation of the heart's response to stress or injury [27, 78]. Intelligent scaffolds that mimic ECM have recently emerged as a way to elucidate the interaction of native ECM molecules with living cells, to further understand how the ECM regulates their environment. Tissue engineering will open new avenues to support regeneration of diseased or damaged tissue.

Study provides novel insight into the endogenous regenerative capacity of the immature human heart, which cannot be investigated in the *in vivo* setting and is consistent with aspects of neonatal heart regeneration observed *in vivo*: lack of fibrosis, lack of hypertrophy, high baseline proliferation rates and functional recovery after injury. This model of study therefore provides an opportunity to interrogate the molecular and cellular mechanism governing regeneration of immature human heart tissue [73, 77].

Attempts to repair the injured myocardium with endogenous regeneration require an organized participation from various cell types including cardiac myocytes and local and peripheral stem cells. It is naive to think that only one cell type can participate in this regenerative process and may actually limit therapeutic opportunities. In summary, these approaches appear to have a sufficient effect on their target, a few is known on the strategies and the tissue regeneration [63, 79]. To date, a human model of acute cardiac injury has not been achieved. Instead, most research into regeneration following cardiac injury has relied on the use of animal models. However, recent studies suggest that reactivation of neonatal cardiac regenerative pathways to drive cardiomyocyte proliferation in the adult heart may be possible. Porrelo et al. found in neonatal mice suggest that therapeutic strategies aimed at restoring the proliferative potential of adult mammalian cardiomyocyte will be an important component of attempts to reactivate the dormant regenerative capacity of the adult mammalian heart after MI [80]. Voges et al. reported that immature human heart tissue possesses an innate ability to regenerate following injury. In addition, myocytes possess an endogenous ability to recover contractile force following injury, which occurs independently of other infiltrating/resident cell types, and that immune cells are not required for functional recovery of immature heart tissue. Furthermore, the addition of monocytes in our studies did not have any impact on cardiomyocyte proliferation. This study suggests that immature human heart muscle may possess an intrinsic ability to mount a regenerative response, which occurs even in the absence of inflammation and angiogenesis [77]. Other studies have emerged documenting complete functional recovery of newborn human hearts following MI [63, 67, 81–83].

Adult cardiac myocytes represent a highly specialized and structured cell type; therefore, it is not surprising that complex and often overlapping systems have evolved to regulate cardiomyocyte growth. Typically, adult cardiac myocytes do not re-enter the cell cycle when exposed to growth signals, and further increases in cardiac mass are achieved through an increase in cell size or hypertrophy [84]. Due to the absence of understanding and the potential of how the heart cells can be migrated from health tissue to necrosis area, an intervention has not yet been developed that can regenerate the damaged myocardium after an infarct.

However, several groups have described the presence of a progenitor or steam cell that can differentiate into cardiac myocytes [15, 36, 61, 74, 85]. Thus, evidence has accumulated over the past 20 years to demonstrate that there is some amount of cardiac turnover, both in healthy and injured tissues [63, 77], and this has prompted efforts to devise cardiomyocyte replacement therapies by the promotion of endogenous regenerative process [60, 62]. Evidence from rodent [86], axolotls, newts, zebrafish [77] and human studies challenges the view of the heart as a terminally differentiated organ and unable to regenerate [67]. The ability of ECM biomaterials to stimulate robust endogenous cardiac regeneration without adding exogenous cells would avoid these issues, today more and more studies show that this utopia is not more the 'holy grail' of heart regeneration. The direct reprogramming of non-muscle cells into cardiomyocyte-like cells by infecting or treating the heart with defined factors has improved. This approach was reviewed by Sadahiro et al., quoted by Foglia et al. [74, 82, 87, 88] that shows evidences about cardiac regeneration without exogenous cells.

We believe that the structure and function of cardiac tissues are regulated by weak nanoscale signals provided by ECM, which are exerting control over the function of cells and tissues. Thus, in the design of scaffolds, it is important to evaluate the effects of dialysis at the cell-biomaterial interface for the creation of truly biomimetic cardiac constructs that replicate the structural and functional aspects of ventricular organization *in vivo*.

4.2. Scaffolds

ECM was commonly viewed as a rather inert scaffold, merely providing structural support for the cells embedded in its environment. However, ECM plays a critical and crucial role in the maintenance of the myocardium function, it is now recognized that the ECM forms in fact a very dynamic and plastic milieu also to regulation of the heart's response to stress and a wide variety of cellular events and [78, 89].

A principal goal of regenerative medicine is the use of scaffolds to replace or repair damaged tissues. The scaffolds are modelled on the natural extracellular matrix, which is a 3D porous hydrogel that provides both mechanical support and biochemical signals to cells. Microscopic and ultra-structural scaffold topography is the key for cellular homing and migration to the target tissue.

For the increasing interest in tissue engineering to create a device for human tissues and organs, innovative techniques have been developed to generate scaffolds that arrange and form micro-architecture mimicking physiological structures. Tissue engineering requires the ability to promote the production and accumulation of ECM component. Alternative therapies with biomimetic scaffold expand the option of adult patient care; efficient means for repairing, reconstructing or regenerating damaged tissues reduce the need for scarce donor organs.

In the field of myocardial tissue engineering, many efforts by the scientific community are dedicated to identify materials possessing specific mechanical properties that play a pivotal role.

Cells, biomaterials, scaffolds and growth-stimulating signals are referred to as the tissue engineering triad. Scaffolds typically made of polymeric biomaterials represent important components for tissue regeneration. They are biopersistent (6 or 8 weeks), biodegradable, provide the structural support for cell attachment and growth and subsequent for tissue development [50, 90]. The stiffness of native heart tissue ranges from 10 to 20 kPa at early diastole and increases to 50 kPa at the end of diastole, which may shoot up 200 kPa or more in infarcted heart [5]. The bioengineered scaffold matches as closely as possible those of the heart ECM in terms of stiffness, since the scaffold should be flexible enough to promote the contraction of the growing cells.

In the scaffolds, nanoscale features must be included that replicate some of the functions of the ECM. In many cases, control of cell alignment and growth direction is essential to obtain functional tissues. Scaffolds 3D are able to control the growing of the cells. Hydrogel scaffold with oriented channels has been used with some success [91]. An injectable myocardial matrix hydrogel, derived from decellularized porcine ventricular ECM increase endogenous cardiomyocytes, preserves cardiac function and is shown to reduce negative LV remodelling in both rat and pig models when delivered weeks after MI [92, 93].

Biomaterial has shown increasing potential as a tool for such procedures, due to its properties favourable for implantation while eliciting minimal side effect. Somewhat ambiguous, the biomaterial needs biocompatibility to interact organically and synergistically with the healthy part of the organ in order to reconstruct channels by cell migration a new and neovascularized tissue over the area damaged by ischemia. If designed appropriately, biomaterials scaffolds can be delivered through minimally invasive approaches and stimulate cardiac repair, while avoiding many of the complications associated with a living product.

Biomaterial scaffolds are expected to provide a compliant and highly hydrated environment, similar to soft tissues having high water contents, thus facilitating diffusion of nutrients and cellular waste. The diverse nature of the organic tissue architecture requires pores sized in specific ranges compatible to each tissue. Since the myocardial tissue is subjected to cyclical and constant deformation, thus scaffolds are requested to show elastomeric properties and possibly long-term elasticity. The dimensions of the cardiac scaffold pores must be compatible with the size of the heart cell phenotype, and the porosity should be above 85%, with pore interconnectivity favouring cell attachment.

Lim et al. related that their results on poly(L-lactic acid)-polystyrene blends reveal that cell adhesion is affected by surface chemistry, topography, and wettability simultaneously and that nanotextured surfaces may be utilized in regulating cell adhesion [94]. Therefore, chemical and physical signals from biomaterial surfaces (chemistry, topography, charge, energy and wettability) are critical extracellular stimulators that have the potential to regulate cell behaviour. The ability to robustly and reproducibly generate uniformly controlled (both structurally and functionally) and precisely defined engineered cardiac tissue will likely be necessary for eventual therapeutic products. This makes fabrication of biomaterials for myocardial tissue engineering an attractive strategy. An effective scaffold for myocardial repair is a critical unmet need, where combining elasticity and strength without compromising heart cell viability and contractility have proved challenging.

Bioengineered cardiac tissue constructs can be divided into two categories: scaffold-cell-based and scaffold-free. A number of 3D devices have been fabricated from natural biological materials, such as collagen, fibrinogen, elastin, hyaluronic acid, fibrin, alginate fibronectin, laminin, from naturally occurring ECM and also fibroin, chitosan, gelatine, fibrin glue, associated with biocompatible synthetic or decellularized heart matrix.

Collagen is the most prevalent extracellular component of the myocardium and has used as a matrix for studying myocardial electrophysiology and contraction. Collagen is the primary force-conducting component of the ECM coupling cells to their force environment [90]. Cardiac myocytes are attached by integrin at specific sites near the Z band to an interconnected collagen network containing other mechanically and biologically active extracellular matrix (ECM) components, including glycoproteins, proteoglycans, growth factors, cytokines and proteases A key property of collagen hydrogel-based engineered heart muscle (EHM) is its contractile performance, which reproduces many, but not all, aspects of contractile performance in native myocardium [95]. Some of the ECM protein collagens contribute the major tensile strength and viscoelasticity. In tissues, cells are anchored in a stationary, 3D network of collagen in a highly characteristic spatial arrangement. This spatial order arises from vectoral deposition of matrix and its subsequent tractional organization by cells. Bhatnagar et al. have shown that P-15 (mimic the bioactivity of type I collagen) bearing matrices mimic the physiological interactions of collagen with cells. For example, the interstitial collagens are subjected to a myriad of post-translational modifications, including covalent and non-covalent cross-linking, leading to haptotactic cell migration, activation of signalling pathways, induction of growth factors, cell differentiation, tissue remodelling and morphogenesis *in vitro* and *in vivo*. Up-regulation of collagen cross-linking because of excess LOX activities can change various cellular behaviours. The authors describe such scaffolds as collagen substitute matrices [90, 96–98].

The plateau phase of the time trace at later times of the coagulation process, the fibrinogen is cleaved and cross-linked by exposure to thrombin and clotting factor XIII to form an insoluble fibrin mesh that captures blood cells to form a blood clot stiffness via thrombosthenin protein. Fibrin meshes can also be used to capture any target cell for tissue engineering [53]. Fibrinogen can also be conjugated with polyethylene glycol (PEG) to generate scaffolds that covalently bind growth factors and other proteins [99].

Elastin is a major fibrous protein in the extracellular matrix of organs and tissues that exhibit stretch and recoil, such as vessel walls. The elastic property maintains heart's structural and biological functions. After an MI, the loss of elastin is the main factor adversely affecting ECM remodelling of the infarcted heart. The restoration of the elastic properties of the infarct region can prevent ventricular dysfunction [54]. The elastic properties of tissues are essential to maintain their structural and biological functions. The lack of elastic recoil contributes to the thinning and expansion of the infarct region, which frequently progresses after a myocardial infarction and results in cardiac enlargement and cardiac failure with time [100]. When the tissue is stretched, the elastin molecule is elongated, and when the stretching force is released, the molecule returns to its more stable random-coil structure and maintains the organ structure [85]. Most cells, such as fibroblasts, endothelial cells and smooth muscle cells, synthesize and secrete glycoproteins to form a microfibrillar scaffold on top of which tropoelastins, the soluble precursors of elastin. Restoration of the elastic properties of the infarct region can prevent progressive cardiac dilation and deterioration of ventricular function following an MI [101]. Expression of elastin in the myocardial scar may be able to change the composition of the extracellular matrix of the infarct region, preserves the elasticity of the infarcted heart, stabilizes an infarct and preserves ventricular function [100]. Many other molecules, though lower in quantity, function as essential components of the extracellular matrix in soft tissues. Besides their basic structures, biochemistry and physiology, their roles in disorders of soft tissues are discussed only briefly.

Fibrous protein, including as well as collagen and elastin the fibronectin, laminin, fibrillin and integrins get a multi-domain structure plays a role of 'master organizer' in matrix assembly as it forms a bridge between cell surface receptors. Then these proteins contribute to the structure of the ECM, and modulate cellular functions such as adhesion, differentiation, migration, stability of phenotype, and resistance towards apoptosis. Integrin also plays an essential role in the assembly of fibrillin-1 into a structured network [78, 89]. Fibulins are tightly connected with basement membranes, elastic fibres and other components of extracellular matrix and participate in formation of elastic fibres [83, 86].

Tenascins are ECM polymorphic glycoproteins found in many connective tissues in the body. Their expression is regulated by mechanical stress both during development and in adulthood [46]. Tenascins mediate both inflammatory and fibrotic processes to enable effective tissue repair and play roles in pathogenesis of Ehlers-Danlos, heart disease, and regeneration and recovery of muscle-tendinous tissue [46, 89]. Increased expression of thrombospondin and TGF-β activity was observed in fibrotic skin disorders such as keloids and scleroderma. Thrombospondin-5 is primarily present in the cartilage. High levels of Thrombospondin-5 are present in fibrotic scars, it plays a role in vascular wall remodelling and has been found in atherosclerotic plaques as well [30].

On the other hand, the extracellular macromolecules, notably glycosaminoglycan, are important mediators of angiogenesis. Hyaluronic acid (hyaluronan) is widely utilized as a biosynthetic biomaterial because of its biocompatibility and diverse array of physiologic functions. In its natural setting, hyaluronan is the only non-sulphated glycosaminoglycan present in the ECM of vertebrates, it is the most abundant and one of more important glycosaminoglycan in the heart development, and there are indications of playing a role in epicardium development [102]. The highly viscous aqueous solutions (1000× its weight in water) thus formed give hyaluronan unique physicochemical and biologic properties that preserve tissue hydration, regulate tissue permeability through steric. In the ECM of connective tissues, hyaluronan forms a scaffold for binding other large glycosilaminoglicans and proteoglycans. Bernanke and Markwald showed that hyaluronan treatment of avian AV cushion explants resulted in an increase of mesenchymal cells invading a 3D collagen matrix [103]. Large hyaluronan molecules are space-filling polymers presenting regulatory as well as structural functions, while small hyaluronan fragments are involved in immune stimulation, angiogenesis and inflammation [104]. 2-iminothiolane-grafted hyaluronan hydrogel and sodium periodate oxidated hyaluronan cross-linked with adipic acid dihydrazide hydrogel were implanted into rat adductor muscles. The degradation tests demonstrated that the hydrogel could maintain the gel matrix over 35 days, depending on the ADH concentration, while inducing low inflammation and dense blood vessel formation in the areas surrounding the implanted hydrogels [105, 106]. Hyaluronan-mediated angiogenic effect *in vivo* is related to its degradation products, which stimulate endothelial cell proliferation and migration [102, 107]. Consequently, hyaluronan plays important roles in maintaining tissue morphologic organization, preserving extracellular space and transporting ions, solutes and nutrients. Along with ECM proteins, hyaluronan binds to specific cell surface receptors such as CD44 and RHAMM [108]. The resulting activation of intra-cellular signalling events leads to cartilage ECM stabilization. Despite the fact that studies in the avian and murine heart clearly show the importance for hyaluronan in heart development, there are, to the best of our knowledge, no published studies that show that hyaluronan insufficiency is associated with congenital heart disease in the human.

Other biomaterials that have been used to engineer myocardial tissue equivalents include alginates, silk fibroin and gelatine. In addition, it seems essential that the exogenous matrix be replaced to prevent interference with tissue formation. Still, vascularization of the matrix occurs quickly in conjunction with the immune response to the presence of foreign materials.

Alginate is a natural polysaccharide characterized by long chains of α-L-glucoronic acid and β-D-mannuronic acid. Alginate is a versatile biocompound and can be obtained with rang of molecular weight (32–400 kDa) and has been utilized as a hydrogel for tissue engineering. Due to its biologically favourable properties including the ease of gelation and its biocompatibility, alginate-based hydrogels have been considered a particularly attractive material for the application in cardiac regeneration and valve replacement techniques [109]. Alginate scaffolds were demonstrated to be safe, biodegradable and feasible to provide a conductive environment to facilitate the 3D culturing of cardiac cells. Scaffold with Ca-alginate enables incorporation and retention of cells and proteins inside the hydrogel scaffold. After implantation into the infarcted myocardium, the biograft stimulated intense neo-vascularization and attenuated LV dilatation and failure in experimental rats compared with controls [29, 110]. Alginates possess several crucial properties, which make them suitable for use in cardiac regeneration. The first is their ability to be dissolved in water to yield aqueous solutions with moderate viscosity, which is particularly important for formulating injectable mixtures for cardiac therapies. The second is their ability to form hydrogels in mild conditions, for example, by adding cationic polyetrolytes salt to an aqueous solution to form the ionically triggered gelation, which is chemo reversible. Another advantage property of alginates is possible to modulate degradation rates and mechanical stiffness by choice of the appropriate cross-linking agents [29, 55, 109]. Ionically, cross-linked, high molecular weight alginate hydrogels degrade in an uncontrolled fashion due to the slow loss of cross-linking cations.

As mentioned previously, one of the limitations for the use of synthetic polymers such as scaffolds is its low biodegradation in biological medium. Ashton et al. describe an alternative approach to design degradable alginate hydrogels based on the enzymatic degradation of alginate polymers. They incorporate poly(lactide-coglycolide) (PLGA) microspheres loaded with the enzyme alginate-lyase (PLGA-AL) into alginate hydrogels; controlling the amount of incorporated enzyme, and its rate of release from the PLGA microspheres, enables the rate of enzymatic degradation of the alginate hydrogels to be tuned [111].

There are two main solid forms of alginate used in cardiac regeneration, namely, hydrogels and porous 3D scaffolds. The hydrogel can contain over 99% water trapped in the network of water-insoluble polymer chains. Depending on the freezing regime, the hydrogels could become 3D scaffold with interconnected pores, up to 200 μm in diameter and 90% of matrix porosity [112]. The current application involving alginate for cardiac regeneration include ventricle restoration

http://dx.doi.org/10.5772/intechopen.69952

by injecting cell-free alginate, treating myocardial infarction by injecting calcium-cross-linked sodium alginate solution, patches for cell transfer in cardiac regeneration [109]. The potential use of the different alginate hydrogels as pharmaceutical excipients has not yet been fully evaluated but alginate is likely to make an important contribution in the development of polymeric delivery systems. The precise chemical degradation mechanism and approaches in the choice of alginate for drug deliveries was previously broadly reviewed [113]. Leor et al. have suggested that intra-myocardial injection of alginate induces neo-vascularization and improved LV function [28, 110]. Alginate-based biomaterials for the treatment of MI is entering the clinical trials stage, therefore understanding the mechanisms by which these therapies affect LV remodelling, cardiac function and cardiac electrophysiology becomes a pivotal issue.

Biocomposite 3D scaffold consisting of two or more polymeric blends is used in order to obtain scaffolds with desired functional and mechanical properties depending on their applications. It is obtained useful the properties of natural and synthetic biomaterial. One such an attempt of using copolymer as poly(l-lactic acid)-co-poly(ε-caprolactone) (PLACL), polyglycerol sebacate, polyethylene glycol, polyvinyl alcohol, silk fibroin and alginate, collagen, hyaluronic acid, laminin and others biopolymers, that contribute to mimic the ECM, for fabricating biocomposite scaffolds for cardiac tissue engineering have been proven to exhibit suitable biodegradable and mechanical properties. Mukherjee et al. showed that the hydrophilic, biocompatible nanofibrous scaffolds made of PLACL/collagen blend provide superior attachment and growth of adult cardiac cells favouring native myocardium-like alignment of newly seeded cardiac cells compared to purely synthetic PLACL counterparts [5]. Biomaterial and biocomposite's scaffold have featured prominently in cardiac regenerative therapy and have been explored to nanofibres, 3D devices, nanoparticles and patches to enhanced cell delivery and, more recently, has been used as acellular therapy.

5. Cardiac patches: *in situ* tissue regeneration

If on one hand to create engineered muscle construct in bioreactor is fascinating, on the other hand, it faces significant difficulties, in particular constructing significant cardiac muscle from scaffold and cells *in vitro*, and poor graft survival. In this approach, acellular scaffolds are implanted on the damaged myocardium and after their vascularization, they create a friendly environment and space for the implanted cardiomyocytes. Bioactive molecules with collagen, fibrinogen, alginate, hyaluronan, integrin, fibronectin and laminin improve viability and survival and may enhance stem cell homing and self-repair [28]. There is accumulating evidence that the heart contains resident progenitor cells that can be induced to develop into cardiac muscle and vascular tissue. These cardiac progenitors could be recruited to repair the infarcted myocardium. The incompleteness of myocardial repair as executed by all endogenous mechanisms collectively, including whatever progenitors exist intrinsic and extrinsic to the heart. At the same time, the identification of these cells opens the tantalizing possibility that they might be coached *in vivo* to home within the damaged myocardium, subsequently promoting functional cardiac repair without the need of introducing exogenous cells [118]. With this approach, the biomaterial itself or its degradation/dissolution products are used to stimulate local tissue repair.

Cardiac patches thought that the microenvironment and architecture provided by such a scaffold can support cellular differentiation and organization, and prevent a form of programmed cell death that occurs in anchorage-dependent cells when they detach from the surrounding ECM [114, 115]. When cells are detached from the ECM, there is a loss of normal cell-matrix interactions, and they may undergo anoikis. Nonetheless, the major drawback with this method remains the inability to generate patches with sizable thickness due to diffusion limitations. The biomaterial surface plays a crucial role as it forms the interface between the scaffold (or cardiac patch) and the cells [58]. The cells once implanted inside the scaffold will help the body heal itself [28, 116]. In animal models for ischemic cardiomyopathy, a variety of biodegradable materials as interventional therapeutic strategies have been investigated, including epicardial patches with and without cellular constituents. This approach might provide an attractive alternative to cellular cardiomyoplasty or larger ventricular restraint devices for cardiac repair.

'Plastic Compression' technique to create a biomimetic cellular dense collagen is used as a scaffold for bone engineering, demonstrating its potential in tissue engineering. Double multiple compressions of hydrated collagenous scaffolds may initially result in enhanced mechanical properties; nonetheless, the double compression process exerted a negative impact on the seeded cell survival [117]. This compression technique can be also combined with other biomaterial to create a patch of collagen hybrid to cardiac regeneration with acellular scaffold.

Dense collagen scaffolds were characterized in terms of gel contraction ratio, morphology, the viability of seeded cells and mechanical properties of the gel and an engineered cardiac patch out of compressed type I collagen that improves recovery of the heart after acute myocardial infarction through several processes, independently from exogenous cells and other factors. The main mechanisms responsible for this effect include the mechanical support of the scaffold, facilitation of cell migration and angiogenesis, and partial preservation of the heart muscle cells within the lesion and the area beneath the patch. Further investigations on therapeutic factors and/or cells that can be seeded within the engineered patch can be used in a new clinical therapy for cardiac repair following severe heart injuries [26].

Cardiac repair using chitosan-hyaluronan/silk fibroin patches in a rat heart model with myocardial infarctions was examined after 8 weeks of study. The patches significantly improved LV functions in MI hearts with markedly reducing the dilation of LVs, increasing the thickness of their walls and improving their fractional shortening (LVFS) in the CHS group.

6. Scientific overview

As a concept, tissue regeneration is intuitively attractive and challenger. For conditions characterized by MI and HF, the basic principle is that non-viable myocardium may be regenerated or repaired by delivery of stem or progenitor cells, including the heart itself. The cardiomyocyte has been considered terminally differentiated, with the response to injury characterized by hypertrophy. Recent evidence increases the possibility that a natural system of myocyte repair exists. Although the concept of myocyte repair is straightforward in theory, realizing the potential of therapeutic strategies based on this concept is extraordinarily complex, and the magnitude of this task has been highlighted recently. The patches using new biomaterial composite have been improved LV function in MI hearts with markedly reducing the dilation of LVs, increasing the thickness of their walls and improving their fractional shortening. Collagen matrices are some of the oldest and most well understood classes of biomaterials. We agree with Chan et al.'s early efforts to simply harness isolated ECM as a substrate for culturing cells have expanded into a wide range of techniques to capitalize on the protein's unique properties to create biomimetic hydrogels with well-defined properties. These materials and the knowledge they generate are enabling the development of new cell and matrix-based therapies that will overcome the limitations of less biologically relevant biomaterials [50]. Cells internally synthesize, modify and assemble the alpha chains into a procollagen form, which is secreted to the extracellular space and then partially cleaved by specific enzymes to form the tropocollagen molecule. These nanoscale subunits further self-assemble into fibrils and are covalently bound to each other in a staggered manner, giving collagen fibrils a distinctive banded pattern when viewed at high magnification. Serpooshan et al. in an approach about the effect of bioengineered acellular collagen patch on cardiac remodelling and ventricular function post-myocardial infarction, brilliantly confirmed that there is integration histological and immunostaining of the scaffold in the patch form with native cardiac cells including fibroblasts, smooth muscle cells, epicardial cells and immature cardiomyocytes [26]. In conclusion, biomaterials synthetic and/or natural can be selected to form a 3D scaffold with controlled porosity, in the dense or laminate form, to recruit and organize native cells to repair or to attenuate remodelling and improve heart function following myocardial infarction.

Author details

Marco V. Chaud[1]*, Thais F. R. Alves[1], Márcia A. Rebelo[1], Juliana F. de Souza[1], Venâncio A. Amaral[1], Cecilia T. Barros[1], Katiusca S. Pontes[1], Carolina Santos[1], Patricia Severino[2] and Lindemberg M. Silveira Filho[3]

*Address all correspondence to: marco.chaud@prof.uniso.br

1 University of Sorocaba, Sorocaba, São Paulo, Brazil

2 University of Tiradentes, Aracaju, Sergipe, Brazil

3 University of Campinas, Campinas, São Paulo, Brazil

References

[1] Maron BJ, Towbin JA, Thiene G, Antzelevitch C, Corrado D, Arnett D, Moss AJ, Seidman CE, Young JB. Contemporary definitions and classification of the cardiomyopathies: An American Heart Association Scientific Statement from the Council on Clinical Cardiology, Heart Failure and Transplantation Committee; Quality of Care and Outcomes Research and Functio. Circulation. 2006;**113**(14):1807-1816

[2] Jefferies JL, Towbin JA, Crespo-Diaz R, Hambrecht R, Holschermann H, et al. Dilated cardiomyopathy. The Lancet. 2010;**375**(9716):752-762

[3] Dickstein K, Cohen-Solal A, Filippatos G, McMurray JJV, Ponikowski P, Poole-Wilson PA, Stramberg A, Van Veldhuisen DJ, Atar D, Hoes AW, Keren A, Mebazaa A, Nieminen M, Priori SG, Swedberg K, Vahanian A, Camm J, De Caterina R, Dean V, Funck-Brentano C, Hellemans I, Kristensen SD, McGregor K, Sechtem U, Silber S, Tendera M, Widimsky P, Zamorano JL, Auricchio A, Bax J, Bauhm M, Correa U, Della Bella P, Elliott PM, Follath F, Gheorghiade M, Hasin Y, Hernborg A, Jaarsma T, Komajda M, Kornowski R, Piepoli M, Prendergast B, Tavazzi L, Vachiery JL, Verheugt FWA, Zannad F. ESC Guidelines for the diagnosis and treatment of acute and chronic heart failure 2008. European Heart Journal. 2008;**29**(19):2388-2442

[4] Ponikowski P, Voors AA, Anker SD, Bueno H, Cleland JGF, Coats AJS, Falk V, Gonzalez-Juanatey JR, Harjola VP, Jankowska EA, Jessup M, Linde C, Nihoyannopoulos P, Parissis JT, Pieske B, Riley JP, Rosano GMC, Ruilope LM, Ruschitzka F, Rutten FH, Van Der Meer P. 2016 ESC Guidelines for the diagnosis and treatment of acute and chronic heart failure. European Heart Journal. 2016;**37**(27):2129-2200m

[5] Mukherjee S, Reddy Venugopal J, Ravichandran R, Ramakrishna S, Raghunath M. Evaluation of the biocompatibility of PLACL/collagen nanostructured matrices with cardiomyocytes as a model for the regeneration of infarcted myocardium. Advanced Functional Materials. 2011;**21**(12):2291-2300

[6] Kraus S, Ogunbanjo G, Sliwa K, Ntusi NAB. Heart failure in sub-Saharan Africa: A clinical approach. South African Medical Journal. 2016;**106**(1):23-31

[7] Olson TM, Michels VV, Ballew JD, Reyna SP, Karst ML, Herron KJ, Horton SC, Rodeheffer RJ, Anderson JL. Sodium channel mutations and susceptibility to heart failure and atrial fibrillation. Journal of the American Medical Association. 2005;**293**(4):447-454

[8] Sanbe A. Dilated cardiomyopathy: A disease of the myocardium. Biological and Pharmaceutical Bulletin. 2013;**36**(1):18-22

[9] Cesselli D, Jakoniuk I, Barlucchi L, Beltrami AP, Hintze TH, Nadal-Ginard B, Kajstura J, Leri A, Anversa P. Oxidative stress-mediated cardiac cell death is a major determinant of ventricular dysfunction and failure in dog dilated cardiomyopathy. Circulation Research. 2001;**89**(3):279-286

[10] Nadal-Ginard B, Kajstura J, Anversa P, Leri A. A matter of life and death: Cardiac myocyte apoptosis and regeneration. Journal of Clinical Investigation. 2003;**111**(10):1457-1459

[11] Weintraub RG, Semsarian C, Macdonald P. Dilated cardiomyopathy. The Lancet. 2017. epub ahead

[12] Li D, Tapscoft T, Gonzalez O, Burch PE, Quiñones MA, Zoghbi WA, Hill R, Bachinski LL, Mann DL, Roberts R. Desmin mutation responsible for idiopathic dilated cardiomyopathy. Circulation. 1999;**100**(5):461-464

[13] Towbin JA, Lowe AM, Colan SD, Sleeper LA, Orav EJ, Clunie S, Messere J, Cox GF, Lurie PR, Hsu D, Canter C, Wilkinson JD, Lipshultz SE. Dilated cardiomyopathy in children. Journal of the American Medical Association. 2006;**296**(15):1867-1876

[14] Zheng QS, Guo WG, Lu ZF, Shi XQ, Su FF, Li H. Dystrophin: From non-ischemic cardiomyopathy to ischemic cardiomyopathy. Medical Hypotheses. 2008;**71**(3):434-438

[15] Gonzalez A, Ravassa S, Beaumont J, Lopez B, Diez J. New targets to treat the structural remodeling of the myocardium. Journal of the American College of Cardiology. 2011;**58**(18):1833-1843

[16] Fan Z, Guan J. Antifibrotic therapies to control cardiac fibrosis. Biomaterials Research. 2016;**20**:13

[17] Spinale FG, Janicki JS, Zile MR. Membrane-associated matrix proteolysis and heart failure. Circulation Research. 2013;**112**(1):195-208

[18] Leask A. Potential therapeutic targets for cardiac fibrosis: TGF-B, angiotensin, endothelin, CCN2, and PDGF, partners in fibroblast activation. Circulation Research. 2010; **106**(11):1675-1680

[19] Gabbiani G. The evolution of the myofibroblast concept: A key cell for wound healing and fibrotic diseases. Giornale di Gerontologia. 2004;**52**(5):280-282

[20] Hinz B. Formation and function of the myofibroblast during tissue repair. Journal of Investigative Dermatology. 2007;**127**(3):526-537

[21] Leask A. Getting to the heart of the matter: New insights into cardiac fibrosis. Circulation Research. 2015;**116**(7):1269-1276

[22] Harvey A, Montezano AC, Lopes RA, Rios F, Touyz RM. Vascular fibrosis in aging and hypertension: Molecular mechanisms and clinical implications. Canadian Journal of Cardiology. 2016;**32**(5):659-668

[23] Gupta N, Liu JR, Patel B, Solomon DE, Vaidya B, Gupta V. Microfluidics-based 3D cell culture models: Utility in novel drug discovery and delivery research. Bioengineering & Translational Medicine. 2016;**1**(1):63-81

[24] Balmert SC, Little SR. Biomimetic delivery with micro- and nanoparticles. Advanced Materials. 2012;**24**(28):3757-3778

[25] Liao R, Pfister O, Jain M, Mouquet F. The bone marrow – cardiac axis for myocardial regeneration. Progress in Cardiovascular Diseases. 2008;**50**(1):18-30

[26] Serpooshan V, Wu SM. Patching up broken hearts: Cardiac cell therapy gets a bioengineered boost. Cell Stem Cell. 2014;**15**(6):671-673

[27] Nursalim A, Katili PA, Santoso T. Cellular cardiomyoplasty for myocardial infarction: A 2014 evidence-based update. Acta Medica Indonesiana. 2014;**46**(2):150-162

[28] Leor J, Amsalem Y, Cohen S. Cells, scaffolds, and molecules for myocardial tissue engineering. Pharmacology & Therapeutics. 2005;**105**(2):151-163

[29] Curtis MW, Russel B. Cardiac tissue engineering. Journal of Cardiovascular Nursing. 2009;**24**(2):87-92

[30] Patterson NL, Iyer RP, de Castro Brás LE, Li Y, Andrews TG, Aune GJ, Lange RA, Lindsey ML. Using proteomics to uncover extracellular matrix interactions during cardiac remodeling. Proteomics – Clinical Applications. 2013;**7**(7-8):516-527

[31] Frey N, Linke A, Süselbeck T, Müller-Ehmsen J, Vermeersch P, Schoors D, Rosenberg M, Bea F, Tuvia S, Leor J. Intracoronary delivery of injectable bioabsorbable scaffold (IK-5001) to treat left ventricular remodeling after ST-elevation myocardial infarction: A first-in-man study. Circulation: Cardiovascular Interventions. 2014;**7**(6):806-812

[32] Alrefai MT, Murali D, Paul A, Ridwan KM, Connell JM, Shum-Tim D. Cardiac tissue engineering and regeneration using cell-based therapy. Stem Cells and Cloning: Advances and Applications. 2015;**8**:81-101

[33] Rosenbloom J, Mendoza FA, Jimenez SA. Strategies for anti-fibrotic therapies. Biochimica et Biophysica Acta (BBA) – Molecular Basis of Disease. 2013;**1832**(7):1088-1103

[34] Murry CE, Soonpaa MH, Reinecke H, Nakajima H, Nakajima HO, Rubart M, Pasumarthi KBS, Virag JI, Bartelmez SH, Poppa V, Bradford G, Dowell JD, Williams DA, Field LJ. Haematopoietic stem cells do not transdifferentiate into cardiac myocytes in myocardial infarcts. Nature. 2004;**428**(6983):664-668

[35] Hudson W, Collins MC, deFreitas D, Sun YS, Muller-Borer B, Kypson AP. Beating and arrested intramyocardial injections are associated with significant mechanical loss: Implications for cardiac cell transplantation. Journal of Surgical Research. 2007; **142**(2):263-267

[36] Feng Y, Wang Y, Cao N, Yang H, Wang Y. Progenitor/stem cell transplantation for repair of myocardial infarction: Hype or hope? Annals of Palliative Medicine. 2012;**1**(1):65-77

[37] Nelson TJ, Ge Z, Van Orman J, Barron M, Rudy D, Hacker TA, Misra R, Auchampach JA, Lough J. Improved cardiac function in infarcted mice after treatment with pluripotent embryonic stem cells. The Anatomical Record Part A Discoveries in Molecular Cellular and Evolutionary Biology. 2008;**288**(11):1216-1224

[38] Meyer GP, Wollert KC, Lotz J, Steffens J, Lippolt P, Fichtner S, Hecker H, Schaefer A, Arseniev L, Hertenstein B, Ganser A, Drexler H. Intracoronary bone marrow cell transfer after myocardial infarction: Eighteen months' follow-up data from the randomized, controlled BOOST (Bone marrow transfer to enhance ST-elevation infarct regeneration) trial. Circulation. 2006;**113**(10):1287-1294

[39] Perin EC, Dohmann HFR, Borojevic R, Silva SA, Sousa ALS, Silva GV, Mesquita CT, Belém L, Vaughn WK, Rangel FOD, Assad JAR, Carvalho AC, Branco RVC, Rossi MID, Dohmann HJF, Willerson JT. Improved exercise capacity and ischemia 6 and 12 months after transendocardial injection of autologous bone marrow mononuclear cells for ischemic cardiomyopathy. Circulation. 2004;**110**(11 Suppl.) 213-218

[40] Schächinger V, Assmus B, Britten MB, Honold J, Lehmann R, Teupe C, Abolmaali ND, Vogl TJ, Hofmann WK, Martin H, Dimmeler S, Zeiher AM. Transplantation of progenitor cells and regeneration enhancement in acute myocardial infarction: Final one-year results of the TOPCARE-AMI trial. Journal of the American College of Cardiology. 2004;**44**(8):1690-1699

[41] Jeevanantham V, Afzal MR, Zuba-Surma EK, Dawn B. Clinical trials of cardiac repair with adult bone marrow-derived cells. Methods in Molecular Biology 2013;**1036**:179-205

[42] Soler-Botija C, Bagó JR, Llucià-Valldeperas A, Vallés-Lluch A, Castells-Sala C, Martínez-Ramos C, Fernández-Muiños T, Chachques JC, Pradas MMP, Semino CE, Bayes-Genis A. Engineered 3D bioimplants using elastomeric scaffold, self-assembling peptide hydrogel, and adipose tissue-derived progenitor cells for cardiac regeneration. American Journal of Translational Research. 2014;**6**(3):291-301

[43] Prat-Vidal C, Gálvez-Montón C, Nonell L, Puigdecanet E, Astier L, Solé F, Bayes-Genis A. Identification of temporal and region-specific myocardial gene expression patterns in response to infarction in swine. PLoS One. 2013;**8**(1):e54785

[44] Iborra-Egea O, Gálvez-Montón C, Roura S, Perea-Gil I, Prat-Vidal C, Soler-Botija C, Bayes-Genis A. Mechanisms of action of sacubitril/valsartan on cardiac remodeling: A systems biology approach. npj Systems Biology and Applications. 2017;**3**(1):12

[45] Van Laake LW, Passier R, Doevendans PA, Mummery CL. Human embryonic stem cell-derived cardiomyocytes and cardiac repair in rodents. Circulation Research. 2008;**102**(9):1008-1010

[46] Lindsey ML, Hall ME, Harmancey R, Ma Y. Adapting extracellular matrix proteomics for clinical studies on cardiac remodeling post-myocardial infarction. Clinical Proteomics. 2016;**13**(19):1-8

[47] Vandenabeele P, Galluzzi L, Vanden Berghe T, Kroemer G. Molecular mechanisms of necroptosis: An ordered cellular explosion. Nature Reviews Molecular Cell Biology. 2010;**11**(10):700-714

[48] Vakili V, Shu LH. Towards biomimetic concept generation. In: Proceedings of DETC'01 ASME 2001 Design Engineering Technical Conferences Design Theory and Methodology; Pittsburgh, Pennsylvania. 2001;**4**:1-9

[49] Rebelo MA, Alves TFR, de Lima R, Oliveira JM, Vila MMDC, Balcão VM, Severino P, Chaud MV. Scaffolds and tissue regeneration: An overview of the functional properties of selected organic tissues. Journal of Biomedical Materials Research Part B: Applied Biomaterials. 2015;**104**(7):1-12

[50] Chan BP, Leong KW. Scaffolding in tissue engineering: General approaches and tissue-specific considerations. European Spine Journal. 2008;**17**(4):S467–S479

[51] Gu Y, Zhu J, Xue C, Li Z, Ding F, Yang Y, Gu X. Chitosan/silk fibroin-based, Schwann cell-derived extracellular matrix-modified scaffolds for bridging rat sciatic nerve gaps. Biomaterials. 2014;**35**(7):2253-2263

[52] Eghbali M, Blumenfeld OO, Seifter S, Buttrick PM, Leinwand LA, Robinson TF, Zern MA, Giambrone MA. Localization of types I, III and IV collagen mRNAs in rat heart cells by in situ hybridization. Journal of Molecular and Cellular Cardiology. 1989;**21**(1): 103-113

[53] Ye L, Zimmermann WH, Garry DJ, Zhang J. Patching the heart: Cardiac repair from within and outside. Circulation Research. 2013;**113**(7):922-932

[54] Lam MT, Wu JC. Biomaterial applications in cardiovascular tissue repair and regeneration. Expert Review of Cardiovascular Therapy. 2012;**10**(8):1039-1049

[55] Mizuno H, Roy AK, Zaporojan V, Vacanti CA, Ueda M, Bonassar LJ. Biomechanical and biochemical characterization of composite tissue-engineered intervertebral discs. Biomaterials. 2006;**27**(3):362-370

[56] O'Brien FJ. Biomaterials & scaffolds for tissue engineering. Materials Today. 2011;**14** (3):88-95

[57] Ozawa T. Optimal biomaterial for creation of autologous cardiac grafts. Journal of the Medical Society of Toho University. 2004;**51**(1):38-40

[58] Tallawi M, Rosellini E, Barbani N, Cascone MG, Rai R, Saint-Pierre G, Boccaccini AR. Strategies for the chemical and biological functionalization of scaffolds for cardiac tissue engineering: A review. Journal of The Royal Society Interface. 2015;**12**(108): 20150254

[59] Markovsky E, Baabur-Cohen H, Eldar-Boock A, Omer L, Tiram G, Ferber S, Ofek P, Polyak D, Scomparin A, Satchi-Fainaro R. Administration, distribution, metabolism and elimination of polymer therapeutics. Journal of Controlled Release. 2012; **161**(2):446-460

[60] Bayomy AF, Bauer M, Qiu Y, Liao R. Regeneration in heart disease – Is ECM the key? Life Sciences. 2012;**91**(17-18):823-827

[61] Kubo H, Jaleel N, Kumarapeli A, Berretta RM, Bratinov G, Shan X, Wang H, Houser SR, Margulies KB. Increased cardiac myocyte progenitors in failing human hearts. Circulation. 2008;**118**(6):649-657

[62] Mouquet F, Pfister O, Jain M, Oikonomopoulos A, Ngoy S, Summer R, Fine A, Liao R. Restoration of cardiac progenitor cells after myocardial infarction by self-proliferation and selective homing of bone marrow-derived stem cells. Circulation Research. 2005;**97**(11):1090-1092

[63] Finan A, Richard S. Stimulating endogenous cardiac repair. Frontiers in Cell and Developmental Biology. 2015;**3**(September):57

[64] Urbanek K, Cesselli D, Rota M, Nascimbene A, De Angelis A, Hosoda T, Bearzi C, Boni A, Bolli R, Kajstura J, Anversa P, Leri A. Stem cell niches in the adult mouse heart. Proceedings of the National Academy of Sciences of the United States of America. 2006;**103**(24):9226-9231

[65] Naqvi N, Li M, Calvert JW, Tejada T, Lambert JP, Wu J, Kesteven SH, Holman SR, Matsuda T, Joshua D, Howard WW, Iismaa SE, Chan AY, Crawford BH, Wagner B, Martin DIK, Lefer DJ, Graham RM, Husain A. NIH public access. A proliferative burst during preadolescence establishes the final cardiomyocyte number. Cell. 2014;**157**(4): 795-807

[66] Kajstura J, Urbanek K, Perl S. Cardiomyogenesis in the adult human heart. Circulation Research. 2010;**107**(2):305-315

[67] Bergmann O, Bhardwaj RD, Bernard S, Zdunek S, Barnabé-Heider F, Walsh S, Zupicich J, Alkass K, Buchholz BA, Druid H, Jovinge S, Frisén J. Evidence for cardiomyocyte renewal in humans. Science. 2009;**324**(5923):98-102

[68] Maraj S, Pressman GS, Figueredo VM. Primary cardiac tumors. International Journal of Cardiology. 2009;**133**(2):152-156

[69] Laflamme MA, Murry CE. Heart regeneration. Nature. 2011;**473**(7347):326-335

[70] Laflamme MA, Murry CE. Regenerating the heart. Nature Biotechnology. 2005;**23**(7): 845-856

[71] Larsen PM, Teerlink JR. Team-based care for patients hospitalized with heart failure. Heart Failure Clinics. 2015;**11**(3):359-370

[72] Nanthakumar CB, Hatley RJD, Lemma S, Gauldie J, Marshall RP, Macdonald SJF. Dissecting fibrosis: therapeutic insights from the small-molecule toolbox. Nature Reviews Drug Discovery. 2015;**14**(10):693-720

[73] Wassenaar JW, Gaetani R, Garcia JJ, Braden RL, Luo CG, Huang D, Demaria AN, Omens JH, Christman KL. Evidence for mechanisms underlying the functional benefits of a myocardial matrix hydrogel for post-MI treatment. Journal of the American College of Cardiology. 2016;**67**(9):1074-1086

[74] Cui Z, Yang B, Li R-K. Application of biomaterials in cardiac repair and regeneration. Engineering. 2016;**2**(1):141-148

[75] Nguyen DT, Ding C, Wilson E, Marcus GM, Olgin JE. Pirfenidone mitigates left ventricular fibrosis and dysfunction after myocardial infarction and reduces arrhythmias. Heart Rhythm. 2010;**7**(10):1438-1445

[76] Wang Y, Wu Y, Chen J, Zhao S, Li H. Pirfenidone attenuates cardiac fibrosis in a mouse model of TAC-induced left ventricular remodeling by suppressing NLRP3 Inflammasome formation. Cardiology. 2013;**126**(1):1-11

[77] Voges HK, Mills RJ, Elliott DA, Parton RG, Porrello ER, Hudson JE. Development of a human cardiac organoid injury model reveals innate regenerative potential. Development. 2017;**144**(6):1118-1127

[78] Valiente-Alandi I, Schafer AE, Blaxall BC. Extracellular matrix-mediated cellular communication in the heart. Journal of Molecular and Cellular Cardiology. 2016;**91**:228-237

[79] Katz MG, Fargnoli AS, Williams RD, Steuerwald NM, Isidro A, Ivanina AV, Sokolova IM, Bridges CR. Safety and efficacy of high-dose adeno-associated virus 9 encoding sarcoplasmic reticulum Ca^{2+} adenosine triphosphatase delivered by molecular cardiac surgery with recirculating delivery in ovine ischemic cardiomyopathy. Journal of Thoracic and Cardiovascular Surgery. 2014;**148**(3):1065-1073

[80] Porrello ER, Mahmoud AI, Simpson E, Johnson BA, Grinsfelder D, Canseco D, Mammen PP, Rothermel BA. Olson EN, Sadek HA. Regulation of neonatal and adult mammalian heart regeneration by the miR-15 family. Proceedings of the National Academy of Sciences of the United States of America. 2013;**110**(1):187-192

[81] Porrello ER, Olson EN. A neonatal blueprint for cardiac regeneration. Stem Cell Research. 2014;**13**(3):556-570

[82] Foglia MJ, Poss KD. Building and re-building the heart by cardiomyocyte proliferation. Development. 2016;**143**(5):729-740

[83] Haubner BJ, Schneider J, Schweigmann U, Schuetz T, Dichtl W, Velik-Salchner C, Stein JI, Penninger JM. Functional recovery of a human neonatal heart after severe myocardial infarction. Circulation Research. 2016;**118**(2):216-221

[84] Ahuja P, Sdek P, Maclellan WR. Cardiac myocyte cell cycle control in development, disease, and regeneration. Physiological Reviews. 2007;**87**:521-544

[85] Hashizume R, Fujimoto KL, Hong Y, Guan J, Toma C, Tobita K, Wagner WR. Biodegradable elastic patch plasty ameliorates left ventricular adverse remodeling after ischemia-reperfusion injury: A preclinical study of a porous polyurethane material in a porcine model. Journal of Thoracic and Cardiovascular Surgery. 2013;**146**(2):391-399

[86] Haubner BJ, Adamowicz-Brice M, Khadayate S, Tiefenthaler V, Metzler B, Aitman T, Penninger JM. Complete cardiac regeneration in a mouse model of myocardial infarction. Aging (Albany NY). 2012;**4**(12):966-977

[87] Ieda M, Fu JD, Delgado-Olguin P, Vedantham V, Hayashi Y, Bruneau BG, Srivastava D. Direct reprogramming of fibroblasts into functional cardiomyocytes by defined factors. Cell. 2010;**142**(3):375-386

[88] Sadahiro T, Yamanaka S, Ieda M. Direct cardiac reprogramming: Progress and challenges in basic biology and clinical applications. Circulation Research. 2015;**116**(8):1378-1391

[89] Lockhart M, Wirrig E, Phelps A, Wessells A. Extracellular matrix and heart development. Birth Defects Research Part A: Clinical and Molecular Teratology. 2011;**91**(6):535-550

[90] Bhatnagar R, Li S. Biomimetic scaffolds for tissue engineering. Conference Proceedings of IEEE Engineering in Medicine and Biology Society 2004;**7**:5021-5023

[91] Deming TJ. Regenerative medicine: Noodle gels for cells. Nature Materials. 2010;**9**(7): 535-536

[92] Singelyn JM, DeQuach JA, Seif-Naraghi SB, Littlefield RB, Schup-Magoffin PJ, Christman KL. Naturally derived myocardial matrix as an injectable scaffold for cardiac tissue engineering. Biomaterials. 2009;**30**(29):5409-5416

[93] Seif-Naraghi SB, Singelyn JM, Salvatore MA, Osborn KG, Wang JJ, Sampat U, Kwan OL, Strachan GM, Wong J, Schup-Magoffin PJ, Braden RL, Bartels K, DeQuach JA, Preul M, Kinsey AM, DeMaria AN, Dib N, Christman KL. Safety and efficacy of an injectable extracellular matrix hydrogel for treating myocardial infarction. Science Translational Medicine. 2013;**5**(173):173ra25

[94] Lim JY, Hansen JC, Siedlecki CA, Hengstebeck RW, Cheng J, Winogard N, Donahue HJ. Osteoblast adhesion on poly(L-lactic acid)/polystyrene demixed thin film blends: Effect of Nanotopography, surface chemistry, and wettability. Biomacromolecules. 2005;**6**(6):3319-3327

[95] Holmes JW, Borg TK, Covell JW. Structure and mechanics of healing myocardial infarcts. Annual Review of Biomedical Engineering. 2005;**7**(1):223-253

[96] Bhatnagar RS, Qian JJ, Wedrychowska A, Sadeghi M, Wu YM, Smith N. Design of biomimetic habitats for tissue engineering with P-15, a synthetic peptide analogue of collagen. Tissue Engineering. 1999;**5**(1):53-65

[97] Nguyen H, Qian JJ, Bhatnagar RS, Li S. Enhanced cell attachment and osteoblastic activity by P-15 peptide-coated matrix in hydrogels. Biochemical and Biophysical Research Communications. 2003;**311**(1):179-186

[98] Lu P, Takai K, Weaver VM, Werb Z. Extracellular matrix degradation and remodeling in development and disease. Cold Spring Harbor Perspectives in Biology. 2011;**3**(12):1-24

[99] Hansen A, Eder A, Bönstrup M, Flato M, Mewe M, Schaaf S, Aksehirlioglu B, Schwörer A, Uebeler J, Eschenhagen T. Development of a drug screening platform based on engineered heart tissue. Circulation Research. 2010;**107**(1):35-44

[100] Mizuno T, Yau TM, Weisel RD, Kiani CG, Li RK. Elastin stabilizes an infarct and preserves ventricular function. Circulation. 2005;**112**(9 SUPPL.):81-89

[101] Mecham RP, Broekelmann T, Davis EC, Gibson MA, Brow-Augsburger P. 'Elastic fibre assembly: Macromolecular interactions.' The molecular biology and pathology of elastic tissues. The Molecular Biology and Pathology of Elastic Tissues. 2008;**773**:172-184

[102] Peattie RA, Nayate AP, Firpo MA, Shelby J, Fisher RJ, Prestwich GD. Stimulation of in vivo angiogenesis by cytokine-loaded hyaluronic acid hydrogel implants. Biomaterials. 2004;**25**(14):2789-2798

[103] Bernanke DH, Markwald RR. Effects of two glycosaminoglycans on seeding of cardiac cushion tissue cells into a Collagen-Lattice culture system. 1984;**31**:25-31

[104] Silva A, Juenet M, Meddahi-Pellé A, Letourneur D. Polysaccharide-based strategies for heart tissue engineering. Carbohydrate Polymers. 2015;**116**:267-277

[105] Taylor P, Su W, Chen K, Chen Y, Tseng C, Lin F. An injectable oxidated hyaluronic Acid/Adipic acid dihydrazide hydrogel as a vitreous substitute.Journal of Biomaterials Science. 2011;**22**(13):1777-1779

[106] Shen X, Tanaka K, Takamori A. Coronary arteries angiogenesis in ischemic myocardium: Biocompatibility and biodegradability of various hydrogels. Artificial Organs. 2009;**33**(10):781-787

[107] Silva C, Ribeiro A, Ferreira D, Veiga F. Administração oral de peptídeos e proteínas: II. Aplicação de métodos de microencapsulação. Revista Brasileira de Ciências Farmacêuticas. 2003;**39**(1):1-20

[108] Slevin M, Krupinski J, Gaffney J, Matou S, West D, Delisser H, Savani RC, Kumar S. Hyaluronan-mediated angiogenesis in vascular disease: Uncovering RHAMM and CD44 receptor signaling pathways. Matrix Biology. 2007;**26**(1):58-68

[109] Liberski A, Latif N, Raynaud C, Bollensdorff C, Yacoub M. Alginate for cardiac regeneration: From seaweed to clinical trials. Global Cardiology Science & Practice. 2015:**4**

[110] Leor J, Aboulafia-Etzion S, Dar A, Shapiro L, Barbash IM, Battler A, Granot Y, Cohen S. Bioengineered cardiac grafts: A new approach to repair the infarcted myocardium? Circulation. 2000;**102**(19 Suppl 3):III56–III61

[111] Ashton RS, Banerjee A, Punyani S, Schaffer DV, Kane RS. Scaffolds based on degradable alginate hydrogels and poly(lactide-co-glycolide) microspheres for stem cell culture. Biomaterials. 2007;**28**(36):5518-5525

[112] Zmora S, Glicklis R, Cohen S, Tailoring the pore architecture in 3-D alginate scaffolds by controlling the freezing regime during fabrication. Biomaterials. 2002;**23**(20):4087-4094

[113] Tønnesen HH, Karlsen J. Alginate in drug delivery systems. Drug Development and Industrial Pharmacy. 2002;**28**(6):621-630

[114] Sarig U, Machluf M. Engineering cell platforms for myocardial regeneration. Expert Opinion on Biological Therapy. 2011;**11**(8):1055-1077

[115] Rane AA, Christman KL. Biomaterials for the treatment of myocardial infarction: A 5-year update. Journal of the American College of Cardiology. 2011;**58**(25):2615-2629

[116] Beltrami AP, Barlucchi L, Torella D, Baker M, Limana F, Chimenti S, Kasahara H, Rota M, Musso E, Urbanek K, Leri A, Kajstura J, Nadal-Ginard B, Anversa P. Adult cardiac stem cells are multipotent and support myocardial regeneration. Cell. 2003;**114**(6):763-776

[117] Bitar M, Salih V, Brown RA, Nazhat SN. Effect of multiple unconfined compression on cellular dense collagen scaffolds for bone tissue engineering. Journal of Materials Science: Materials in Medicine. 2007;**18**(2):237-244

Chapter 11

Biodegradable Scaffolds for Gastric Tissue Regeneration

Yaser Greish, Sunitha Pulikkot,
Abdel-Hamid I. Mourad and Sherif M. Karam

Additional information is available at the end of the chapter

http://dx.doi.org/10.5772/intechopen.70525

Abstract

Tissue engineering has been viewed as a valid approach toward the partial or total replacement of defective tissues and organs. Recent advances in nanotechnology have made it possible to develop biocompatible materials at the micro- and nano-scales to be used as scaffolds for cellular growth and regeneration of defective tissues. Gastric mucosal lining is an example of soft tissues that are highly susceptible to damage due to various reasons including cancer or ulcer development. Current therapeutic approaches to these diseases have some limitations. This chapter describes the basis for development of a novel modality combining nanotechnology, stem cells, and tissue engineering for the replacement of defective gastric tissues using synthetic biocompatible scaffolds. These microfibrous scaffolds are seeded with gastric stem cells, which are studied for their proliferation and differentiation into functional gastric mucous cells.

Keywords: gastric stem cells, mucous neck cells, electrospinning, porous microfibrous scaffold, tissue regeneration

1. Introduction

Gastric cancer remains the second or third largest cause of cancer-related mortality worldwide [1, 2]. The standard operation for early or advanced gastric cancer is partial or radical gastrectomy. Patients after gastrectomy suffer from various complications such as dumping syndrome and pernicious anemia. These conditions are attributed to the loss of storage, digestive, and exocrine glandular functions of the stomach such as secretion of gastric enzymes, acid and intrinsic factor [3]. More than two-third of gastric cancer cases are unresectable and their response rate to chemotherapy is very low. Cases subjected to gastrectomy have less than 30% chance of 5-year survival [4, 5].

The stomach is the most dilated part of the digestive tube, which connects the esophagus with the small intestine (**Figure 1**). The wall of the stomach comprises four coats: serosa, musculosa, submucosa, and mucosa. The outermost serosa layer represents the peritoneal covering of the stomach. The luminal surface of the stomach has innumerable micro-openings for gastric pits (foveolae) continuous with the isthmus, neck and base of the gastric glands. In the cardia and pyloric regions, these glands are mostly populated by mucous cells and some enteroendocrine cells. However, the corpus glands are populated by two different types of mucous cells (pit and neck cells), pepsinogen-secreting chief or zymogenic cells, acid-secreting parietal cells, and hormone-secreting enteroendocrine cells (**Figure 1**). All these cells originate from stem cells residing in the isthmus region of the gastric glands. In addition to the role of these stem cells in physiological maintenance of the heterogeneous epithelium of the gastric gland, these cells are necessary for regeneration, healing, and repair of the gastric mucosa.

Although some gastric replacement approaches have been proposed to improve the quality of life of patients after gastrectomy, the optimal reconstruction procedure remains controversial [6]. Recent advances in the field of tissue engineering allowed fabrication of many tissues and organs. As an alternative remedy to the post-gastrectomy complications, gastric mucosal tissue engineering has been proposed. It has long been believed that the stomach never regenerates once it has been resected [7]. However, tissue-engineered stomach is an attractive solution post-gastrectomy to restore an adequate food digestion and appropriate gastric physiology.

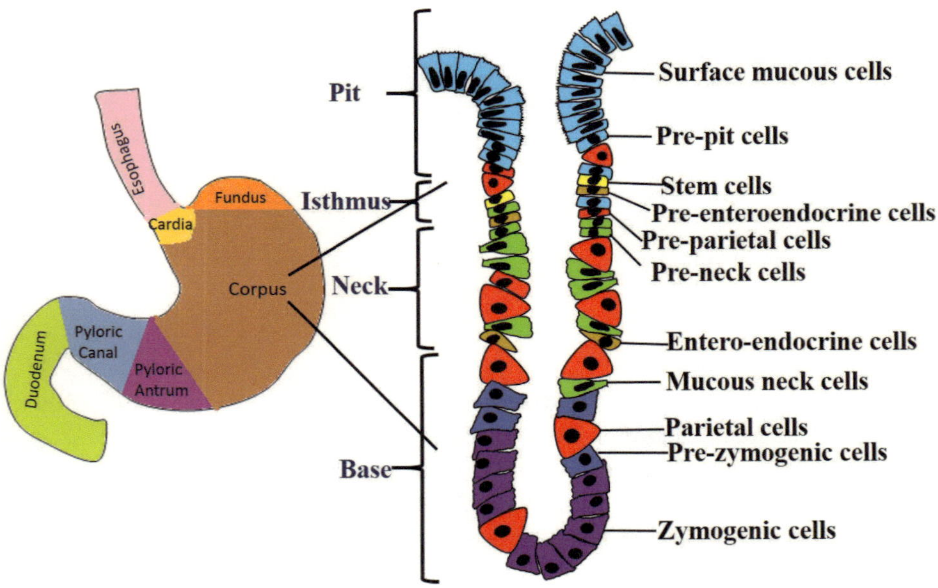

Figure 1. Diagrams depicting the structure of the stomach and gastric epithelial unit including the pit and three glandular regions: isthmus, neck and base. Note that stem cells and their immediate descendants are located in the isthmus. Surface mucous, mucous neck, and zymogenic cells are located in the pit, neck, and base, respectively. Parietal and enteroendocrine cells are scattered throughout.

Studies on tissue engineering of the stomach are very limited. This may be attributed to the unique geometry and biomechanics of the stomach compared to other soft tissues. There are some reports on syngeneic and autologous tissue-engineered stomachs in Lewis rats [8–11] and Yorkshire swine [12], respectively. Although these studies demonstrated a regenerated epithelium organized into gastric glands, epithelial differentiation, and proliferation were not comprehensively analyzed. In addition, investigations involving the mechanism of formation of these tissue-engineered gastric glands are lacking.

Tissue engineering is an interdisciplinary field that combines the knowledge and technology of materials design and optimization, cell cultures, and appropriate use of growth or biochemical factors to create artificial tissues, and regenerate damaged organs [13–22]. Examples of organs that have been tissue engineered include urinary bladder [23], trachea [24, 25], urethra [26], heart [27], liver [28–30], and lung [31, 32].

For tissue engineering, stem cells are derived from a patient, cultured to increase their number, seeded onto a certain carrier or a "scaffold," and then incubated *in vitro* to cellular integration and differentiation. The appropriate factors are also added to the culture system and over a relatively short time, a new tissue is formed. This newly developed tissue is finally ready for implantation to restore the function of a defective organ in a patient [33]. By definition, a scaffold is a three-dimensional (3D) porous construct with pre-tailored architecture and internal morphology that serves as a template for tissue regeneration [34]. It also serves as a carrier for cells, growth factors, or other biomolecular signals. Being porous in nature, a scaffold directs the growth of cells either seeded within its porous structure or migrating from surrounding tissue. Many studies demonstrated the fabrication of scaffolds with different structure and topography.

Electrospinning is a versatile processing technique that can be used for the fabrication of random and aligned fibrous scaffolds with fibers of average diameter in the nm- or μm-scale. The technique utilizes an applied voltage (up to 30 kV) to overcome the surface tension forces of a polymer solution, hence causing it to stretch into fibers that are deposited on a grounded metallic substrate (**Figure 2**) [35, 36]. Various materials have been electrospun into micro- and nano-fibers using electrospinning for a wide range of applications. In tissue engineering, a fibrous scaffold made by electrospinning closely matches the morphology of the extracellular matrix; hence, this increases its potential for tissue regeneration applications. Moreover, the interconnectivity of the pores in 3D-fibrous scaffolds made by electrospinning enhances the communication between the cells in culture, and facilitates their proliferation and differentiation. In contrast, cells grown on 2D platform can proliferate but their differentiation potential would be limited [37].

Various studies developed 3D porous scaffold systems that were either cells-free (acellular) or containing different types of cells. In the acellular approach, cells-free porous scaffolds are implanted inside the internal lining of the stomach to promote the formation of new gastric tissues. Examples included a porous poly(glycolic acid)-reinforced collagen scaffold with a silicone sheet covering the luminal side of the stomach [7, 38]. These batches incompletely supported tissue regeneration of the muscular layer and the patch grafts contracted significantly over time. An alternative scaffold composition was developed by Araki et al. and

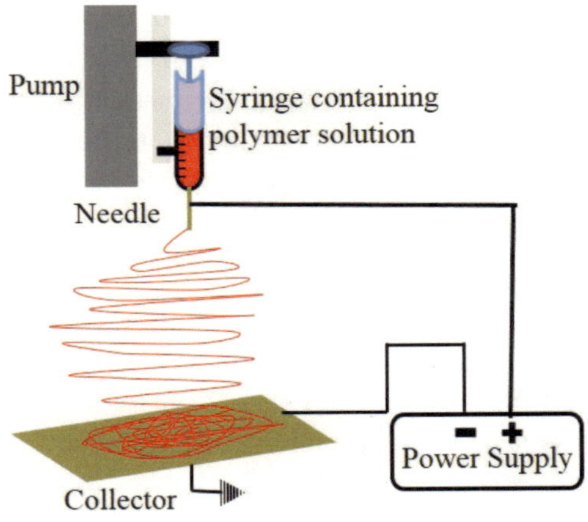

Figure 2. A schematic diagram of the electrospinning process.

was composed of poly(D,L-lactide), poly(ε-caprolactone) (PCL), collagen, and poly(glycolic acid) nonwoven fabric [39]. This multi-layer scaffold was used to repair a large stomach wall defect in a dog without infection or anastomotic dehiscence, and showed sufficient mechanical strength for suturing and better biocompatibility. However, early shrinkage of the implanted scaffold was eventually observed and regeneration of the muscle layer did not occur. In another study, Lourenco et al. developed a model comprising a gastric stromal cell line (NST-20) embedded in a 3D alginate-RGD hydrogel prepared on the basolateral side of a Transwell insert. This assembly closely mimicked the extracellular matrix of the gastric mucosa. It was used for the growth of a moderately differentiated gastric adenocarcinoma cell line (MKN28). This cell-containing scaffold was capable of reproducing the physiological conditions of the gastric barrier [40]. Lourenco et al. proved the closer similarity of this model to the native structure of the gastric mucosa, in which stromal cells appeared to have a role in the establishment of mucosal architecture. This was further confirmed by the production of extracellular matrix. In a different study, isolated gastric epithelial units were seeded onto the inner luminal surface of microporous biodegradable polymer tubes. These tubes were made from a fibrous, nonwoven mesh made of polyglycolic acid and coated with 5% poly-L-lactic acid. The seeded polymer tubes were completely wrapped and sutured into the omentum of adult Lewis rat. These gastric unit/scaffold constructs formed cyst-like structures, which were called "tissue-engineered stomachs" [41]. Recently, Noguchi et al. developed a method to induce the formation of stomach organoids from mouse embryonic stem cells. In this regard, gastric primordial epithelium and underlying mesenchyme were developed using a Matrigel-based 3D culture system. The differentiated organoid were found to contain both corpus- and antrum-specific mature gastric epithelial cells [42]. Finally and in another recent study, scientists grew tissues from the stomach's corpus/fundus region in a petri dish, which were able to produce hydrochloric acid and digestive enzymes [43]. This took place through differentiation of human pluripotent stem cells into gastric organoids containing fundic epithelium [43].

Recently, in our lab, a 3D scaffold made of a biodegradable PCL was fabricated by an electrospinning technique and was used for the growth and differentiation of gastric stem cells [44]. The main objective was to develop the basis for a new modality for the regeneration of defective gastric tissue as a result of gastric cancer or severe gastric ulcer. The experimental procedures and results of this 3D model system are summarized below.

2. Experimental procedures

2.1. Scaffolds preparation and characterization

PCL with an average molecular weight (M_n) of 70,000–90,000 by GPC was used for preparing scaffolds for gastric stem cell culture. An electrospinning technique operating at an applied voltage of 12 kV, a spinning distance of 14 cm, and a feeding rate of 0.16 mL/min, was used in making porous PCL scaffolds. Scaffolds were characterized for their microstructure using a scanning electron microscopy (SEM) technique operating at 15 kV. Tensile properties of the PCL fibrous scaffolds were measured using a universal testing machine MTS with a load cell of 100 kN with an overhead speed of 5 mm/min at room temperature. The measurement of the tensile strength was done in triplicate according to published protocol [45]. For comparison, the stomach wall of 6-month-old mice ($n = 3$) was used after washing in cold phosphate buffered saline (PBS). Tissues were immediately tested for their tensile strength. Before and after the tensile tests, SEM examination was conducted to investigate the effect of applied load and deformation on the surface topography of the scaffolds.

2.2. Culture of mouse gastric stem (mGS) cells

PCL microporous scaffolds were sterilized with various degrees of ethanol solutions and completely dried prior cell culture experiments. The mGS cells were seeded (2.5×10^5 cells) onto scaffolds (15 mm diameter and 0.9 mm thickness) placed in a 12-well plate, and allowed to grow in a 37°C incubator containing 5% CO_2 and 95% O_2 for up to 12 days. The serum-containing RPMI culture medium was changed every 48 hours. After 3 days in culture, cells were processed for initial screening of cell viability and for microscopic examinations using an inverted Olympus microscope, and Phillips SEM. For SEM, fixation was in 4% paraformaldehyde for 15 min and post-fixation, in 1% osmium tetroxide for 10 min. Following dehydration in ethanol, cells were processed for gold-palladium coating, and finally examined with SEM.

After 3, 6, 9, and 12 days, cultured cells were processed for the quantification of DNA. Cells were washed with PBS and stored at −80°C in 1 ml of Milli-Q water. DNA was extracted from the samples by repeated freeze-thaw cycles followed by ultrasonication. The Quant-iT PicoGreen dsDNA kit was used to quantify DNA according to the manufacturer's instructions. The fluorescence intensity was measured at 520 nm by using the PerkinElmer reader. Scaffolds without cells were used as blank samples. For the measurement of statistical significance, a one way ANOVA with Tukey Multiple Comparison Test was used.

2.3. Immuno- and lectin-cytochemical analysis

The mGS cells cultured on scaffolds for 3 and 9 days were fixed for 15 min in 4% paraformaldehyde. Following PBS wash, cell-containing scaffolds were processed for overnight incubation in 20% buffered sucrose at 4°C. The scaffolds were then mounted on an aluminum stalk using Shandon Cryomatrix. Frozen sections (10–30 micron-thick) were obtained using Cryostome FSE cryostat and immediately mounted on gelatin-coated slides. To visualize cellular morphology and orientation, a few sections were stained with hematoxylin and eosin. The adjacent sections were processed for lectin cytochemistry and immunoprobing of cellular-specific biomarkers.

Following incubation with blocking solution, cryosections were incubated for 60 min with different fluorophore-conjugated lectins: *Ulex europaeus* agglutinin (UEA) I (specific for surface mucous cells), *Griffonia simplicifolia* (GS) II (for mucous neck cells), or *Dolichos biflorus* agglutinin (DBA, for parietal cells) [46, 47]. Cryosections of cell-containing scaffolds were also incubated overnight with several antibodies specific for H,K-ATPase alpha and beta subunits (for parietal cells), TFF1 (for surface mucous cells), TFF2 (for mucous neck), chromogranin-A (for enteroendocrine cells), and ghrelin (for a subtype of enteroendocrine cells). Probed sections were washed in PBS and the appropriate fluorophore-conjugated secondary antibody was added. Finally, cells were visualized using Olympus fluorescence microscope or Nikon confocal microscope.

3. Results and discussion

PCL scaffolds made of microfibers in the range of 2–3 μm in diameter were prepared by electrospinning (**Figure 3**). The scaffold is characterized by a homogeneous fiber size distribution and interconnected porosity. These features are strongly recommended for tissue engineering applications, where the small size fibers provide high surface area for better cell adhesion, while the interconnected porosity provides pathways for cell interaction and new tissue formation. **Figure 4** shows a comparison between the tensile strength of a pure PCL microfibrous scaffold and that of a mouse stomach tissue. On one hand, synthetic scaffold showed a variable degree of tensile strengths at a higher range (0.35–0.6 MPa) than that of natural gastric tissues (0.22 MPa). On the other hand, both synthetic scaffold and natural stomach tissues showed a high degree of elasticity. These properties indicate the suitability of the PCL fibrous scaffold to replace the natural gastric tissues. More importantly, the higher strain of the PCL fibers makes them more durable to expansion and contraction, as dictated by the stomach biomechanical properties. While electrospun PCL fibers appeared nonwoven with random distribution (**Figure 3**), they re-align upon applying tensile forces. This feature is attributed to the interconnected porosity that allows the re-orientation of the fibers during tensile strength measurement.

Microscopic examination of the toluidine blue-stained mGS cells revealed their variable appearance on the PCL microfibrous scaffold [44]. When examined with SEM, they tend to

Figure 3. A scanning electron micrograph showing the PCL microfibrous scaffold (insert: a higher magnification of the fibers).

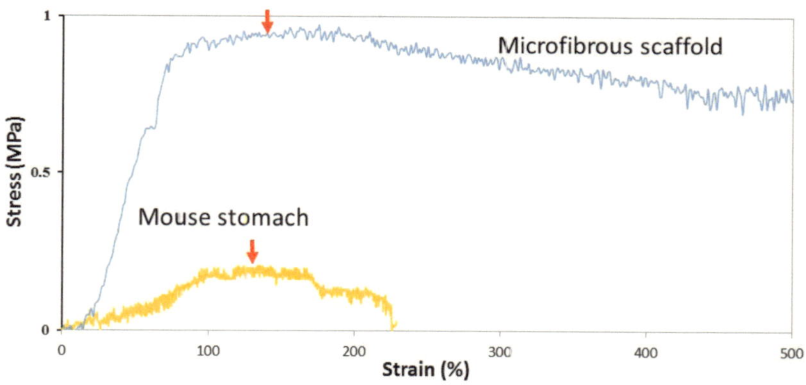

Figure 4. Stress-strain curves showing the tensile strengths of a synthetic PCL microfibrous scaffold (a), and a natural mouse stomach issue (b).

appear flattened [44] with cytoplasmic processes spanning the space between microfibers. Therefore, mGS cells were attached to more than one microfiber and integrated into the pores of the scaffold to grow in 3D (**Figure 5**).

Studies on the surface topography and porosity of scaffolds demonstrated their role on cellular adhesion, growth, and differentiation. Inducing abrasions on the surface of polyvinyl alcohol improved orientation/elongation of fibroblasts and cardiac muscle cells [48]. Different types of scaffolds with variable geometries have been tested for cell culture and adequate growth.

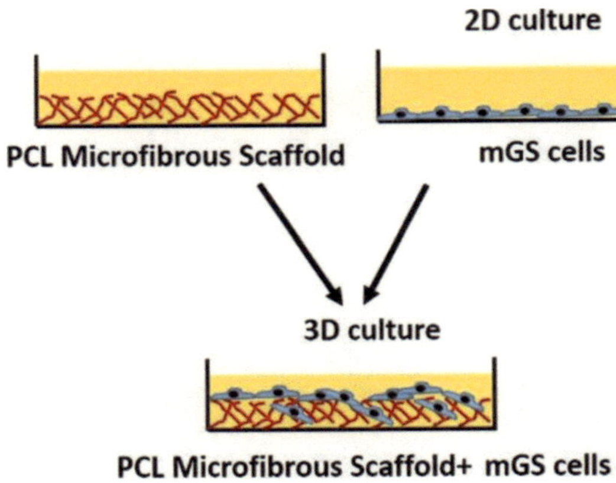

Figure 5. Diagrammatic representation of the microfibrous PCL scaffold immersed in the culture media before and after seeding with mGS cells. Note that the fibrous structure of the scaffold allows the cells to integrate through its interconnected pores and grow in 3D.

The electrospun scaffolds with microfibers seem to be the most suitable because of their 3D architecture, large surface area, and interconnected porosity. It was shown previously that PCL fibrous scaffolds support proliferation and differentiation of mesenchymal stem cells extracted from periodontal ligament [49, 50] and oligodendrocyte precursor cells [51]. Some studies showed that the fiber diameter could influence cell function and behavior on the scaffold [52–57]. Porosity is also an important factor for transport of nutrients and metabolites [58, 59].

Based on the above information, the mechanical and topographical characteristics of the microfibrous scaffold, cell viability and DNA quantification assays were conducted. Both Calcein and MTT cell viability assays showed that microfibrous scaffold support mGS cell growth. The DNA PicoGreen assay was also used to estimate the amount of cells and confirmed the advantage of the fibrous nature of the scaffold in promoting mGS cell binding and growth. It is known that cells interact with the extracellular matrix via integrin binding and sense difference in mechanical stresses through integrin signaling pathway. It was shown that increasing porosity is associated with increasing the expression of integrins [60]. This could partly explain the results obtained in the present study and the value of high porosity of microfibrous scaffold and their significant support to mGS cell growth and attachment.

The mGS cells were seeded on the PCL microfibrous scaffolds for 3, 6, 9, and 12 days to determine their pattern of growth as a function of time. The DNA was extracted from attached cells at different time points and quantified by using PicoGreen assay. The data reflected the number of attached cells on the scaffolds at 3 to 12 days of culture. **Figure 5** in [44] showed that the amount of DNA increased from day 3 to day 6 indicating proliferation of mGS cells on the scaffolds. However, when the cells were cultured for 9 days, the amount of DNA was significantly reduced. A reduction in the amount of DNA was also observed in cells cultured for 12 days [44]. This decrease in the number of cells is either due to down-regulation of cell proliferation or cell death. This effect could be regulated by integrins cytoplasmic domains that are known to affect cell proliferation [61]. Integrin's extracellular domain is also involved in adhesion through interactions with laminin [62]. Targeted deletion of the cytoplasmic domain of integrin

http://dx.doi.org/10.5772/intechopen.70525

induced reduction in cell proliferation and cell cycle arrest [63, 64]. The microfibers of scaffolds are made of inert material and lack the integrin binding sites. Therefore, the modification in the mGS cell cycle signaling is not expected for mGS cells growing on microfibrous PCL scaffolds. It was interesting to find that the reduction in cell proliferation after 9-day culture was associated with an increase the size of mGS cells [44] which could suggest cell differentiation with even loss of some of the differentiated cells. To test whether the decrease of cell number and the associated cellular enlargement were due to differentiation, cryostat sections of mGS cell-containing scaffolds of 3 and 9 days were processed for antibody probing and lectin binding. Expressions of lineage-specific proteins and glycoconjugates were taken as an indication of cell differentiation. Cryosections stained with hematoxylin and eosin demonstrated general morphology [44]. Adjacent sections probed with anti-TFF2 antibodies revealed that after 9 days of culture of mGS cells on scaffolds, some cells synthesized TFF2 specific for mucous neck cells [65, 44]. When adjacent sections were incubated with a lectin specific for mucous neck cells (GSII), the results revealed positive binding to GSII lectin as seen with fluorescence [44] and confocal [44] microscopes. Therefore, these findings demonstrate that PCL microfibrous scaffolds are suitable for growth and differentiation of mGS cells into mucous neck cells.

4. Conclusions

A synthetic biocompatible microfibrous scaffold made of PCL and fabricated by an electrospinning technique has been used for the culture of mGS cells. The scaffold is characterized by its high surface area and interconnectivity of its 3-dimensional porosity. These factors were shown to provide suitable construct for the proliferation and differentiation of mGS cells. Results showed the continued growth of the mGS cells for 6 days, followed by differentiation at 9 days. Histo- and immunocytochemistry measurements combined with SEM analysis showed multiple evidences in support of the differentiation of the gastric stem cells to mucous neck cells. These results provide the basis for a valid potential application of tissue engineering for regeneration of gastric tissues.

Acknowledgements

Data presented in this chapter are supported by research grants from UAE University.

Author details

Yaser Greish[1*], Sunitha Pulikkot[1,2], Abdel-Hamid I. Mourad[3] and Sherif M. Karam[2]

*Address all correspondence to: y.afifi@uaeu.ac.ae

1 Department of Chemistry, United Arab Emirates University, Al Ain, UAE

2 Department of Anatomy, United Arab Emirates University, Al Ain, UAE

3 Department of Mechanical Engineering, United Arab Emirates University, Al Ain, UAE

References

[1] Parkin DM. Global cancer statistics in the year 2000. The Lancet Oncology. 2001;**2**(9): 533-543

[2] Ferlay J, Soerjomataram I, Dikshit R, Eser S, Mathers C, Rebelo M, Parkin DM, Forman D, Bray F. Cancer incidence and mortality worldwide: Sources, methods and major patterns in GLOBOCAN 2012. International Journal of Cancer. 2015;**136**(5):E359-E386

[3] Kim HH, Hyung WJ, Cho GS, et al. Morbidity and mortality of laparoscopic gastrectomy versus open gastrectomy for gastric cancer. An interim report – A phase III multicenter, prospective, randomized trial (KLASS trial). Annals of Surgery. 2010;**251**:417

[4] Lordick F, Siewert JR. Recent advances in multimodal treatment for gastric cancer: A review. Gastric Cancer: Official Journal of the International Gastric Cancer Association and the Japanese Gastric Cancer Association. 2005;**8**(2):78-85

[5] Wöhrer SS, Raderer M, Hejna M. Palliative chemotherapy for advanced gastric cancer. Annals of Oncology: Official Journal of the European Society for Medical Oncology / ESMO. 2004;**15**(11):1585-1595

[6] Speer AL, Sala FG, Matthews JA, Grikscheit TC. Murine tissue-engineered stomach demonstrates epithelial differentiation. The Journal of Surgical Research. 2011;**171**(1):6-14

[7] Hori Y, Nakamura T, Matsumoto K, Kurokawa Y, Satomi S, Shimizu S. Experimental study on in situ tissue engineering of the stomach by an acellular collagen sponge scaffold graft. ASAIO Journal. 2001;**47**:206-210

[8] Grikscheit TC, Srinivasan A, Vacanti JP. Tissue-engineered stomach: A preliminary report of a versatile in vivo model with therapeutic potential. Journal of Pediatric Surgery. 2003;**38**:1305

[9] Maemura T, Shin M, Sato M, et al. A tissue-engineered stomach as a replacement of the native stomach. Transplantation. 2003;**76**:61

[10] Maemura T, Shin M, Ishii O, et al. Initial assessment of a tissue engineered stomach derived from syngeneic donors in a rat model. ASAIO Journal. 2004;**50**:468

[11] Maemura T, Ogawa K, Shin M, et al. Assessment of tissue-engineered stomach derived from isolated epithelium organoid units. Transplantation Proceedings. 2004;**36**:1595

[12] Sala FG, Kunisaki SM, Ochoa ER, et al. Tissue-engineered small intestine and stomach form from autologous tissue in a preclinical large animal model. The Journal of Surgical Research. 2009;**156**:205

[13] van Es JH, van Gijn ME, Riccio O, van den Born M, Vooijs M, Begthel H, Cozijnsen M, Robine S, Winton DJ, Radtke F, Clevers H. Notch/gamma-secretase inhibition turns proliferative cells in intestinal crypts and adenomas into goblet cells. Nature. 2005 Jun 16;**435**(7044):959-963

[14] Langer R, Vacanti JP. Tissue engineering. Science (New York, N.Y.). 1993;**260**(5110):920-926

[15] Mason C, Dunnill P. A brief definition of regenerative medicine. Regenerative Medicine. 2008;**3**(1):1-5

[16] Orlando G, Wood KJ, Stratta RJ, Yoo JJ, Atala A, Soker S. Regenerative medicine and organ transplantation: Past, present, and future. Transplantation. 2011;**91**(12):1310-1317

[17] Hibino N, McGillicuddy E, Matsumura G, Ichihara Y, Naito Y, Breuer C, Shinoka T. Late-term results of tissue-engineered vascular grafts in humans. The Journal of Thoracic and Cardiovascular Surgery. 2010;**139**(2):431-436

[18] L'Heureux N, McAllister TN, de la Fuente LM. Tissue-engineered blood vessel for adult arterial revascularization. The New England Journal of Medicine. 2007;**357**(14):1451-1453

[19] Matsumura G, Hibino N, Ikada Y, Kurosawa H, Shin'oka T. Successful application of tissue engineered vascular autografts: Clinical experience. Biomaterials. 2003;**24**(13):2303-2308

[20] McAllister TN, Maruszewski M, Garrido SA, Wystrychowski W, Dusserre N, Marini A, Zagalski K, et al. Effectiveness of haemodialysis access with an autologous tissue-engineered vascular graft: A multicenter cohort study. Lancet. 2009;**373**(9673):1440-1446

[21] Shin'oka T, Imai Y, Ikada Y. Transplantation of a tissue-engineered pulmonary artery. The New England Journal of Medicine. 2001;**344**(7):532-533

[22] Shin'oka T, Matsumura G, Hibino N, Naito Y, Watanabe M, Konuma T, Sakamoto T, et al. Midterm clinical result of tissue-engineered vascular autografts seeded with autologous bone marrow cells. The Journal of Thoracic and Cardiovascular Surgery. 2005;**129**(6):1330-1338

[23] Atala A, Bauer SB, Soker S, Yoo JJ, Retik AB. Tissue-engineered autologous bladders for patients needing cystoplasty. Lancet. 2006;**367**(9518):1241-1246

[24] Baiguera S, Birchall MA, Macchiarini P. Tissue-engineered tracheal transplantation. Transplantation. 2010;**89**(5):485-491

[25] Macchiarini P, Jungebluth P, Go T, Asnaghi MA, Rees LE, Cogan TA, Dodson A, et al. Clinical transplantation of a tissue-engineered airway. Lancet. 2008;**372**(9655):2023-2030

[26] Raya-Rivera A, Esquiliano DR, Yoo JJ, Lopez-Bayghen E, Soker S, Atala A. Tissue-engineered autologous urethras for patients who need reconstruction: An observational study. Lancet. 2011;**377**(9772):1175-1182

[27] Ott HC, Matthiesen TS, Goh S-K, Black LD, Kren SM, Netoff TI, Taylor DA. Perfusion-decellularized matrix: Using nature's platform to engineer a bioartificial heart. Nature Medicine. 2008;**14**(2):213-221

[28] Baptista PM, Orlando G, Mirmalek-Sani S-H, Siddiqui M, Atala A, Soker S. Whole organ decellularization – A tool for bioscaffold fabrication and organ bioengineering. In: Conference Proceedings. Annual International Conference of the IEEE Engineering

in Medicine and Biology Society. IEEE Engineering in Medicine and Biology Society. Annual Conference. Vol. 2009; 2009. p. 6526-6529

[29] Soto-Gutierrez A, Zhang L, Medberry C, Fukumitsu K, Faulk D, Jiang H, Reing J, et al. A whole-organ regenerative medicine approach for liver replacement. Tissue Engineering. Part C, Methods. 2011;**17**(6):677-686

[30] Uygun BE, Soto-Gutierrez A, Yagi H, Izamis M-L, Guzzardi MA, Shulman C, Milwid J, et al. Organ reengineering through development of a transplantable recellularized liver graft using decellularized liver matrix. Nature Medicine. 2010;**16**(7):814-820

[31] Ott HC, Clippinger B, Conrad C, Schuetz C, Pomerantseva I, Ikonomou L, Kotton D, et al. Regeneration and orthotopic transplantation of a bioartificial lung. Nature Medicine. 2010;**16**(8):927-933

[32] Petersen TH, Calle EA, Zhao L, Lee EJ, Gui L, Raredon MB, Gavrilov K, et al. Tissue-engineered lungs for in vivo implantation. Science (New York, N.Y.). 2010;**329**(5991):538-541

[33] Fre S, Huyghe M, Mourikis P, Robine S, Louvard D, Artavanis-Tsakonas S. Notch signals control the fate of immature progenitor cells in the intestine. Nature. 2005;**435**(7044): 964-968

[34] Griffith LG. Emerging design principles in biomaterials and scaffolds for tissue engineering. Annals of the New York Academy of Sciences. 2002;**961**:83-95

[35] Laurencin CT, Nair LS, Bhattacharyya S, Allcock HR, Bender JD, Brown PW, et al. Polymeric nanofibers for tissue engineering and drug delivery. US Patent No. 7235295. 2006

[36] Bhattacharyya S, Nair LS, Singh A, Krogman NR, Greish YE, Brown PW, et al. Electrospinning of poly[bis(ethyl alanato) phosphazene] nanofibers. Journal of Biomedical Nanotechnology (**2**, 36-45. Netherlands). 2006;**12**(1):81-84

[37] Knight E, Przyborski S. Advances in 3D cell culture technologies enabling tissue-like structures to be created in vitro. Journal of Anatomy. 2014, 2015;**227**:746-756

[38] Hori Y, Nakamura T, Kimura D, et al. Functional analysis of the tissue-engineered stomach wall. Artificial Organs. 2002;**26**:868

[39] Araki M, Tao H, Sato T, et al. Development of a new tissue-engineered sheet for reconstruction of the stomach. Artificial Organs. 2009;**33**:816

[40] Gjorevski N, Sachs N, Manfrin A, Giger S, Bragina ME, Ordóñez-Morán P, Clevers H, Lutolf MP. Designer matrices for intestinal stem cell and organoid culture. Nature. 2016 Nov 24;**539**(7630):560-564. doi: 10.1038/nature20168. Epub 2016 Nov16. PubMed PMID: 27851739

[41] Maemura T, Shin M, Kinoshita M. Tissue engineering of the stomach. The Journal of Surgical Research. 2013;**183**:285-295

[42] Noguchi TK, Kurisaki A. Formation of Stomach Tissue by Organoid Culture Using Mouse Embryonic Stem Cells. Methods Mol Biol. 2017;**1597**:217-228. doi: 10.1007/978-1-4939-6949-4_16. PubMed PMID: 28361321

[43] McCracken KW, Aihara E, Martin B, Crawford CM, Broda T, Treguier J, Zhang X, Shannon JM, Montrose MH, Wells JM. Wnt/β-catenin promotes gastric fundus specification in mice and humans. Nature. 2017;**541**:182-187

[44] Pulikkot S, Greish YE, Mourad AH, Karam SM. Establishment of a three-dimensional culture system of gastric stem cells supporting mucous cell differentiation using microfibrous polycaprolactone scaffolds. Cell Proliferation. 2014;**47**:553-563

[45] Mourad AHI. Thermo-mechanical characteristics of thermally aged polyethylene/polypropylene blends. Materials and Design. 2010;**31**:918-929

[46] Falk P, Roth KA, Gordon JI. Lectins are sensitive tools for defining the differentiation programs of mouse gut epithelial cell lineages. The American Journal of Physiology. 1994;**266**(6 Pt 1):G987-1003

[47] Karam SM, John R, Alpers DH, Ponery AS. Retinoic acid stimulates the dynamics of mouse gastric epithelial progenitors. Stem Cells (Dayton, Ohio). 2005;**23**(3):433-441

[48] Au HTH, Cheng I, Chowdhury MF, Radisic M. Interactive effects of surface topography and pulsatile electrical field stimulation on orientation and elongation of fibroblasts and cardiomyocytes. Biomaterials. 2007;**28**(29):4277-4293

[49] Hess R, Jaeschke A, Neubert H, Hintze V, Moeller S, Schnabelrauch M, Scharnweber D. Synergistic effect of defined artificial extracellular matrices and pulsed electric fields on osteogenic differentiation of human MSCs. Biomaterials. 2012;**33**(35):8975-8985

[50] Kim SE, Yun Y-P, Han Y-K, Lee D-W, Ohe J-Y, Lee B-S, Choi B-J. Osteogenesis induction of periodontal ligament cells onto bone morphogenic protein-2 immobilized PCL fibers. Carbohydrate Polymers. 2014;**99**:700-709

[51] Li Y, Ceylan M, Shrestha B, Wang H, Lu QR, Asmatulu R, Yao L. Nanofibers support oligodendrocyte precursor cell growth and function as a neuron-free model for myelination study. Biomacromolecules. 2014;**15**(1):319-326

[52] Badami AS, Kreke MR, Thompson MS, Riffle JS, Goldstein AS. Effect of fiber diameter on spreading, proliferation, and differentiation of osteoblastic cells on electrospun poly(lactic acid) substrates. Biomaterials. 2006;**27**(4):596-606

[53] Sun T, Norton D, Ryan AJ, MacNeil S, Haycock JW. Investigation of fibroblast and keratinocyte cell-scaffold interactions using a novel 3D cell culture system. Journal of Materials Science. Materials in Medicine. 2007;**18**(2):321-328

[54] Christopherson GT, Song H, Mao H-Q. The influence of fiber diameter of electrospun substrates on neural stem cell differentiation and proliferation. Biomaterials. 2009;**30**(4):556-564

[55] Liu Y, Ji Y, Ghosh K, Clark RAF, Huang L, Rafailovich MH. Effects of fiber orientation and diameter on the behavior of human dermal fibroblasts on electrospun PMMA scaffolds. Journal of Biomedical Materials Research. Part A. 2009;**90**(4):1092-1106

[56] Yao L, O'Brien N, Windebank A, Pandit A. Orienting neurite growth in electrospun fibrous neural conduits. Journal of Biomedical Materials Research. Part B, Applied Biomaterials. 2009;**90**(2):483-491

[57] Daud MFB, Pawar KC, Claeyssens F, Ryan AJ, Haycock JW. An aligned 3D neuronal-glial co-culture model for peripheral nerve studies. Biomaterials. 2012;**33**(25):5901-5913

[58] Freed LE, Guilak F, Guo XE, Gray ML, Tranquillo R, Holmes JW, Vunjak-Novakovic G. Advanced tools for tissue engineering: Scaffolds, bioreactors, and signaling. Tissue Engineering. 2006;**12**(12):3285-3305

[59] Pham QP, Sharma U, Mikos AG. Electrospun poly(epsiloncaprolactone) microfiber and multilayer nanofiber/microfiber scaffolds: Characterization of scaffolds and measurement of cellular infiltration. Biomacromolecules. 2006;**7**(10):2796-2805

[60] Knudson W, Loeser RF. CD44 and integrin matrix receptors participate in cartilage homeostasis. Cellular and Molecular Life Sciences: CMLS. 2002;**59**(1):36-44

[61] Mainiero F, Murgia C, Wary KK, Curatola AM, Pepe A, Blumemberg M, Giancotti FG. The coupling of alpha6beta4 integrin to Ras-MAP kinase pathways mediated by Shc controls keratinocyte proliferation. The EMBO Journal. 1997;**16**(9):2365-2375

[62] Simon-Assmann P, Kedinger M, De Arcangelis A, Rousseau V, Simo P. Extracellular matrix components in intestinal development. Experientia. 1995;**51**(9-10):883-900

[63] Fang F, Orend G, Watanabe N, Hunter T, Ruoslahti E. Dependence of cyclin E-CDK2 kinase activity on cell anchorage. Science (New York, N.Y.). 1996;**271**(5248):499-502

[64] Zhu X, Ohtsubo M, Böhmer RM, Roberts JM, Assoian RK. Adhesion-dependent cell cycle progression linked to the expression of cyclin D1, activation of cyclin E-cdk2, and phosphorylation of the retinoblastoma protein. The Journal of Cell Biology. 1996;**133**(2):391-403A

[65] Karam SM, Tomasetto C, Rio M-C. Trefoil factor 1 is required for the commitment programme of mouse oxyntic epithelial progenitors. Gut. 2004;**53**(10):1408-1415

Chapter 12

Polymeric Scaffolds for Bioartificial Cardiovascular Prostheses

Marcel Ricklefs, Sotiris Korossis, Axel Haverich and Tobias Schilling

Additional information is available at the end of the chapter

http://dx.doi.org/10.5772/intechopen.71846

Abstract

The reconstruction or replacement of diseased heart valves, the revascularisation of coronary arteries by coronary artery bypass grafting, the replacement of the central or peripheral blood vessels, and the reconstruction of the irreversibly damaged heart muscle represent the most common fields of application of cardiovascular surgery. In such cases, the diseased tissue is replaced by either a synthetic (metallic or polymeric) or a biological (xenograft, homograft, or autograft) prosthesis, or tissue engineered constructs. The aim of this book chapter is to give an overview over the most frequently used synthetic and biologic polymers as scaffold material in cardiovascular surgery.

Keywords: tissue engineering, polymer, scaffold, heart valve, cardiothoracic surgery

1. Introduction

Cardiovascular disease is the leading cause of death worldwide. In 2015 alone, 17.7 million people died due to cardiovascular disease, accounting for 31% of all deaths worldwide [1]. Beyond limiting the risk factors that could potentially lead to cardiovascular disease and the administration of a pharmaceutical regime, surgical treatment represents a life-saving option for severe forms of cardiovascular disease. The reconstruction or replacement of diseased heart valves, the revascularisation of coronary arteries by coronary artery bypass grafting, the replacement of the central or peripheral blood vessels, and the reconstruction of the irreversibly damaged heart muscle represent the most common fields of application of cardiovascular surgery. In such cases, the diseased tissue is replaced by either a synthetic (metallic or polymeric) or a biological (xenograft, homograft, or autograft) prosthesis, or tissue engineered constructs.

1.1. Limitations of currently available implants

The limitations of currently available cardiovascular prostheses necessitate further research and development of improved materials for diseased tissue replacement. For example, the implantation of synthetic prostheses, such as mechanical heart valves, requires lifelong anticoagulation treatment. On the other hand, glutaraldehyde-treated biological prostheses (xenografts) tend to calcify [2–5], whereas homograft prostheses have been reported to initiate an immunogenic reaction in the recipient, which leads to rejection responses. Also, most implants that are in direct contact with the bloodstream show thrombogenicity [6]. Especially small calibre vessels are prone to stenosis due to thrombus formation, intimal hyperplasia, or neointima formation. Synthetic implants are very susceptible to infections, which like any graft failure is at the very most an indication for high-risk and expensive revision surgery [7]. Moreover, cardiovascular implants need to meet biomechanical requirements for appropriate long-term *in vivo* function. The development of aneurysms is not a rare case with prostheses in the high-pressure arterial system. Also, the characteristics of the prosthesis should support its integration with the surrounding tissues. Accordingly, grafts with high porosity can lead to good tissue integration, but high porosity could also potentially cause bleeding complications due to the insufficient sealing. Covering the prosthesis with anti-thrombogenic substances such as albumin, collagen, or gelatine predisposes to the same complication [8]. Owing to the ageing of the population and the subsequent need for cardiovascular implants, optimised materials and prostheses are necessary in the future to avoid high-risk, stressful and expensive revision surgery, which induces a threatening economic pressure on healthcare systems.

1.2. Requirements for ideal cardiovascular prostheses

Based on previous experience and large-scale clinical studies with implants and the long-term postoperative medical aftercare, there are some characteristics of an ideal cardiovascular prosthesis. The prosthesis should demonstrate high long-term durability and should have the capacity for regeneration and growth. Grafts that have the potential to regenerate would be especially beneficiary for the paediatric population since they would be capable of growing in the host, avoiding reoperations. Prosthetic vessels and valves should also present physiological hemodynamics, allow for unrestricted blood flow without any volume losses, turbulences, or stasis, and present no thrombogenicity [9] or interference with the blood constituents [10]. Since anticoagulation or platelet inhibition therapy might lead to bleeding complications, these therapeutic regimes should be avoided [11]. Furthermore, there should be no interference with other cardiac structures [12, 13]. In case of *in vitro*–seeded prostheses, the utilised cells need to be non-immunogenic (autologous or HLA-silenced homologous) and functional. Homologous donor cells in homografts can cause immunogenic reactions and subsequent graft degeneration, scar tissue formation, loss of flexibility, and, ultimately, graft rejection [14]. Moreover, any degradation products of the implanted prosthesis should be non-toxic and metabolise physiologically [15, 16]. Preferably, the ideal prosthesis should also possess a physiological structure, resembling the tissue it replaces [17, 18]. The prosthesis should be available for all patients in need, easily storable, transportable, and implantable.

1.3. Tissue engineering of regenerative grafts

Tissue engineering allows for the development of viable implants, with a regenerative and growth potential after implantation in the patient. The fundamental approach of tissue engineering comprises five steps [19, 20] (**Figure 1**):

A. Cell sourcing, cultivation, and expansion *in vitro.*

B. Development of a scaffold of either synthetic or biological origin (e.g. collagen or decellularised tissue), which could either be implanted for subsequent endogenous cell repopulation (*in vivo* tissue engineering) or seeded with appropriate cells *in vitro* prior to implantation (*in vitro* tissue engineering).

C. Cell seeding of the scaffold *in vitro,* followed by physical conditioning and maturation (regeneration and neo-tissue formation) in a functional simulation system (bioreactor).

D. Implantation of the scaffold or the tissue-engineered construct.

E. *In vivo* remodelling of the graft (guided tissue engineering).

Scaffolds should contain as many physiological characteristics of the natural extracellular matrix (ECM) as possible, including histoarchitecture and composition. Numerous complex interactions between the ECM and cells are necessary for inducing appropriate cellular function, including adhesion, migration, proliferation, and differentiation, as well as ECM degradation and synthesis [21–25]. The closer the scaffold's substrate for cell seeding is to its natural equivalent, the more the seeded cells develop their physiologic phenotype [26] to produce a viable, metabolically active prosthesis. Mainly interstitial cells and endothelial cells are used in cardiovascular tissue engineering. Seeding with endothelial cells (ECs) fulfils two main functions especially for cardiovascular tissue–engineered constructs; the scaffold is a foreign body and, therefore, a target not only for thrombocytes but also for immune cells. An EC layer shields the scaffold material and reduces both thrombogenicity and immunogenicity, the latter being the reason for the prostheses' degradation.

Mimicking natural ECM synthesis and degradation processes, the scaffold of the tissue-engineered construct would be degraded and regenerated by the seeded connective tissue cells. To maintain structural and mechanic integrity, the degradation rate of the scaffold should match the formation of new tissue [27]. Because of these remodelling, the tissue-engineered prosthesis would not be indistinguishable from the native organ after a while.

Decellularised xenogeneic or allogeneic tissue ECMs or polymers are commonly utilised substrates for engineering of cardiovascular scaffolds. The main drawbacks of xenogeneic or allogeneic tissue ECMs are the potential risk of disease transmission, variability in production, and limited availability (allogeneic) [28]. On the other hand, polymers made of biological or synthetic substances can be produced in unlimited amounts under sterile conditions. The aim of this chapter is to give an overview of polymer-based synthetic or biologic cardiovascular prosthesis.

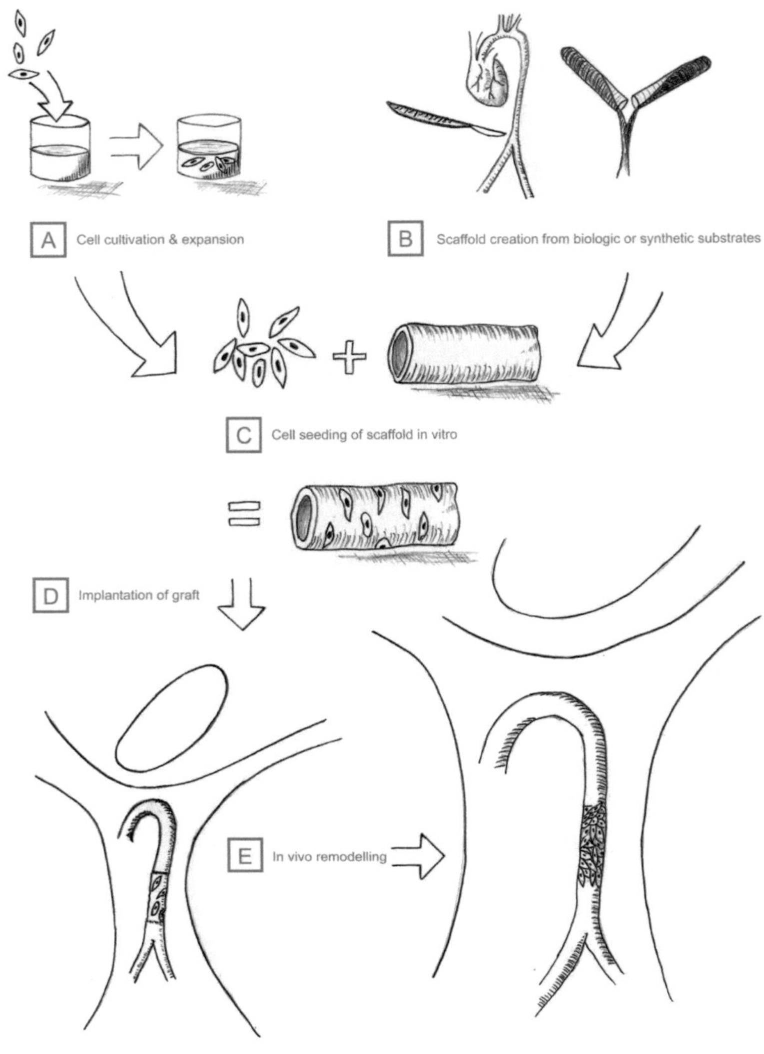

Figure 1. Drawing of the principles of tissue engineering. A: Cell sourcing, cultivation, and expansion *in vitro*. B: Development of a synthetic or biological scaffold. C: Cell seeding of the scaffold *in vitro*. D: Implantation of the scaffold. E: *In vivo* remdelling of the graft.

2. Indications for the use of polymer-based cardiovascular implants

2.1. Aetiology, pathophysiology, and surgical therapy of heart valve disease

There is a huge diversity of aetiologies of heart valve defects requiring surgical intervention. **Table 1** gives an overview of heart valve diseases requiring surgical intervention. Tricuspid or pulmonary valve dysfunction has a lower incidence than pathologies of the aortic or mitral valve [29, 30] since the left heart bears higher blood pressures and the valves are, therefore, subjected to a higher grade of mechanical stress. In general, heart valve dysfunction involves regurgitation,

http://dx.doi.org/10.5772/intechopen.71846

Congenital	Immunological	Infectious	Degenerative	Traumatic	Others
• Prolapse • Aortic aneurysm • Bicuspid valves • Primary cardiomyopathy • Stenosis • Atresia • Regurgitation • Marfan syndrome • Ehlers-Danlos syndrome	• Rheumatic fever • Scarlet fever • SLE • Scleroderma	• Endocarditis • Endomyocarditis	• Prolapse of mitral valve • Calcification	• Aortic dissection • Aortectasis • Blunt thoracic trauma	• Ischemia • Mechanic (HOCM) • Idiopathic

Table 1. Overview of heart valve diseases with the potential need for surgery (mitral- and aortic valve).

stenosis, or a combination of both. Left untreated, valvular dysfunction, which initiates either a concentric (stenosis) or an eccentric (regurgitation) hypertrophy, results in congestive heart failure—a considerable economic burden and the most common reason for hospitalization of the elderly [31]. Since the early days of valve replacement, the procedure is being carried out by applying either mechanical (**Figure 2A**) or biological (**Figure 2B**) prostheses, the latter being xenogeneic (made from treated animal valves or animal-derived pericardium) or allogeneic.

There are different access paths and principles to implant heart valve prostheses. Since the 1960's, the standard surgical approach, via full sternotomy, is the therapy of choice for most patients. In surgery of the elderly and severely co-morbid patients, it is beneficial to reduce the time of the intervention to a minimum, which can be achieved by implanting sutureless heart valve prostheses (**Figure 2C**). In contrast to the standard surgical approach, these valves do not require the time-consuming suturing of the valve into the annulus, as those grafts keep their position by an expandable metal stent. The third approach is the percutaneous, catheter-based application of a heart valve transcatheter aortic valve implantation (TAVI). This intervention is performed in the operating room, cardiac catheter laboratory, or hybrid operating room (**Figure 2D**). Innovative materials for polymer-based valve prostheses, the aforementioned different procedures, and the resulting options and restrictions should always be considered.

2.2. Aetiology, pathophysiology, and surgical therapy of vascular disease

Arteriosclerosis is an arterial disease, in which the lumen of the vessel is occluded by calcification [32]. In case of a complete obstruction, the area perfused by the affected artery undergoes critical hypoperfusion. A complete acute obstruction of the vessels leads to myocardial infarction, apoplexy, or insufficient blood circulation in the extremities. An overview of the aetiology of aortic disease with the need for interventions is presented in **Table 2**. The ischemic cardiovascular disease carries the highest mortality rate worldwide [1]. The gold standard for the treatment of severe cases of occlusive vascular disease is surgical revascularisation. In these cases, the affected vessels are replaced, or the obstruction is bridged by a vascular graft (i.e. "bypass-surgery"). Synthetic polymer grafts made from alloplastic materials, such as Dacron,

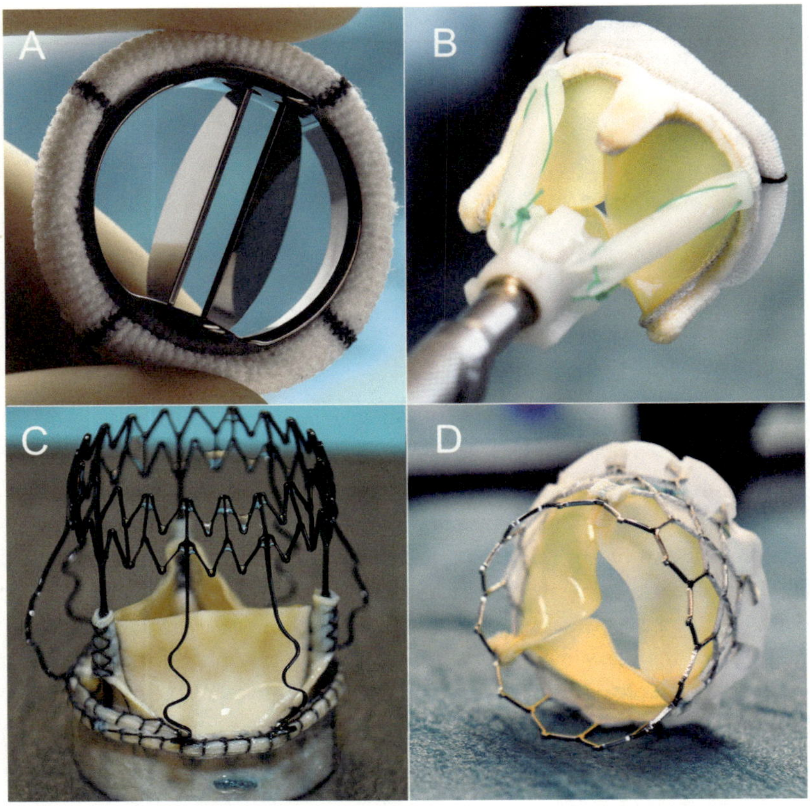

Figure 2. Currently used valve prostheses: (A) mechanical valve, (B) biological, xenogenic valve, (C) sutureless, biological valve, and (D) valve for TAVI procedure.

are routinely used clinically for the replacement of large vessels, such as the aorta (**Figure 3**). However, synthetic polymer grafts for small vessels still show problems regarding patency and development of an intimal hyperplasia or thromboembolic events [6]. Furthermore, a life-threatening adverse effect of long-distance replacement of large vessels is therapy-refractory hypertension with subsequent end-organ damage, since the currently available aortic grafts lack the function of Windkessel. Owing to this, the identification of suitable, anti-proliferative, regenerative, and mechanically functional materials, manufacturing processes, and coatings still are the primary objectives of current research in the field of vascular prostheses.

2.3. Aetiology, pathophysiology, and surgical therapy of myocardial disease

Ischemic heart conditions, such as chronic coronary artery disease, myocardial infarction or infection, or immunologic diseases such as sarcoidosis or amyloidosis, lead to a loss of viable heart muscle cells (cardiomyocytes). In contrast to other cells, adult human cardiomyocytes cannot proliferate. Therefore, the necrotic myocardium is replaced by functionless scar tissue. This weakens the heart muscle pump and eventually causes heart failure. The heart is not able to generate an adequate blood flow. Congestive heart failure is characterised by a high mortality and recurrent, long hospitalisations [33]. In 2013, congestive heart failure was the reason for every

Deformation	Occlusive diseases	Aneurysm	Trauma	Others
• Aneurysm	• Arteriosclerosis	• Arteriosclerosis	• Blunt thoracic trauma	• Aortic insufficiency
• Aortic coarctation	• Stenosis	• Familial thoracic aortic aneurysm	• Penetrating wounds	• Shunts (dialysis)
• Aortic arch abnormalities	• Thrombosis (acute vs. chronic)	• Marfan syndrome	• Iatrogenic trauma	
• Aberration	• Embolism	• Trauma	• erosion (tracheostomy)	
• Stenosis		• AV fistula		
		• Inflammatory (e.g. SLE)		
		• Infection (e.g. rheumatic fever)		
		• Iatrogenic (e.g. puncture)		
		• Pregnancy		

Table 2. Overview of large vessel diseases with the potential need for surgery.

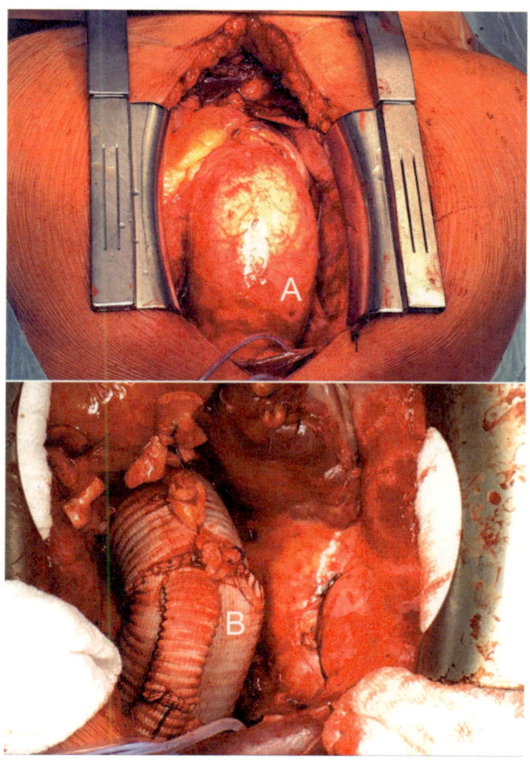

Figure 3. Operative situs of aortic surgery: (A) severe aneurysm of the ascending aorta and (B) aortic prostheses made of Dacron following surgical replacement of the dilated segment.

third death in Germany [34], producing a cost of 3.2 billion Euro [35]. The mortality of congestive heart failure has been reported to reach rates up to 36% per year. In addition to pharmacological therapy, there are surgical options for the treatment of end-stage heart failure, including heart transplantation, implantation of ventricular assist devices, or myocardial reconstruction (Dor procedure) [36]. In a ventriculoplasty, non-functional, dilated tissue is resected and the physiologic geometry and size of the heart chamber are reconstructed utilising a repair patch. For this purpose, a large variety of patch materials and biologic substrates have been assessed.

3. Polymers as a substrate in engineering innovative cardiovascular prostheses

Although the first work on polymeric heart valves goes back to the late 1950s [37–39], polymeric heart valves are not implanted in patients on a regular basis, since they have not yet shown satisfactory results. Depending on the type of polymer, the main issues with polymeric heart valves were low durability due to the tendency to become stiff with subsequently regurgitation or even complete shredding of the leaflets [40], poor long-term survival and a high mortality rate due to perioperative complications [41, 42], and thrombotic and calcific degeneration of the valve leaflets, leading to severe complications (thromboembolic events with end-organ damage, regurgitation, and heart failure) [43]. Thus, extensive research has been done on different polymers. Although there are many ways to classify polymeric scaffolds, the materials used to develop heart valve scaffolds can be classified as natural or synthetic. Examples of synthetic polymers include polycaprolactone (PCL), polytetrafluoroethylene (PTFE), polyethylene glycol (PEG), and polyurethanes (PU) [44–46], whereas examples of natural materials include, amongst others, hyaluronic acid, fibrin and collagen [47–51], and the combination of them [52].

The major drawbacks of natural polymers are their poor mechanical properties and fast degradation rate. Several studies have reported on mixing natural polymers with biologic and synthetic materials to combine the high biocompatibility of the natural polymers with the increased mechanical properties of the co-material with encouraging results [53–55]. For example, Stamm et al. used enzymatically decellularised porcine aortic valves and impregnated them with biodegradable polyhydroxybutyrate, since decellularisation leads to the exposure of the ECM collagen and by that to a high thrombogenicity. In that study, impregnating the valves with polyhydroxybutyrate had two positive effects: (a) the collagen matrix was covered and, therefore, was no longer thrombogenic and (b) the biomechanical properties of the valve were improved since decellularisation using enzymatic digestion weakened the mechanical properties of the valve. In large animal testing, the aforementioned valves functioned well for up to three months and partially developed the morphological characteristics of the native aortic valve [56].

The work of Stamm et al. is a good example for the feasibility of implanting polymeric scaffolds *in vivo*. However, the implantation of these valves is still not an option in the clinical setting. On the one hand, the lack of long-term results by means of safety and effectiveness, in comparison to regularly implanted valves, does not allow for clinical use; only long-term *in vivo* studies can reveal, for example, the toxicity and side effects of degradation products. On the other hand, a major hurdle for the clinical use is calcific degeneration. Although tissue-engineered polymer-based heart valves show a lower tendency to calcify *in vivo* compared to standard bioprostheses, there is again a lack of long-term clinical studies, since the standardly used sheep model is less thrombogenic than the human coagulation system. Experience with the utilisation and application of the most frequently used synthetic and natural polymers as graft material will be introduced below. An overview of the pros and cons of these polymers is given in **Table 3**.

3.1. Polycaprolactone-based polymers (PCL)

Thermoplastics based on polycaprolactone have been largely used in biomedical applications, as they show excellent biocompatibility and a controlled biodegradation. Examples of these types of polymers are polylactide (PLA) or polyglycolide (PGA), which are used in a mixture with other natural or synthetic materials, or alone. Both PLA and PGA are classified as biocompatible and non-toxic and are FDA-approved for human implantation [57, 58]. Degradation of these polymers *in vivo* is facilitated through a hydrolytic splitting of the ester bonds that leads to the formation of the natural metabolite lactic acid and glycolic acid. The risk for acidosis through local accumulation of these degradation products is low [59]. Owing to the presence of the additional methyl group, PLA is more hydrophobic than PGA and has

Polymer	Pros	Cons	References
Synthetic polymers			
Polycaprolactone	Excellent biocompatibility, controlled biodegradation	High stiffness, limited ability for cell adhesion	[57, 58, 61, 62]
Polyhydroxyalkanoate	High flexibility, occurs naturally, thermoplasticity	High production costs	[66–69]
Polyurethane	Resistant to degradation, high durability	Limited biocompatibility	[75, 76]
Polyglycerol sebacate	Low stiffness, high elastic properties	Low porosity, poor cell adhesion	[82–84]
Polyethylene glycol	Good mechanical properties, potential for functionalization	Poor cell adhesion	[98]
Biological polymers			
Collagen	High biocompatibility, low immunogenicity	Very high thrombogenicity, poor mechanical properties	[106–108]
Fibrin	High biocompatibility, low immunogenicity, easily derived	Poor mechanical characteristics	[48, 112]
Hyaluronic acid	High biocompatibility, low immunogenicity	Poor mechanical characteristics	[104, 118]

Table 3. Overview of pros and cons of most frequently used synthetic and natural polymers.

a lower degradation rate; while PGA is generally completely degraded in 2–3 months, the degradation of PLA in its preferred metabolic D-configuration takes an average 2 years [57]. A possibility for the regulation of the degradation rate is the copolymerization of PLA with PGA [60]. However, PCL scaffolds have a high stiffness and have a limited ability for cell adhesion and proliferation [61, 62]. Hence, the production of hybrid compound scaffolds is only indicated for myocardial grafts, vascular prostheses, and bioartificial heart valves.

Shinoka et al. used a scaffold made of a PLA mesh between two layers of randomly orientated PGA fibres and seeded it *in vitro* with ovine fibroblasts, smooth muscle cells (SMCs), and ECs. The autologous implantation as a replacement of the posterior leaflet of the pulmonary valve showed an adequate functionality, although the flexibility was reduced in comparison to the native leaflet. Furthermore, the cellular architecture and synthesis of collagen were indicative of a delicate expression of ECM [63]. Sodian et al. reported on the production of a tricuspid heart valve made of a PGA structure that unfortunately demonstrated insufficient structural integrity under physiological flow and pressure conditions [64]. To improve both the plasticity and mechanical properties, Hoerstrup et al. coated the fibre structure above with poly-4-hydroxybutyrate, which resulted in a thermoplastically editable composite [65].

3.2. Polyhydroxyalkanoate (PHA)

Linear polyesters of hydroxy fatty acids are generally pooled as polyhydroxyalkanoates. PHAs show a high biocompatibility and occur naturally as reserve substances in a variety of bacteria [66, 67]. The widespread use as scaffold materials in valvular tissue engineering, in combination with other materials, or as a standalone substrate is due to their higher flexibility in comparison to PLA and PGA, as well as their thermoplasticity, which allows for moulding with different thermal procedures [68, 69]. The degradation rates, besides the specific molecular weight, are dependent on their crystallinity [70]. Thus, poly-4-hydroxybutyric acid degrades rather fast *in vivo* [71], whereas polyhydroxyoctanoate is still detectable after 24 weeks [72]. The mechanism of degradation is based on a hydrolytic splitting of ester bonds and distinguishes through a delayed loss of mass, as the loose chain fragments only start to diffuse at a certain length [16]. Furthermore, it is expected that these degradation products are non-toxic and show a lower acidity compared to PLA [67, 73]. Sodian et al. seeded ovine vascular cells onto a tricuspid heart valve scaffold made of porous polyhydroxyoctanoate and implanted it in autologous pulmonary position. After 17 weeks, the collagen content was above the reference value of native tissue, whereas after explantation, the valves showed a characteristic non-linear stress-strain behaviour. Although there was no confluent endothelium on the grafts' surface, there was no sign of material-induced thrombogenicity [74].

3.3. Polyurethane (PU)

Polyurethanes are polymers of organic units joined by carbamate links and similar to PCL-based polymers; they are widely used in biomedical applications. PUs show a good long-term durability since they are resistant to degradation. However, modifications of the original structure of PUs have been conducted to allow for a controlled biodegradation, mechanical stability, and improved cell colonisation in PU-based heart valve prostheses, vascular grafts, and myocardial patches [75, 76]. PUs often were used in combination with other materials or as an unblended polymer. Scaffolds made of PU are a promising option for tissue engineering of myocardial replacement tissue, especially after seeding with mesenchymal cells [77] and in combination with other polymers for improving cell adhesion, porosity, and mechanic stability [78]. Fromstein et al. seeded PU scaffolds with embryonic stem cells in a bioreactor and investigated the effect of the macro-architecture on the adhesion, viability, and morphology on the seeded cells [79]. The authors found cells with the typical morphology of cultured cardiomyocytes on electrospun fibrous scaffolds, whereas there were no cardiomyocyte-like cells on scaffolds made through thermally induced phase separation (TIPS).

In vascular prostheses, McCarthy et al. investigated PGA, PLA, PCL, and PU as supportive materials of the elastica interna of a decellularised murine aorta [55]. The developed grafts were comparable to the native saphenous vein regarding burst pressure and wall diameter. In a direct comparison of their hybrid grafts, polyurethane grafts showed better burst pressure and tensile properties than the other polymeric scaffolds. Furthermore, Nieponice et al. reported that in vitro seeding of the PU grafts with muscular stem cells improved stability and functionality [80]. The authors implanted seeded and non-seeded grafts in the aortic position of a rat model and found a substantial lower graft failure in the cell-seeded group. In developing vascular prostheses, not only the superior mechanical properties of the polyurethane grafts should be pursued, but also the potential neointima formation and graft stenosis should be considered in the production of these grafts, which could be addressed with drug release mechanisms [81].

3.4. Polyglycerol sebacate (PGS)

Firstly produced in 2002 by a polycondensation reaction of glycerol sebacic acid, PGS has been widely studied in the context of biomedical applications. Owing to the low stiffness and high elastic properties, PGS has been reported to be a promising scaffold material for tissue engineering [82–84]. Depending on the production parameters and structural conditions, the elastic modulus of PGS can vary between 0.025 and 12 MPa, which corresponds to the modulus of human myocardial tissue [85]. The mechanical properties can be altered by adjusting the duration of the cross-linking and the concentration of the educts, which have been FDA-approved regarding their biocompatibility [84]. To improve the mechanical properties, Xu et al. combined PGS with PLA, and they reported a stiffness that was comparable to the native myocardial tissue [86].

One major challenge in the production of PGS-based scaffolds is the fabrication of a porous structure [87]. To face this challenge, Masoumi et al. used laser ablation to produce a diamond-shaped porous microstructure [88]. Sant et al. produced a fibrous scaffold using electrospinning; however, they needed to adapt the viscosity of the fluid by adding polycaprolactone. Following the seeding of these scaffolds with human valvular intermediate cells (VICs), the authors found a high cell viability, as well as the expression of a dense collagen network [27]. Jeffries et al. compared electrospun fibrous PGS matrices to porous PGS foam and found a fivefold higher torque strength and better suture retention of the former [89]. The porosity of the PGS scaffolds is crucial for *in vitro* and *in vivo* cell seeding [90], as well as for mechanical stability. This is of particular importance in the development of cardiac patches and myocardial prostheses since these grafts need to have a reliable mechanical integrity, which normally is achieved by increasing the wall thickness of synthetic grafts. However, the increase of the wall thickness limits oxygen supply by diffusion to the seeded cells, posing the need for vascularisation. Therefore, PGS grafts are produced with a high porosity to mimic native vascularisation [90, 91]. Furthermore, cell adhesion can be improved by simulating epitope sequences of laminin and fibronectin through connecting specific amino acid sequences with the PGS scaffolds [92].

In an *in vitro* haemocompatibility study with human blood, PGS showed a lower adhesion of thrombocytes and release of inflammation markers, compared to polytetrafluoroethylene, which is indicative of improved haemocompatibility [93]. Guler et al. investigated the ECM of decellularised sheep aorta connected to PGS *in situ* and showed that there was no additional impairment on the smooth muscle cells of a human aorta [94]. In this fashion, Guler et al. augmented the regenerative potential of allogeneic prostheses with the superior mechanical properties of PGS elastomers. To investigate *in vivo* degeneration, Pomerantseva et al. implanted disc-shaped PGS samples subcutaneously in rats, which only generated a minor tissue reaction. They reported a superficial degradation process by enzyme-mediated hydrolytic splitting of ester bonds [83]. The PGS samples were maintained *in vivo* for several weeks and were characterised by a linear loss of mass [83, 95]. The degradation products were glycerine and a metabolic intermediate of the ω-oxidation (sebacic acid) [84]. Moreover, Stuckey et al. implanted myocardial PGS patches in rodent hearts. They found a significant faster degeneration than in previous *in vitro* experiments [96]. Khosravi et al. have found aneurysms after infrarenal implantation of aortic grafts made from electrospun PGS [54]. The authors suggested that the reason of the aneurysms was the suspected degradation of the PGS without a sufficient remodelling process simultaneously. However, there was no aortic dissection, since the grafts were covered with a thin PCL layer. In contrast to these findings, Wu et al. reported on an early remodelling process after implantation of a heparin-covered PGA graft in the abdominal aortic position [97].

3.5. Polyethylene glycol (PEG)

PEG is a hydrophilic polyether bond of the divalent alcohol ethylene glycol, which is neither toxic nor immunogenic and FDA-approved for implantation in humans. Also, it has been used on a regular basis in the field of tissue engineering [98] due to its potential for functionalisation of the terminal hydroxyl groups by means of an adaption for cell adhesion and degradation, as well as its mechanical properties [98]. For example, a hydrolytically degradable copolymer results from an integration of PLA or PGA into the polyether structure [99]. Benton et al. cross-linked the PEG chains with peptide sequences, which contained a proteolytic sensitivity towards matrix metalloproteinases, therefore, allowing for a cell-controlled degradation. Also, a specific amino acid sequence (arginine-glycine-asparagine) was integrated into the hydrogel matrix. The encapsulated porcine VICs showed an increasing elongated morphology, as well as an improved occurrence of integrin binding as a sign for increased cell adhesion [100]. Moreover, porcine VICs cultivated in a copolymer of methacrylised hyaluronic acid and PEG molecules produced both collagen and elastin [101]. Hockaday et al. combined photopolymerisation of PEG with 3D printing and reproduced an anatomically precise aortic valve [102]. Zhang et al., however, produced an anisotropic scaffold structure by using an

http://dx.doi.org/10.5772/intechopen.71846

aperture that allowed for local varying light irradiation and, subsequently, a varying grade of cross-linking [103].

3.6. Collagen

Collagen is the most prevalent structure protein of the human body and is also a substantial component of the valvular extracellular matrix [104, 105]. Subsequently, collagen is highly biocompatible and shows only low immunogenicity [106]. Moreover, the specific peptide sequences promote cell adhesion, which could drive the cellular population of collagenous matrices [107]. However, the conventional isolation from animal tissue bears the risk of zoonosis [108]. Furthermore, collagen has a high thrombogenic potential [109]. Therefore, its use in the engineering of tissues with contact to the blood flow requires coating or masking of the surface. This is often realised by endothelialisation [104].

Further limitations of collagen-derived hydrogels as a scaffold material are its poor mechanical properties in comparison to native valve tissue as well as the rapid and hardly predictable degradation process in vivo [104, 106]. The degradation happens enzymatically through collagenases [110]. The degradation rate varies depending on the implantation site [109]. Cross-linking of peptide chains may improve the stability and influence the degradation rate [98]. For this, both chemical agents (glutaraldehyde and formaldehyde) and physical procedures (UV radiation and heat treatment), as well as insertion of polymers, are applied [110].

To improve the compressive strength, Flanagan et al. integrated chondroitin sulphate into a collagen matrix and subsequently seeded it with porcine VIC and valvular endothelial cells (VEC). During the total cultivation time of 28 days, the cells were mitotic active and kept their initial phenotype. Histochemical examination showed both collagen synthesis and, most likely because of chondroitin sulphate, the production of elastin. Nevertheless, they observed a significant contraction of the matrix through cellular interaction [111], which was in line with the findings of Benton et al. who also reported on halving the initial matrix dimension due to contraction in *in vitro* studies [100].

3.7. Fibrin

Fibrin represents the fundamental substrate for ubiquitous tissue repair mechanisms and, therefore, is a natural scaffold material, whose precursors are relatively easy and can be derived from the patients' blood plasma. Thus, allogenic and autologous scaffolds can be generated by fibrin, which contains only a low risk of a graft-versus-host disease. The degradation products also stimulate the production of extracellular matrix angiogenesis [112]. Hence, fibrin displays favourable properties for tissue engineering of cardiovascular scaffolds. On the other hand, cardiovascular implants are in direct contact with the bloodstream and its inherent lytic enzymes such as plasmin, which may rapidly degrade fibrin [48].

In a study done by Ye et al. fibrinolysis was not evident at a concentration of fibrin at a level of 20 μg/ml in a surrounding culture medium, whereas the fibrin hydrogel degraded in two days without adding the plasmin antagonist aprotinin. The modified-release degradation also led to an increased collagen expression in human myofibroblasts [113]. However, the cellular production of collagen fibres led to a contraction of the fibrin matrix. Jockenhövel et al. inhibited the contraction through fixation of the fibrin scaffold margin with a poly-l-lysine-solution [48]. They also introduced an injection moulding process to produce and to seed cardiac valve–shaped scaffolds. The polymerisation of fibrin took place in a mould and was triggered by a gradual injection of a fibrin solution into a thrombin-containing cell suspension. The produced fibrin gel showed a homogenous distribution of cells [48]. Robinson et al. set up a mould, in which the fibrin matrix compacted in a determined direction. In this way, an anisotropic collagen expression similar to a native heart valve's configuration was initiated [114].

However, the mechanical stability of fibrin-based scaffolds is not sufficient to withstand the hemodynamic loads of the high-pressure zones of a human circulation. Flanagan et al. performed a pulsatile preconditioning of fibrin-based heart valve prostheses to face this problem and achieved an improved synthesis of extracellular matrix with a subsequently higher stability of the fibrin structure [115].

After that, they populated a cardiac valve scaffold with fibroblasts, smooth muscle cells, and endothelial cells of an ovine carotid artery for 28 days. They implanted this preconditioned, autologous scaffold into pulmonary valve position. Three months following implantation, the scaffolds showed a biological tissue-like consistency as well as a functional endothelium, whereas no fibrin could be detected in the grafts anymore. However, the remodelling led to a contraction of the leaflets with a subsequent valve insufficiency [50].

Keeping up the structural integrity of fibrin-based scaffolds was of lesser importance in a study by Chi et al. They seeded a fibrin–hyaluronic acid matrix with bone marrow mesenchymal stem cells and used the patches in a rat infarction model [116]. However, the patch was not used as a prosthesis, but as a vehicle to deposit stem cells in the infarction area.

For the production of fibrin-based vessel prostheses, Aper et al. condensed blood-derived fibrin by centrifugation [117]. The resulting cross-linking of the fibrin fibres led to a higher mechanical stability of the vessel segments up to a pressure of 230 mmHg. Six months after the replacement of carotid arteries of sheep, the authors reported on an almost complete physiologic remodelling of the grafts.

3.8. Hyaluronic acid

Hyaluronic acid is an unsulphated glycosaminoglycan, whose repetitive disaccharide units consist of d-glucuronic acid and n-acetylglucosamine. Hyaluronic acid occurs in mammals mainly as an extracellular part of the connective tissue, has a water-binding as well as a texture-priming function [118], and is to be found in large extent in heart valve tissue (up to 50% of the glycosaminoglycan content) [104]. Hyaluronic acid is also an essential component of the embryonic development of the heart [119]. Furthermore, the immunogenic potential of commercially available hyaluronic acid (mainly made from bacterial fermentation) is low due to the cross-species structural homology [120]. Finally, the presence of hyaluronic acid promotes the expression of extracellular matrix proteins: VIC cultivated on a hyaluronic acid coated surface produced significantly higher amounts of ECM proteins than VIC on uncoated polystyrene [49].

However, the weak mechanical properties of hyaluronic acid and the high *in vivo* degradation rate advise against its utilisation in tissue engineering of cardiovascular scaffolds [118]. The process of degradation is based on the enzyme hyaluronidase, and its half-life period lasts from several hours up to a few days as a function of the local enzyme concentration.

To manufacture a hydrogel employable as a scaffold for cardiovascular prostheses, cross-linking of hyaluronic acid is possible. For this purpose, methacrylation is an often used tool [121]. The resulting methyl acrylate hyaluronic acid is photo–cross-linkable, whereby the mild conditions of a photopolymerisation allow for the encapsulation of cells into the manufactured hydrogel [119]. Hyaluronan benzyl ester can be processed in manifold ways to serve as an appropriate substrate for tissue engineering [122]. By seeding neonatal rat cardiomyocytes on knitted hyaluronan benzyl ester, Boublik et al. produced hybrid myocardial prostheses with sufficient mechanical properties for an in vivo implantation in a rat model [123].

4. Conclusions

Currently, no optimal polymer for manufacturing an ideal cardiovascular prosthesis has been identified yet. There are rather specific requirements, such as functionality, biocompatibility, and regenerative potential to be taken into account while selecting a substrate for a particular application, implantation site, and host species. Beyond promising synthetic and biologic elastomers, their combination with natural, cell-free matrices can be employed to develop cardiovascular scaffolds as well. However, the huge variety of the currently available materials, coatings, and manufacturing processes, the differences of investigation techniques in vitro and in vivo, as well as the inconsistent selection of evaluation criteria and test parameters impede the comparability of the currently conducted investigations. Consistent analytical standards would facilitate a clear, valid, and comparable perspective on polymer research and thereby would increasingly lead to a faster translation of this important field of research into clinical reality.

The current overview of developments in the investigation of cardiovascular implants warrants the hope for innovative grafts, which are manufactured according to the principles of tissue engineering and which can be used clinically to make sustainable and regenerative therapies available for all patients in the future.

Acknowledgements

We are grateful to Anna Junge for providing excellent photographs of the cardiac valve prostheses and to Sven Gitte for significantly contributing to the literature research necessary for this chapter. We also thank the German Ministry Federal Ministry of Education and Research for funding parts of our polymer-based scaffolds project within the RESPONSE research collaboration.

Author details

Marcel Ricklefs*, Sotiris Korossis, Axel Haverich and Tobias Schilling

*Address all correspondence to: ricklefs.marcel@mh-hannover.de

Department of Cardiac-, Thoracic-, Transplantation and Vascular Surgery, Hannover Medical School, Hannover, Germany

References

[1] World Health Organization. Cardiovascular Diseases (CVDs). World Health Organization, Geneva, Switzerland; 2017 [updated May 2017; cited 2017 27.09.2017]. Available from: http://www.who.int/mediacentre/factsheets/fs317/en/

[2] Zilla P, Fullard L, Trescony P, Meinhart J, Bezuidenhout D, Gorlitzer M, et al. Glutaraldehyde detoxification of aortic wall tissue: a promising perspective for emerging bioprosthetic valve concepts. The Journal of Heart Valve Disease. 1997;**6**(5):510-520

[3] Vincentelli A, Latremouille C, Zegdi R, Shen M, Lajos PS, Chachques JC, et al. Does glutaraldehyde induce calcification of bioprosthetic tissues? The Annals of Thoracic Surgery. 1998;**66**(6 Suppl):S255-S2S8

[4] Liao K, Frater RW, LaPietra A, Ciuffo G, Ilardi CF, Seifter E. Time-dependent effect of glutaraldehyde on the tendency to calcify of both autografts and xenografts. The Annals of Thoracic Surgery. 1995;**60**(2 Suppl):S343-S3S7

[5] Angell WW, Angell JD. Porcine valves. Progress in Cardiovascular Diseases. 1980;**23**(2): 141-166

[6] Aper T, Haverich A, Teebken O. New developments in tissue engineering of vascular prosthetic grafts. VASA. 2009;**38**(2):99-122

[7] Carrel T, Englberger L, Schmidli J. How to treat aortic graft infection? With a special emphasis on xeno-pericardial aortic tube grafts. General Thoracic and Cardiovascular Surgery. 2017. Epub ahead of print

[8] Kogel H. Postoperative Komplikationen in der Gefasschirurgie. In: Hepp W, Kogel H, editors. Gefaesschirurgie. Muenchen: Urban&Fischer Verlag; 2001. pp. 645-703

[9] Meyer BJ, Beer JH. Prevention of thrombosis in heart valve diseases. Therapeutische Umschau. 1998;**55**(12):756-761

[10] Ward RP, Sugeng L, Weinert L, Korcarz C, Verdino RJ, Spencer KT, et al. Images in cardiovascular medicine. Hemolysis after mitral valve repair. Circulation. 2000;**101**(6): 695-696

[11] Cannegieter SC, Rosendaal FR, Briet E. Thromboembolic and bleeding complications in patients with mechanical heart valve prostheses. Circulation. 1994;**89**(2):635-641

[12] Pai GP, Ellison RG, Rubin JW, Moore HV, Kamath MV. Disc immobilization of Bjork-Shiley and Medtronic Hall valves during and immediately after valve replacement. The Annals of Thoracic Surgery. 1987;**44**(1):73-76

[13] Moreno R, Dobarro D, Lopez de Sa E, Prieto M, Morales C, Calvo Orbe L, et al. Cause of complete atrioventricular block after percutaneous aortic valve implantation: Insights from a necropsy study. Circulation. 2009;**120**(5):e29-e30

[14] Armiger LC. Postimplantation leaflet cellularity of valve allografts: are donor cells beneficial or detrimental? The Annals of Thoracic Surgery. 1998;**66**(6 Suppl):S233-S2S5

[15] O'Brien FJ. Biomaterials & scaffolds for tissue engineering. Materials Today. 2011;**14**(3): 88-95

[16] Laurencin CT, Nair LS. Nanotechnology and tissue engineering: The scaffold. Boca Raton: CRC Press; 2008. xvii, 359 p

[17] Bell E. Tissue engineering in perspective. In: Lanza RP, Langer R, Vacanti JP, editors. Principles of Tissue Engineering. 2nd. San Diego, London: Academic Press; 2000. p. XXXV-XIi

[18] Vacanti JP, Langer R. Tissue engineering: the design and fabrication of living replacement devices for surgical reconstruction and transplantation. Lancet. 1999;**354**(Suppl 1): SI32-SSI4

[19] Cheung DY, Duan B, Butcher JT. Current progress in tissue engineering of heart valves: Multiscale problems, multiscale solutions. Expert Opinion on Biological Therapy. 2015;**15**(8):1155-1172

[20] Mendelson K, Schoen FJ. Heart valve tissue engineering: Concepts, approaches, progress, and challenges. Annals of Biomedical Engineering. 2006;**34**(12):1799-1819

[21] Sakata N, Kawamura K, Takebayashi S. Effects of collagen matrix on proliferation and differentiation of vascular smooth muscle cells in vitro. Experimental and Molecular Pathology. 1990;**52**(2):179-191

[22] Meredith JE Jr, Fazeli B, Schwartz MA. The extracellular matrix as a cell survival factor. Molecular Biology of the Cell. 1993;**4**(9):953-961

[23] Ingber D. Extracellular matrix and cell shape: Potential control points for inhibition of angiogenesis. Journal of Cellular Biochemistry. 1991;**47**(3):236-241

[24] Hubbell JA. Matrix effects. In: Lanza RP, Langer R, Vacanti JP, editors. Principles of Tissue Engineering. 2nd ed. San Diego, London: Academic Press; 2000. pp. 237-250

[25] Hoerstrup SP, Zund G, Ye Q, Schoeberlein A, Schmid AC, Turina MI. Tissue engineering of a bioprosthetic heart valve: Stimulation of extracellular matrix assessed by hydroxyproline assay. ASAIO Journal. 1999;**45**(5):397-402

[26] Lalka SG, Oelker LM, Malone JM, Duhamel RC, Kevorkian MA, Raper BA, et al. Acellular vascular matrix: A natural endothelial cell substrate. Annals of Vascular Surgery. 1989;**3**(2):108-117

[27] Sant S, Iyer D, Gaharwar AK, Patel A, Khademhosseini A. Effect of biodegradation and de novo matrix synthesis on the mechanical properties of valvular interstitial cell-seeded polyglycerol sebacate-polycaprolactone scaffolds. Acta Biomaterialia. 2013;**9**(4):5963-5973

[28] Blusch JH, Patience C, Martin U. Pig endogenous retroviruses and xenotransplantation. Xenotransplantation. 2002;**9**(4):242-251

[29] Bruckenberger E. Deutscher Herzbericht 2016. Hannover: Bruckenberger E; 2016

[30] Maganti K, Rigolin VH, Sarano ME, Bonow RO. Valvular heart disease: Diagnosis and management. Mayo Clinic Proceedings. 2010;**85**(5):483-500

[31] Salem K, ElKhateeb O. Gender-adjusted and age-adjusted economic inpatient burden of congestive heart failure: Cost and disability-adjusted life-year analysis. ESC Heart Failure. 2017;**4**(3):259-265

[32] Haverich A. A surgeon's view on the pathogenesis of atherosclerosis. Circulation. 2017;**135**(3):205-207

[33] Kannel WB. Incidence and epidemiology of heart failure. Heart Failure Reviews. 2000;**5**(2):167-173

[34] Bundesamt S. Gesundheit—Todesursachen in Deutschland 2011. Fachserie 12 Reihe 42012

[35] Bundesamt S. Krankheitskostenrechnung: Krankheitskosten: Deutschland, Jahre, Krankheitsdiagnosen (ICD10); 2009. [Available from: https://www-genesis.destatis.de/genesis/online/logon?language=de&sequenz=tabellen&selectionname=23631*

[36] Dor V, Saab M, Coste P, Kornaszewska M, Montiglio F. Left ventricular aneurysm: A new surgical approach. The Thoracic and Cardiovascular Surgeon. 1989;**37**(1):11-19

[37] Braunwald NS, Cooper T, Morrow AG. Complete replacement of the mitral valve. Successful clinical application of a flexible polyurethane prosthesis. The Journal of Thoracic and Cardiovascular Surgery. 1960;**40**:1-11

[38] Roe BB, Moore D. Design and fabrication of prosthetic valves. Experimental Medicine and Surgery 1958;**16**(2-3):177-82

[39] Akutsu T, Dreyer B, Kolff WJ. Polyurethane artificial heart valves in animals. Journal of Applied Physiology. 1959;**14**:1045-1048

[40] Roe BB. Late follow-up studies on flexible leaflet prosthetic valves. The Journal of Thoracic and Cardiovascular Surgery. 1969;**58**(1):59-61

[41] Jansen J, Reul H. A synthetic three-leaflet valve. Journal of Medical Engineering & Technology. 1992;**16**(1):27-33

[42] Roe BB, Kelly PB Jr, Myers JL, Moore DW. Tricuspid leaflet aortic valve prosthesis. Circulation. 1966;**33**(4 Suppl):I124-I130

[43] Jansen J, Willeke S, Reiners B, Harbott P, Reul H, Rau G. New J-3 flexible-leaflet polyurethane heart valve prosthesis with improved hydrodynamic performance. The International Journal of Artificial Organs. 1991;**14**(10):655-660

[44] Puperi DS, Balaoing LR, O'Connell RW, West JL, Grande-Allen KJ. 3-Dimensional spatially organized PEG-based hydrogels for an aortic valve co-culture model. Biomaterials. 2015;**67**:354-364

[45] Rahmani B, Tzamtzis S, Ghanbari H, Burriesci G, Seifalian AM. Manufacturing and hydrodynamic assessment of a novel aortic valve made of a new nanocomposite polymer. Journal of Biomechanics. 2012;**45**(7):1205-1211

[46] Ando M, Takahashi Y. Ten-year experience with handmade trileaflet polytetrafluoroethylene valved conduit used for pulmonary reconstruction. The Journal of Thoracic and Cardiovascular Surgery. 2009;**137**(1):124-131

[47] Lee CH, Singla A, Lee Y. Biomedical applications of collagen. International Journal of Pharmaceutics 2001;**221**(1-2):1-22

[48] Jockenhoevel S, Zünd G, Hoerstrup SP, Chalabi K, Sachweh JS, Demircan L, et al. Fibrin gel—Advantages of a new scaffold in cardiovascular tissue engineering. European Journal of Cardio-Thoracic Surgery. 2001;**19**(4):424-430

[49] Masters KS, Shah DN, Walker G, Leinwand LA, Anseth KS. Designing scaffolds for valvular interstitial cells: Cell adhesion and function on naturally derived materials. Journal of Biomedical Materials Research Part A. 2004;**71**(1):172-180

[50] Flanagan TC, Sachweh JS, Frese J, Schnoring H, Gronloh N, Koch S, et al. In vivo remodeling and structural characterization of fibrin-based tissue-engineered heart valves in the adult sheep model. Tissue Engineering. Part A. 2009;**15**(10):2965-2976

[51] Tedder ME, Liao J, Weed B, Stabler C, Zhang H, Simionescu A, et al. Stabilized collagen scaffolds for heart valve tissue engineering. Tissue Engineering. Part A. 2009;**15**(6):1257-1268

[52] Nazir R. Collagen—Hyaluronic acid based interpenetrating polymer networks as tissue engineered heart valve. Materials Science and Technology-London. 2016;**32**(9):871-882

[53] Duan B, Kapetanovic E, Hockaday LA, Butcher JT. Three-dimensional printed trileaflet valve conduits using biological hydrogels and human valve interstitial cells. Acta Biomaterialia. 2014;**10**(5):1836-1846

[54] Khosravi R, Best CA, Allen RA, Stowell CE, Onwuka E, Zhuang JJ, et al. Long-term functional efficacy of a novel electrospun poly(glycerol sebacate)-based arterial graft in mice. Annals of Biomedical Engineering 2016;**44**(8):2402-16

[55] McCarthy CW, Ahrens DC, Joda D, Curtis TE, Bowen PK, Guillory RJ 2nd, et al. Fabrication and short-term in vivo performance of a natural elastic lamina-polymeric hybrid vascular graft. ACS Applied Materials & Interfaces. 2015;**7**(30):16202-16212

[56] Stamm C, Khosravi A, Grabow N, Schmohl K, Treckmann N, Drechsel A, et al. Biomatrix/polymer composite material for heart valve tissue engineering. The Annals of Thoracic Surgery. 2004;**78**(6):2084-2092. discussion 92-3

[57] FONG P, Shin'oka T, Lopez-Soler RI, Breuer C. The use of polymer based scaffolds in tissue-engineered heart valves. Progress in Pediatric Cardiology. 2006;**21**(2):193-199

[58] Sewell-Loftin MK, Chun YW, Khademhosseini A, Merryman WD. EMT-inducing biomaterials for heart valve engineering: Taking cues from developmental biology. Journal of Cardiovascular Translational Research. 2011;**4**(5):658-671

[59] Gunatillake PA, Adhikari R. Biodegradable synthetic polymers for tissue engineering. European Cells & Materials. 2003;**5**(16):1-16

[60] Ravi S, Chaikof EL. Biomaterials for vascular tissue engineering. Regenerative Medicine. 2010;**5**(1):107-120

[61] Reddy CS, Venugopal JR, Ramakrishna S, Zussman E. Polycaprolactone/oligomer compound scaffolds for cardiac tissue engineering. Journal of Biomedical Materials Research. Part A. 2014;**102**(10):3713-3725

[62] Li WJ, Cooper JA, Jr., Mauck RL, Tuan RS. Fabrication and characterization of six electrospun poly(alpha-hydroxy ester)-based fibrous scaffolds for tissue engineering applications. Acta Biomaterialia 2006;**2**(4):377-85

[63] Shinoka T, Breuer CK, Tanel RE, Zünd G, Miura T, Ma PX, et al. Tissue engineering heart valves: Valve leaflet replacement study in a lamb model. The Annals of Thoracic Surgery. 1995;**60**:S513-S5S6

[64] Sodian R, Hoerstrup SP, Sperling JS, Martin DP, Daebritz S, Mayer JE, et al. Evaluation of biodegradable, three-dimensional matrices for tissue engineering of heart valves. ASAIO Journal. 2000;**46**(1):107-110

[65] Hoerstrup SP, Sodian R, Daebritz S, Wang J, Bacha EA, Martin DP, et al. Functional living trileaflet heart valves grown in vitro. Circulation. 2000;**102**(Suppl. 3):III-44-III-9

[66] Jana S, Tefft BJ, Spoon DB, Simari RD. Scaffolds for tissue engineering of cardiac valves. Acta Biomaterialia. 2014;**10**(7):2877-2893

[67] Ying TH, Ishii D, Mahara A, Murakami S, Yamaoka T, Sudesh K, et al. Scaffolds from electrospun polyhydroxyalkanoate copolymers: Fabrication, characterization, bioabsorption and tissue response. Biomaterials. 2008;**29**(10):1307-1317

[68] Wu Q, Wang Y, Chen G-Q. Medical application of microbial biopolyesters polyhydroxyalkanoates. Artificial Cells, Blood Substitutes, and Immobilization Biotechnology. 2009;**37**(1):1-12

[69] Sodian R, Sperling JS, Martin DP, Egozy A, Stock U, Mayer JE, et al. Fabrication of a trileaflet heart valve scaffold from a polyhydroxyalkanoate biopolyester for use in tissue engineering. Tissue Engineering. 2000;**6**(2):183-188

[70] Valappil SP, Misra SK, Boccaccini AR, Roy I. Biomedical applications of polyhydroxyalkanoates: An overview of animal testing and in vivo responses. Expert Review of Medical Devices. 2006;**3**(6):853-868

[71] Williams SF, Martin DP. Applications of polyhydroxyalkanoates (PHA) in medicine and pharmacy. In: Doi Y, Steinbüchel A, editors. Biopolymers Online. Weinheim, Germany: Wiley-VCH Verlag GmbH & Co. KGaA; 2005

[72] Morsi YS. Bioengineering strategies for polymeric scaffold for tissue engineering an aortic heart valve: An update. The International Journal of Artificial Organs. 2014;**37**(9):651-667

[73] Chen G-Q, Wu Q. The application of polyhydroxyalkanoates as tissue engineering materials. Biomaterials. 2005;**26**(33):6565-6578

[74] Sodian R, Hoerstrup SP, Sperling JS, Daebritz S, Martin DP, Moran AM, et al. Early in vivo experience with tissue-engineered trileaflet heart valves. Circulation. 2000;**102**(Suppl. 3): III-22-III-9

[75] Silvestri A, Sartori S, Boffito M, Mattu C, Di Rienzo AM, Boccafoschi F, et al. Biomimetic myocardial patches fabricated with poly(varepsilon-caprolactone) and polyethylene glycol-based polyurethanes. Journal of Biomedical Materials Research. Part B, Applied Biomaterials 2014;**102**(5):1002-13

[76] Herrmann FE, Lehner A, Hollweck T, Haas U, Fano C, Fehrenbach D, et al. In vitro biological and mechanical evaluation of various scaffold materials for myocardial tissue engineering. Journal of Biomedical Materials Research. Part A. 2014;**102**(4):958-966

[77] Blumenthal B, Golsong P, Poppe A, Heilmann C, Schlensak C, Beyersdorf F, et al. Polyurethane scaffolds seeded with genetically engineered skeletal myoblasts: A promising tool to regenerate myocardial function. Artificial Organs. 2010;**34**(2):E46-E54

[78] Baheiraei N, Yeganeh H, Ai J, Gharibi R, Ebrahimi-Barough S, Azami M, et al. Preparation of a porous conductive scaffold from aniline pentamer-modified polyurethane/PCL blend for cardiac tissue engineering. Journal of Biomedical Materials Research. Part A. 2015;**103**(10):3179-3187

[79] Fromstein JD, Zandstra PW, Alperin C, Rockwood D, Rabolt JF, Woodhouse KA. Seeding bioreactor-produced embryonic stem cell-derived cardiomyocytes on different porous, degradable, polyurethane scaffolds reveals the effect of scaffold architecture on cell morphology. Tissue Engineering. Part A. 2008;**14**(3):369-378

[80] Nieponice A, Soletti L, Guan J, Hong Y, Gharaibeh B, Maul TM, et al. In vivo assessment of a tissue-engineered vascular graft combining a biodegradable elastomeric scaffold and muscle-derived stem cells in a rat model. Tissue Engineering. Part A. 2010;**16**(4):1215-1223

[81] Punnakitikashem P, Truong D, Menon JU, Nguyen KT, Hong Y. Electrospun biodegradable elastic polyurethane scaffolds with dipyridamole release for small diameter vascular grafts. Acta Biomaterialia 2014;**10**(11):4618-28

[82] Sales VL, Engelmayr GC, Johnson JA, Gao J, Wang Y, Sacks MS, et al. Protein precoating of elastomeric tissue-engineering scaffolds increased cellularity, enhanced extracellular matrix protein production, and differentially regulated the phenotypes of circulating endothelial progenitor cells. Circulation 2007;**116**(11 Suppl):I55-63.

[83] Pomerantseva I, Krebs N, Hart A, Neville CM, Huang AY, Sundback CA. Degradation behavior of poly(glycerol sebacate). Journal of Biomedical Materials Research Part A. 2009;**91**(4):1038-1047

[84] Rai R, Tallawi M, Grigore A, Boccaccini AR. Synthesis, properties and biomedical applications of poly(glycerol sebacate) (PGS): A review. Progress in Polymer Science. 2012;**37**(8):1051-1078

[85] Chen QZ, Bismarck A, Hansen U, Junaid S, Tran MQ, Harding SE, et al. Characterisation of a soft elastomer poly(glycerol sebacate) designed to match the mechanical properties of myocardial tissue. Biomaterials 2008;**29**(1):47-57

[86] Xu B, Li Y, Fang X, Thouas GA, Cook WD, Newgreen DF, et al. Mechanically tissue-like elastomeric polymers and their potential as a vehicle to deliver functional cardiomyocytes. Journal of the Mechanical Behavior of Biomedical Materials 2013;**28**:354-65

[87] Sant S, Hwang CM, Lee S-H, Khademhosseini A. Hybrid PGS-PCL microfibrous scaffolds with improved mechanical and biological properties. Journal of Tissue Engineering and Regenerative Medicine. 2011;**5**(4):283-291

[88] Masoumi N, Johnson KL, Howell MC, Engelmayr GC. Valvular interstitial cell seeded poly(glycerol sebacate) scaffolds: Toward a biomimetic in vitro model for heart valve tissue engineering. Acta Biomaterialia. 2013;**9**(4):5974-5988

[89] Jeffries EM, Allen RA, Gao J, Pesce M, Wang Y. Highly elastic and suturable electrospun poly(glycerol sebacate) fibrous scaffolds. Acta Biomaterialia. 2015;**18**:30-39

[90] Chen QZ, Ishii H, Thouas GA, Lyon AR, Wright JS, Blaker JJ, et al. An elastomeric patch derived from poly(glycerol sebacate) for delivery of embryonic stem cells to the heart. Biomaterials. 2010;**31**(14):3885-3893

[91] Radisic M, Park H, Chen F, Salazar-Lazzaro JE, Wang Y, Dennis R, et al. Biomimetic approach to cardiac tissue engineering: Oxygen carriers and channeled scaffolds. Tissue Engineering. 2006;**12**(8):2077-2091

[92] Rai R, Tallawi M, Barbani N, Frati C, Madeddu D, Cavalli S, et al. Biomimetic poly(glycerol sebacate) (PGS) membranes for cardiac patch application. Materials Science & Engineering. C, Materials for Biological Applications. 2013;**33**(7):3677-3687

[93] Motlagh D, Yang J, Lui KY, Webb AR, Ameer GA. Hemocompatibility evaluation of poly(glycerol-sebacate) in vitro for vascular tissue engineering. Biomaterials. 2006;**27**: 4315-4324, 4324

[94] Guler S, Hosseinian P, Aydin HM. Hybrid aorta constructs via in situ crosslinking of poly(glycerol-sebacate) elastomer within a decellularized matrix. Tissue Engineering. Part C, Methods 2017;**23**(1):21-9

[95] Masoumi N, Annabi N, Assmann A, Larson BL, Hjortnaes J, Alemdar N, et al. Trilayered elastomeric scaffolds for engineering heart valve leaflets. Biomaterials. 2014; **35**(27):7774-7785

[96] Stuckey DJ, Ishii H, Chen QZ, Boccaccini AR, Hansen U, Carr CA, et al. Magnetic resonance imaging evaluation of remodeling by cardiac elastomeric tissue scaffold biomaterials in a rat model of myocardial infarction. Tissue Engineering. Part A. 2010;**16**(11):3395-3402

[97] Wu W, Allen RA, Wang Y. Fast-degrading elastomer enables rapid remodeling of a cell-free synthetic graft into a neoartery. Nature Medicine 2012;**18**(7):1148-53

[98] Drury JL, Mooney DJ. Hydrogels for tissue engineering: scaffold design variables and applications. Biomaterials. 2003;**24**(24):4337-4351

[99] Ma PX. Scaffolds for tissue fabrication. Materials Today. 2004;**7**(5):30-40

[100] Benton JA, Fairbanks BD, Anseth KS. Characterization of valvular interstitial cell function in three dimensional matrix metalloproteinase degradable PEG hydrogels. Biomaterials. 2009;**30**(34):6593-6603

[101] Shah DN, Recktenwall-Work SM, Anseth KS. The effect of bioactive hydrogels on the secretion of extracellular matrix molecules by valvular interstitial cells. Biomaterials. 2008;**29**(13):2060-2072

[102] Hockaday LA, Kang KH, Colangelo NW, Cheung PYC, Duan B, Malone E, et al. Rapid 3D printing of anatomically accurate and mechanically heterogeneous aortic valve hydrogel scaffolds. Biofabrication. 2012;**4**(3):035005

[103] Zhang X, Xu B, Puperi DS, Yonezawa AL, Wu Y, Tseng H, et al. Integrating valve-inspired design features into poly(ethylene glycol) hydrogel scaffolds for heart valve tissue engineering. Acta Biomaterialia. 2015;**14**:11-21

[104] Zhang X, Xu B, Puperi DS, Wu Y, West JL, Grande-Allen KJ. Application of hydrogels in heart valve tissue engineering. Journal of Long-Term Effects of Medical Implants 2015;**25**(1-2):105-134.

[105] Gentleman E, Lay AN, Dickerson DA, Nauman EA, Livesay GA, Dee KC. Mechanical characterization of collagen fibers and scaffolds for tissue engineering. Biomaterials. 2003;**24**(21):3805-3813

[106] Patel H, Bonde B, Srinivasan G. Biodegradable polymer scaffold for tissue engineering. Trends in Biomaterials and Artificial Organs. 2011;**25**(1):20-29

[107] Fivola E, Straka F, Mirejovsky T, Masin J, Bacakova L. Tissue-engineered heart valves. Physiological Research. 2009;**58**(Suppl. 2):141-158

[108] Taylor PM, Sachlos E, Dreger SA, Chester AH, Czernuszka JT, Yacoub MH. Interaction of human valve interstitial cells with collagen matrices manufactured using rapid prototyping. Biomaterials. 2006;**27**(13):2733-2737

[109] Chevallay B, Herbage D. Collagen-based biomaterials as 3D scaffold for cell cultures: Applications for tissue engineering and gene therapy. Medical & Biological Engineering & Computing. 2000;**38**(2):211-218

[110] Parenteau-Bareil R, Gauvin R, Berthod F. Collagen-based biomaterials for tissue engineering applications. Materials. 2010;**3**(3):1863-1887

[111] Flanagan TC, Wilkins B, Black A, Jockenhoevel S, Smith TJ, Pandit AS. A collagen-glycosaminoglycan co-culture model for heart valve tissue engineering applications. Biomaterials. 2006;**27**(10):2233-2246

[112] Barsotti MC, Felice F, Balbarini A, Di Stefano R. Fibrin as a scaffold for cardiac tissue engineering. Biotechnology and Applied Biochemistry. 2011;**58**(5):301-310

[113] Ye Q, Zünd G, Benedikt P, Jockenhoevel S, Hoerstrup SP, Sakyama S, et al. Fibrin gel as a three dimensional matrix in cardiovascular tissue engineering. European Journal of Cardio-Thoracic Surgery. 2000;**17**(5):587-591

[114] Robinson PS, Johnson SL, Evans MC, Barocas VH, Tranquillo RT. Functional tissue-engineered valves from cell-remodeled fibrin with commissural alignment of cell-produced collagen. Tissue Engineering Parts A. 2008;**14**(1):83-95

[115] Flanagan TC, Cornelissen C, Koch S, Tschoeke B, Sachweh JS, Schmitz-Rode T, et al. The in vitro development of autologous fibrin-based tissue-engineered heart valves through optimised dynamic conditioning. Biomaterials. 2007;**28**(23):3388-3397

[116] Chi NH, Yang MC, Chung TW, Chen JY, Chou NK, Wang SS. Cardiac repair achieved by bone marrow mesenchymal stem cells/silk fibroin/hyaluronic acid patches in a rat of myocardial infarction model. Biomaterials 2012;**33**(22):5541-51

[117] Aper T, Wilhelmi M, Gebhardt C, Hoeffler K, Benecke N, Hilfiker A, et al. Novel method for the generation of tissue-engineered vascular grafts based on a highly compacted fibrin matrix. Acta Biomaterialia 2016;**29**:21-32

[118] Xu X, Jha AK, Harrington DA, Farach-Carson MC, Jia X. Hyaluronic acid-based hydrogels: From a natural polysaccharide to complex networks. Soft Matter. 2012; **8**(12):3280-3294

[119] Masters KS, Shah DN, Leinwand LA, Anseth KS. Crosslinked hyaluronan scaffolds as a biologically active carrier for valvular interstitial cells. Biomaterials. 2005;**26**(15): 2517-2525

[120] Ramamurthi A, Vesely I. Evaluation of the matrix-synthesis potential of cross-linked hyaluronan gels for tissue engineering of aortic heart valves. Biomaterials. 2005;**26**(9):999-1010

[121] Burdick JA, Prestwich GD. Hyaluronic acid hydrogels for biomedical applications. Advanced Materials (Deerfield Beach, Fla). 2011;**23**(12):H41-56

[122] Vindigni V, Cortivo R, Iacobellis L, Abatangelo G, Zavan B. Hyaluronan benzyl ester as a scaffold for tissue engineering. International Journal of Molecular Sciences. 2009;**10**(7):2972-2985

[123] Boublik J, Park H, Radisic M, Tognana E, Chen F, Pei M, et al. Mechanical properties and remodeling of hybrid cardiac constructs made from heart cells, fibrin, and biodegradable, elastomeric knitted fabric. Tissue Engineering. 2005;**11**(7-8):1122-32

Chapter 13

Bioengineering the Pancreas: Cell-on-Scaffold Technology

Andrea Peloso, Antonio Citro, Graziano Oldani,
Szandra Brambilla, Lorenzo Piemonti and
Lorenzo Cobianchi

Additional information is available at the end of the chapter

http://dx.doi.org/10.5772/intechopen.70990

Abstract

Nowadays, type I diabetes mellitus is a pathology afflicting millions of people globally with a dramatic assessment in the next future. Current treatments including exogenous insulin, pancreas transplantation and islets transplantation, are not free from important lifelong side effects. In the last decade, tissue engineering and regenerative medicine have shown encouraging results about the possibility to produce a functional bioengineered pancreas. Among many technologies, decellularization offers the opportunity to produce an organ-specific acellular matrix that could subsequently repopulate with endocrine cellular population. Herein, we aim to review the state-of-art and this technology highlighting the diabetes burden for the healthcare system and the major achievements toward the manufacturing of a bioengineered pancreas obtained by cell-on-scaffold technology.

Keywords: regenerative medicine, pancreas, extracellular matrix, scaffold, bioengineered pancreas

1. Introduction: definition of diabetes and diabetes' impact on the economy and future projections

In 2014, the World Health Organization (WHO) has estimated diabetes as a worldwide disease afflicting 422 million people with an increasing, globally, heavy burden both for the health-care system and the economic policies [1]. As stated in 2014 by National Diabetes Statistics Report [2] T1D, the so-called diabetes mellitus, is a chronic metabolic disorder, which is afflicting around 9.3% of worldwide adult population widely counting 29.1 million people worldwide. Moreover this data could even be worse if considering prediabetes or borderline diabetes global prevalence of around 7.8% (343 M) [3]. In this regards, according to ADA expert panel, in 20 years, without any kind of therapeutic intervention, it is reasonable to estimate that up to 70% of prediabetic individuals will develop a "real" diabetic status [4], whereas, individuals with lifestyle modifications and drug intervention, normoglycemic conversion will take place ranging from 35 to 50% of cases [5] in a 10-years follow-up cohort. Starting from these data, the prospective future scenario is dramatic.

In the 7th Edition of the Diabetes Atlas, the International Diabetes Federation (IDF) has recently estimated that by 2040, 1 out 10 adult individuals will suffer of diabetes (actually the estimation is set to 1 out 11 adults) with a global picture of more than 640 M of people (raised from actual 415 M) [5]. This scenario will also have huge impact on economic health-care policies. Nowadays, the direct cost of diabetes hits 825 billion dollars a year [6] to whom should be added all the indirect costs deriving from diabetes consequences.

Diabetes is currently recognized as a group of different metabolic disorders, which led to the same final outcome: the organism incapacity to manage intracorporeal glucose levels.

This inability can be referred to two main physiological alterations and consequently to two principal diabetes categories: type 1 and type 2 diabetes.

Type 1 diabetes (T1D) (also recognized as autoimmune or juvenile diabetes) is a lifelong chronic autoimmune disease characterized by insulin deficiency due to pancreatic β cells loss and, therefore to a subsequent hyperglycemia status [7].

Type 2 diabetes (T2D) (formerly called non-insulin-dependent or adult-onset) represents the most common type of diabetes grossly counting 90% of all cases worldwide. T2D pathophysiology is related to genetic and epigenetic factors, environmental conditions and lifestyle (obesity, physical activity and diet) that bring to a mixture of hyper-insulinemia, insulin resistance and pancreatic β cells failure [8, 9].

If insulin resistance cannot be treated by bioengineering approach, insulin deficiency, typical of T1D, could surely benefit bioengineered pancreas.

T1D is a chronic autoimmune metabolic disease characterized by hyperglycemia, which is secondary to insulin deficiency that develops as consequence of the numerical loss of the pancreatic islet β cells. This loss is due in 70–90% of cases by an autoimmunity disorder that brings to a cascade characterized by β cells destruction, dysglycemia and finally hyperglycemia. It represents 5–10% of whole diabetes cases with more than 20 million people worldwide [10]. This disease is also marked by numerous short- or long-term complications. Acute complications include life-threatening crisis ranging from severe hypoglycemic episodes to diabetic ketoacidosis. Long-term complications comprise chronic micro- and macro-vascular diseases including retinopathy, nephropathy and neuropathy due to stroke and ischemic heart attack. This panorama leads, in almost all cases, to lifelong disabilities reducing, at the same time, quality and expectancy of life and involving immense health-care expenditures.

As deeply analyzed by Dall et al. in 2007, at that time, the estimated cost for T1D patient for year was around 3000 pounds (versus around 2000 pounds estimated for T2D patient for year), which includes direct medical costs and indirect costs (such as productivity loss or premature mortality) [11]. These data dramatically changed in recent years accompanying to a huge increase both to the worldwide number of diabetic patients as well as the deriving costs. The World Health Organization [12] has estimated for the 2014 a total of 422 million diabetic people which, compared to 108 millions counted in 1980, have more than triplicated the total global health-care spending for diabetes reaching 827 USD billion/year. [13, 14]. Forecasting models and projections define a situation where diabetes will be pandemic. Recently Krohe et al. has published their future projection for 2030 indicating an increase of the American prevalence (type 1 and type 2 diabetes) of around 54% with a 54.9 M people afflicted by the disease and a total cost increase of 53%, 622 billion USD just for the US population [15].

2. Historical development and current therapeutic options for diabetes' treatment

2.1. Insulin approach

Diabetes was born in 1910 when the English physiologist Sir Edward Albert Sharpey-Schafer discovers a substance normally produced by non-diabetic patients: insulin. The name derives from the Latin word "insula" means island, referring to the islets of Langerhans (www.diabetes.org), is a specific structure that is able to produce insulin in the pancreas. Since then, all the efforts have focused on the possibility to isolate this substance in order to replace for diabetic patients (so far described as simple, non-producing insulin patients) [16]. This need was driven by the fact that, without a substitute, diabetes was a pathologic condition that would surely have led to death. In 1921, Banting and Macleod (Nobel prize in Physiology or Medicine in 1923) succeeded in isolating insulin from a canine diabetic model, literally revolutionizing the entire health world. One year later, with this pioneering finding, the first diabetic patient was treated. Since then, several modifications have been completed, making the insulin replacement more refined, from slower acting insulin (the first form was introduced in 1936) to genetically engineered artificial "human" insulin that was produced in 1978 exploiting *Escherichia coli* bacteria able to avoid frequent allergic reactions derived from the use of cattle or porcine insulin. Nowadays, exogenous insulin administration is considered as an essential therapy for patients affected by T1D, being able to avoid, when accurately managed, acute metabolic compensations.

It is said that insulin is not a "perfect therapy" as it is being accompanied by several side effects, both in short- and long period and finally considering that just 40% of treated patients achieve and maintain a satisfying glucose range [17]. In fact, there are different factors influencing the effectiveness of insulin therapy such as daily stress, the food intake and the physical activity [18] highlighting the vital importance of a baseline good lifestyle management [19]. Insulin therapy for T1D is based on multiple daily subcutaneous injections of insulin trying to follow a patient-tailored scheme created by considering patient carbohydrate assumption, pre-meal glucose levels and anticipated physical activity [20]. Recently, a continuous intra-venous insulin administration has been considered as a simpler way for exogenous insulin administration in order to match insulin requirements. This approach can be effectively considered useful for perioperative period of patients who has undergone surgery [21, 22] but it could not be used as a lifelong option due to the modest differences between the two delivering systems [23] in terms of final outcomes.

Even with some limitations, insulin treatment represents, as of now, the most widespread therapy to face T1D diabetes.

2.2. Pancreas transplantation

Pancreas transplantation has been successfully established for the first time in 1966 by the surgical team driven by Prof. W. Kelly [24]. That ground-breaking success has opened a new chapter in the transplantation field bringing almost 50,000 transplants performed worldwide by 1996 [25, 26]. The worldwide diffusion of this procedure has been strengthened by the improvement of surgical technic but, overall, by the discovery around 1980, of immunosuppressive drugs that prevent the otherwise unavoidable immunological rejection. Cyclosporine was the first immunosuppressive drug tested, which led to very significant improvements for long-term patient survival (91% at 1 year and 84% at 3 years, respectively) [25]. Later tacrolimus and mycophenolate mofetil have been developed [27] and then also T cell depleting agents (Alemtuzumab, OKT3 or Minnesota ALG), until the modern immunosuppressive regimes, have permitted to abandon the initial immunosuppressive drugs (tacrolimus and prednisone) and their side effects (hypertension, hyperlipidemia and nephrotoxicity).

When available, nowadays, pancreas transplantation represents the best treatment to offer T1D patient, yielding higher rates of insulin independence compared to insulin treatment or islet encapsulation (which will be discussed below) [28]. Additionally, the quality of life and the complications, deriving from diabetic status, have more benefits from pancreas transplantation than from insulin therapeutic plan or from islets encapsulation [29, 30].

For the above-explained reasons, pancreas transplantation is known as the only definitive long-term treatment for insulin-dependent patients with survival rates of >95% and graft survival rate at almost 85% at 1 year [31]. The major problem that strongly limits a wider diffusion of pancreas transplantation currently is the lack of viable transplantable pancreas. Recently, the total of pancreas transplants has been reduced. Between 2005 and 2014, the number of pancreas transplants declined by more than 30% as well as the number of candidates on the waiting list (−48.0%) [32] with a total of just 954 pancreas transplants performed in the USA [33]. This is secondary to several factors including above all shortage of a primary referral source, difficulties of acceptance by the diabetes scientific society and developments in diabetes management. The final consequence concerning the effective clinical application is that, in the USA, just 3 out 10.000 T1D patients are treated with pancreas or islets transplantation [17].

2.3. Islets transplantation

Islets cell transplantation is one of the most powerful weapons to treat selected T1D patients. It is based on the transplantation of a single selected cluster of cells (islets cells) that mainly contain β cells responsible for the physiological production of insulin. This procedure is offered only to selected diabetic recipient who did not benefit "standard" insulin treatment with severe hypoglycemic episodes (brittle diabetes) and unstable glycemic profile proposing the whole pancreas transplantation only in case of very poor metabolic control [34, 35]. Islet transplantation has been a solid road to follow just from 1972 via islet isograft performed by Lacy and Ballinger who transplanted pancreatic islet in streptozocin-induced diabetic rats [36]. These revolutionary data have shown how a cell therapy could be effective in the treatment of diabetes even if the translation to the large animal and, finally, to the clinic was still far and full of hurdle to overcome. During the evolution of the whole purification process, Ricordi has furnished a significant improvement [37]. He built an automated chamber to optimize and standardize the isolation procedure thus becoming an indispensable tool both for animal and human models [38] until becoming an internationally accepted procedure for selected candidates. As is done in whole organ pancreas transplantation, the immunosuppressive drugs evolution has been crucial also for the development of islet transplantation. Till date, immunosuppressive regimes are mandatory after islet transplantation due to the source of islets, which need one or more multiple deceased donors in order to be numerically sufficient (it has been calculated that 265,000 islets can be grossly sufficient to achieve and maintain an insulin-independence status [39] up to 7½ years after transplantation). Later, the desired number of transplantable islets cells has been refined and tailored on recipient features, up to 5000–10,000 islets equivalents (IEQ) per kg/body mass that is actually recommended as the minimal β cells mass [40] to transplant. The necessity to obtain this number of viable islets to transplant faces the organ shortage for whole pancreas transplantation. The clinical islet use follows rigid pre-transplant preparation rules, which depend on multi-step approach with the final goal to extract, from the whole pancreas, only the islet fraction (that represents just 1–2% of the total cellular volume).

The entire process is based on enzymatic digestion, controlled mechanical shear and final purification until to enrich a satisfying amount of pure pancreatic islets ready to be injected into the recipient [41]. Even if much less invasive compared to pancreas transplantation (mainly because of its intravascular portal approach), islets transplantation is not free from limitations [42]. Some acute effects comprehend bleeding [43], portal vein thrombosis [44] or a transient increase of hepatic inflammatory markers [45]. All these issues most of the times are transient and much more important long-term side effects related to immunosuppressive regimes (sharing in this way the immunological problems of pancreas transplant).

3. Biomaterials

Biomaterials science is an interdisciplinary field focused on the physical and biological activity of materials and, furthermore, on their interactions if used in a biological environment.

Conventionally, the most intense development and investigation have been oriented toward biomaterials synthesis, optimization, characterization, testing and the biology of host-material relations [46]. The global health care has enormously benefited this field through the creation of heart valve prostheses, artificial hip joints, dental implants, intraocular lenses and many other dispositive, becoming the fundamental cornerstones of many modern therapies.

As reported by the first Consensus of the European Society for Biomaterials in 1976, a biomaterial can be defined as "a nonviable material used in a medical device, intended to interact with biological system" but, alongside the evolution of the field, this definition has been evolved into "a material intended to interface with biological system to evaluate, treat, augment or replace any tissue, organ or function of the body" [46].

Typically we can recapitulate three important big types of biomaterials:

- Ceramics
- Synthetic polymers
- Natural polymers

The correct choice depends on which physiologic function is aimed to augment/replace. For example, in case of bone part replacement, a ceramic scaffold (with major strength and higher mechanical stiffness) will be preferred, otherwise, for cellular therapies, synthetic/polymers will be chosen, favored by their major sustainability in cell behavior.

Scaffolds can be considered as the missing link between biomaterial science and the tissue engineering approach and are defined as biomaterial-based tridimensional structures that should be able to support cellular viability.

According to the biomaterial from which they are manufactured, scaffolds can be divided as

- Ceramics scaffolds
- Synthetic scaffolds
- Natural scaffolds

Regardless of the material, a scaffold must have precise characteristics and important requirements. Biocompatibility is the first mandatory characteristic for every scaffold intended to be used in a biological environment. Generally defined as the capacity of a material to be in contact with a living tissue or integrated in a living environment by not being toxic, injurious or physiologically reactive and not causing immunological rejection, biocompatibility's definition has been modified into "the ability of materials to locally trigger and guide wound healing, reconstruction and tissue integration" [47].

This new definition moves the focus from an outlook where biomaterials must be simply inherent and not causing damages, to a new perspective where they have an active role being a dynamic element. Practically, this is exploited by the capacity of biomaterials to receive cells, allow their attachment and guarantee their proliferation and differentiation at the same time.

Obviously, after implantation the scaffold must be well integrated and not being the target for immunologic reaction.

Mechanical properties are a second important aspect to analyze.

The "ideal" scaffold must present specific mechanical properties suitable both for the anatomical site of implantation (they have also kept mechanical characteristics that make the scaffold handle for surgical implantation) and for the type of cells intended to use. In particular, using natural scaffolds, mechanotransduction regards the transformation of cellular stress into electrochemical responses, crucial for the survival and the right function of both cells and higher organisms [48]. Additionally, it has been shown how mechanosensitivity could as well facilitate mesenchymal stem cells differentiation [49]. Finally, biological features and biological activity include all the cues that the scaffold could reciprocally interchange with seeded cells and the biological environment. These kinds of properties, moreover in terms of growth factors, chemokines and cytokines, are typical of natural scaffolds suitable to regulate cellular functions.

According to these properties four major categorizes of scaffolds are known:

1. Pre-made scaffold for cell seeding [50].
2. Cell sheets with new-secreted extracellular matrix.
3. Cell encapsulation and hydrogel scaffold.
4. Organ extracellular matrix-derived scaffold.

3.1. Pre-made scaffold for cell seeding.

As deeply reviewed by Chan et al., pre-made scaffolds represent the first structure that have been seeded with the birth of "tissue engineering" [51]. The idea is to produce a tridimensional structure that are able to furnish a non-toxic environment for seeded cells providing a gentle transition by which the scaffold degrades together with the functional enhancement and engraftment of seeded cells. Initial pre-made scaffolds were made in the attempt to overcome limitations due to the "classic" 2D cell culturing, offering all the advantages deriving from the third dimension adding a more physiological environment and more predictive data.

The porosity of the scaffold is always been considered as a crucial property for cellular vitality for the possibility to guarantee the effective delivering of oxygen and nutrients to cells.

Porosity of the scaffold is also today seen as paradigmatic for an ideal-scaffold. In the review of "Porous scaffold design for tissue engineering" [52], the author paraphrased the state of an important architect, Robert leRicolais "The art of structure is where to put the holes" transforming it into "The art of scaffolding is where to put the holes and the biofactors".

This approach has numerous advantages: the choice of biomaterials to use is wide (both considering natural or synthetic polymers) and it can exploit a relative precise and repetitive assembling to micrometer size and the material used can be loaded or cross-linked with numerous molecules in the attempt to augment cellular functionality [53, 54].

3.2. Cellular sheets approach for tissue engineering

Cell sheets approach, even if considered as scaffold-free technology, must be included in the tissue-engineered approach.

Cell sheets technology has been recently developed by tissue engineering and regenerative medicine community has a potential technology able to manufacture layer-by-layer transplantable cellular constructs [55, 56]. The entire principle is based on a particular temperature-reversible polymer, the poly(N-isopropylacrylamide) (PIPAAm) able to change from an hydrophilic to an hydrophobic state at 37 and 32°C, respectively [57].

Cell sheet approach has been tested as a potential approach for pancreatic islets and β cell transplantation exploring alternative site of transplantation as well as the intraportal injection. Subcutaneous site is one of the new sites of islets injection under evaluation (in animal model) [58] and the most important difference is the absence of the vascular connection between transplanted cells and the blood flow. Moreover, in this regard, other advantages can be numbered including less-invasiveness (the hypothetical subcutaneous implantation could be performed under local anesthesia), the possibility of repeated-procedures in case of immunological rejection and, finally, the opportunity to safely remove transplanted islets.

In 2009, the group driven by Okano has manufactured a transplantable cell-sheet made by rat islets [59]. In their research, rat β cells have been cultured on temperature-responsive dishes (with or without adding extracellular matrix) to form a transplantable β cells layer. The resulting cellular structure was subcutaneously implanted in streptozotocin-induced diabetic immunodeficient mice achieving a euglicemia state after 1 week, which last over 100 days. Subcutaneous space is still under analysis in order to be translated to clinics principally due a reduced quantity of oxygen and nutrients that can reduce cell viability at long-term [60]. Recently Pepper et al. gave place to a newly born use of the subcutaneous site for islet implantation. In order to overcome the limitation of the avascular condition of the subcutaneous site, the authors have previously vascularized the implantation site to later infuse in a well-oxygenated site the islet mass. The final described result showed the possibility to revert the hyperglycemic state of severely diabetic induced mice. This innovative approach is recently proposed as alternative vascularized site of implantation that can be combined with device or scaffolds for beta cell replacement in type 1 diabetes.

3.3. Cell encapsulation and hydrogel scaffold

As previously described, islets transplantation procured from deceased donors is one of the most used technology in order reverse a T1D state. Although this procedure is able to produce a euglicemia state, is not free from side effects and limited principally by donor shortage, lifelong immunosuppression and the immunological rejection of transplanted islets. The use of encapsulation device, able to separate the transplanted β cells from the surrounding recipients' environment, has emerged as a promising approach with the attempt to eliminate the immunological issue and, consequently, the need of immunosuppression [61]. The first attempt for encapsulation of human insulinoma tissue dated 1933 by Bisceglie et al. who encapsulated human tissue into membranous bags and transplanted into rats [62]. From that moment, several encapsulation technologies have been developed and implemented and actually they can be sorted into micro- and macroencapsulation technologies depending on the size [63] with the recent addition of the nano-encapsulation technology.

Regardless of the capsule size, the whole approach has a common rationale: enveloping pancreatic islets in biocompatible membrane able to permit the diffusion inside the capsule of molecules including oxygen and nutrition and, at least in theory, to shield it from larger molecules such as antibodies or immune cells [64].

In this way, if successful, an immunological physical barrier, through a perm-selective coating, could prevent the systemic administration of immunosuppressive drugs, which are actually essential to avoid graft rejection.

Materials for islets encapsulation are deeply studied always searching for better performances. It is recognized that two important properties must be developed: firstly, capsule must permit the admission inside the capsule (and so on in contact with encapsulated islets) of small molecules and the diffusion out of waste material and insulin; secondly, they must isolate the content from immune competent cells (B or T cells or macrophages).

These abilities are secondary to the material/s used to produce the capsule.

Alginate, a colloidal substance derived from brown seaweed, is the most famous and the first biomaterial suitable to produce capsules. The main advantage of using alginate relies on its capacity of not interfering with islets (and with insulin release) while guaranteeing a good stability [65]. Several materials are then be added as multiple layers in order to improve alginate functionality. On this subject, poly(ethylene glycol) (PEG) and poly-L-lysine (PLL) are the more established aiming to reduce plasma absorption and increasing long-term capsule stability. In 1999 Chandy et al. reported a modified encapsulation technique with alginate and PEG and showing an improved stability [66]. One year later, Desai et al. showed islets good viability and insulin release with the same encapsulation protocol [67]. Alginate/PLL is the most utilized combination for cell encapsulation in a multi-layer composition. A three layer encapsulation protocol was proposed by Goosen et al., which was based on a alginate/PLL/alginate composition providing a good shield from immune system and a limited diffusion of serum immunoglobulin albumin and hemoglobin [68].

3.3.1. *Microencapsulation*

In matter of dimension, the size belonging to the microencapsulation group has been the first to be developed and examined. In 1984, O'Shea et al. manufactured an alginate-based microcapsule [69] open up a new era in islets encapsulation that, after almost 40 years counts more than 100 studies in preclinical model (approximately 96% involving small animals). In all these settings several strategies have been tested including capsule customization with three particular molecules such as the alginate, the glucoronic and mannuronic acids, which if used (or combined with other mono/polymer(s)) [70] can confer more strength and stability. Rodent preclinical models are the most studied, as reviewed by Souza et al. who have evaluated more than 60 encapsulation strategies and have founded that the most effective approach is based on the use of intraperitoneal alginate-base microencapsulation without immunosuppressive strategies (islets mean survival rate 100 days) and intraportal injection with immunosuppression (islets mean survival rate 164 days) [71]. Indeed, high mannuronic acid, as biomaterial for islets encapsulation, has shown prolonged survival rate for more than 350 days [72].

3.3.2. Macroencapsulation

Macroencapsulation (or transplantation systems) principle depends on macro-extravascular chambers containing the transplanted tissue. This approach has been developed by Algire et al. and exploits the presence of a semi-permeable membrane to block immune cells but preserving the diffusion of oxygen, nutrients, glucose, insulin, glucagon and somatostatin [73]. Algire's study has paved the way for the creation of a commercially available device that can be transplanted [74]. The peculiarity of this chamber resides in the porosity dimension, 450 nm, able to avoid the direct contact between islets and immunocompetent cells. Follow-up of this device reported a euglicemia state up to 8 months in rodents [75]. Macroencapsulation devices have also shown same interesting results for large animal also even if still not consistent [76]. Baxter Healthcare is the author of one of the first prototype of macroencapsulation device. Two sealed membranes constituted the designed structure, with an autonomous inlet gate. The outside was defined in order to have a gradient membrane to immune-isolate the transplanted cells but also to allow the vascular growth. Small animal experiments displayed a significantly high level of device vascularization in the subcutaneous site 1 year after the implant. This promising approach could be considered the first generation of the actually known TheraCyte Device.

3.4. Organ extracellular matrix-derived scaffold

Extracellular matrix (ECM) can be defined as a heterogeneous, connective network composed of several fibrous glycoproteins able to coordinate cellular functions providing a physical architecture, mechanical stability and biochemical cues necessary for tissue morphogenesis and homeostasis [77]. These qualities will make ECM a potential successful strategy in order to obtain a bioactive scaffold supporting *in-vivo* cellular viability. Recently decellularization protocols have been implemented in order to achieve an ECM-based scaffold that recapitulate the tridimensional architecture of the organ from which, cell intended to be seeded, had been harvested, as well as its biological-specific features [78]. The importance of using an organ-specific ECM relies on the opportunity of working with a natural, biocompatible, bioactive and structural scaffold. Specific properties of extracellular matrix and its application in the diabetes treatment will be discussed below.

4. Extracellular matrix as a template for cell culturing

The extracellular matrix ECM is an organ-specific complex system composed by a plethora of molecules (structural and non-structural) created by the tissue-/organ-specific cells, and deposited into the nearby medium to provide biophysical and biochemical support to the surrounding cells. Originally identified as an inert and passive structural architecture, today its role has been completely transformed and the entire scientific community indicates ECM as an active environment able to support viability, growth and differentiations both stem [79] and differentiated cells. In particular it has been reported how ECM plays a crucial position, being essential part of micro environmental stem cell niches [80] and being a substantial part of any given tissue. ECM is composed of a multitude of supporting bioactive molecules that are secreted, in different quantities and composition by resident cells in order to support tissue-/organ-specific cellular function(s) specific of each tissue or organ.

The most important constituents of ECM are represented by macromolecules: polypeptide chain of collagen, laminin, fibronectin and glycosaminoglycans. It is important to highlight how this composition could change not only during physiological aging but also under pathological conditions that are able to really upset the whole ECM structure (e.g. liver fibrosis). In native pancreas, ECM lives a sort of "environmental dynamic reciprocity" with pancreatic cells (belonging both to exocrine and the endocrine part).

Islets isolation is a process where Langerhans islets are detached from ECM by the progressive enzyme-based ECM degradation [81]. In other words, the entire cellular compartment is stripped of its surrounding environment leading, in case of islets transplantation, to a very low survival rate around 10% of grafted islets [82]. Obviously the low graft success is not just the consequence of ECM absence being also related to the transplant site characteristics and to a hypoxic state in short- and long-term period, as well as to immunological rejection. Matrix composition is particular important for islets homeostasis and stability.

Pancreatic ECM is an intricate tridimensional network enriched by multiple carbohydrate and protein which, through a system protein-integrin orchestras islets stability, growth, differentiation and death. In this regards, the presence of a specific structure, known as "basement membrane", is essential. This cytological structure, strictly attached to cells is the final activator of specific pathways via interactions between ECM proteins and dedicated integrins. With all these intrinsic properties it is easy to understand how ECM could represent the "ideal" scaffold recapitulating most of the qualities needed such as presence of bioactive molecules, mechanical strength and tridimensional environment.

Moreover these potentials have the additional advantages to be organ specific.

With these bases, the use of an extracellular matrix scaffold as a template for cell seeding can be perfectly in line with the paradigm of tissue engineering.

4.1. Decellularization technology

Decellularization is defined as a multi-step process able to separate the organ/tissue extracellular matrix from its inhabitant cellular component, leaving ECM relatively intact with regard to tridimensional characteristics and biological properties. Resultant structure is an acellular scaffold, which recapitulates the native tissue/organ features furnishing an ideal template to be seeded with new cell families. In 2011 Crapo et al. has profoundly examined all the techniques that can be applied to remove cells from the surrounding ECM [83].

They have divided decellularization techniques in three major types that are still valid today:

1. Chemical decellularization
2. Biological decellularization
3. Physical decellularization

4.1.1. Chemical decellularization

Chemical decellularization is based on the use of chemical agents in order to detach the cells from ECM, catalyzing hydrolytic biomolecular degradation.

Usually, these substances can be acids or bases.

Most common acids used for this purpose are acetic and paracetic acids that have demonstrated good capacity for cellular removal but they also seemed to be too aggressive to ECM structure with an excessive loss of ECM mechanical properties [84]. Calcium hydroxide, sodium sulfide and sodium hydroxide are the most common bases utilized as decellularization agents. It is generally accepted that their use, during decellularization, leads to the elimination of growth factors that enrich the ECM resulting in a loss of bioactivity [85]. For above described reasons, acids and bases are not yet universally used.

Detergents (ionic, nonionic and zwitterionic) may represent the most important chemical agents used for decellularization. Solubilizing cell membranes (cytoplasmatic and nuclear) [86] and separating DNA from proteins, they are therefore effective in removing cellular material from the tissue or the organ treated [77].

Sodium dodecyl sulfate (SDS) and Triton X-100 are the most common used detergents present in decellularization protocols [87].

SDS is an ionic, synthetic, organic compound with an established experience in tissue engineering of decellularized tissues. Dedicated scientific literature offers many protocol based on the use of SDS for organ or tissue decellularization [88–90].

Triton X-100 (TNX-100) is a nonionic surfactant composed by a hydrophilic polyethylene oxide chain and an aromatic hydrocarbon lipophilic or hydrophobic group [91]. By its chemical properties, Triton X-100 can effectively remove cells from tissue but seems to be less aggressive compared to SDS (even bringing to an ECM degradation) and so more useful for thicker tissue [92]. Furthermore, several reports [93, 94] demonstrated how SDS is more effective in removal nuclear material (and consequently, shortening the risk related to residual presence of immunological material).

Finally zwitterionic detergents including 3-[(3-cholamidopropyl) dimethylammonio]-1-proppanesulfonate (CHAPS), sulfobetaine-10 (SB-10) and SB-16 have shown encouraging results, with an important preservation of ECM biological cues but final results need to be more studied [95].

4.1.2. Biological decellularization

Biological decellularization involves the use of biologic enzymatic and non-enzymatic agents able to specifically remove of cell residues or undesirable ECM constituents. Enzyme such nucleases (DNases or RNases) are the perfect archetype centered on their capacity to cleave nucleic acid sequences aiding, in such this way, to eliminate nucleotides after cell lysis [96]. Non-enzymatic agents are mainly represented by chelating agents, such as ethylenediaminetetraacetic acid (EDTA) and ethylene glycol tetraacetic acid (EGTA). Through metal ions, segregating EDTA and EGTA can separate cells from ECM. Unfortunately these agents are not particularly effective if used alone. For this, most of the time, they are added in a multi-step protocol [97–99].

4.1.3. Physical decellularization

Physical decellularization protocols include several procedures that exploit physical strategies in order to remove cell from ECM, counting temperature-protocols (freeze-thaw cycles), mechanical-protocols (via the agitation and the immersion of samples) and pressure-based protocols [100]. Temperature-based decellularization protocols are relatively simple procedures that only necessitate of multiple freeze-thaw cycles to be effective in cellular removal [101]. If, on one hand, these protocols seem to be satisfying for cellular removal, on the other one, multiple temperature changes cause important damages on final tridimensional ECM ultrastructure [102]. Freeze-thaw cycles technology is especially appealing for the decellularization of simple structure such as tendon or cartilage-base organs but results hardly applicable on more structural complex architectures (pancreas, kidney or liver). Samples shaking and immersion have instead a strategic role and they are often used for tissue engineering. Most of the times these protocols are used when samples to decellularized are small and without dedicated vascular inlet of outlet such as pre-cut parenchyma cubes of tissue, blood vessels [103] or bone fragments [104]. These techniques provide numerous benefits, such as the possibility to simply change the duration of protocol (and therefore the time-of-contact between detergent and sample) or the shaking force. The choice of the liquid immersion, the total duration time or the shaking force (expressed by rpm in case of orbicular shaking) permit a variety of different protocols targeted on the density of native tissue.

However, the achievement of an homogeneous decellularization state remains one of the main limit of immersion and shacking techniques, also considering that sample external surfaces will be more treated (with a major probability of ECM degradation) compared to the inner parts (that could contain cellular residuals at the end of the process).

4.2. Pancreas-specific ECM

Pancreas ECM results composed by proteins belonging to the basement membrane (collagen, laminin, fibronectin, nidogen/entactin, vitronectin and perlecan) and by proteoglycans (HS proteoglycan, syndecan, glypican, betaglycan and chondroitin sulfate proteoglycans).

4.2.1. Basement membrane ECM components

Collagen fibers are responsible both for structural strength of pancreatic ECM as well as for biological action including cellular adhesion and morphogenesis: collagen I, II, III, IV V and VI are specifically represented in the islets ECM [104, 105]. Their functions are still not completely clear moreover if considering their role on the pancreatic endocrine pathways. As acutely reviewed by Poole-Warren et al. [106], Collagen I molecules seem to promote islets survival [107] but decreasing at the same time insulin release. Furthermore collagen properties can be exploited only with the presence of specific proteins of surface: the integrins

(Int) [108, 109]. These small particles, classified as Int-α1β1, Int-α2β1, Int-α10β1 and Int-α11β1 have been shown to act as receptor supporting matrix-cellular interactions via the subunit α of collagen 1 fibers [110].

Beyond the collagen, laminin proteins are plenty involved in the structural ECM pancreatic architecture. Laminin takes part in the structural integrity of the extracellular matrix by bounding to collagen, nidogen and glycosaminoglycan fibers [111]. Laminin proteins have been localized in the islets cells permeating their islet's microvasculature, but their actions are still under debate. Laminin-islets relationship is controlled by the presence of following integrins: Int-α1β1, Int-α2β1, Int-α3β1, Int-α6β1, Int-α7β1, Int-α9β1, Int-αvβ3, Int-αvβ5, Int-αvβ8 and Int-α6β4 with islets morphogenesis functions (α3 and β1) [112] and pancreas development (α6) [113]. Finally, vitronectin is an ECM protein expressed during human fetal islets development. Its correct pathway is essential for the appropriate growth of islets to "mature" β cells and moreover for their migration acting as mobility promoter [114].

4.2.2. Proteoglycans ECM components

Proteoglycans are heavily glycosylated proteins that take ubiquitously part in most of all organs and tissues both with a structural and deposit activity. Their composition is based on a core protein with several different, negatively charged, protein chains (GAGs) attached. Different protein chains bring to different proteoglycans [115]. Negative charge is crucial for the deposit role of this protein being able to conserve specific growth factors, cytokines and chemokines (that differ from tissue to tissue in terms of composition) releasing them just if necessary [116].

4.3. Whole organ decellularization and regenerative medicine

Whole organ perfusion is the most commonly used technology, if the target sample is an entire organ. This technique exploits the organ native vascular system to homogenously perfuse the organ with selected detergent(s) [117]. Perfusion-decellularization is the most consistent method to obtain decellularized whole organ scaffolds due to the native vascular architecture, which is naturally designed to permit the delivery of oxygen and nutrients by blood flux. This allows the same time-to-contact between detergent(s) and cells in the entire organ by an anterograde or a retrograde perfusion. Likewise, the presence of a vascular outlet (most of the times represented by the native venous outflow system) allows an efficient strategy to eliminate cellular debris deriving from decellularization. Whole organ perfusion has been already established for several organs or tissues that differ in terms of size, origin (small or large animal model and human) and physiopathological conditions [118]. Obviously, as described for immersion/shaking technology, the appropriateness of the perfusion protocol must be pointed on the organ properties (moreover size and resident cell density).

This technology has been also applied in the attempt of creating a bioengineered pancreas.

Basically the idea is to customize clinical-relevant size acellular pancreas that can be secondly repopulated with patient-own endocrine cellular population and then transplanted in the same patient. If successful, this approach could address the limitations that today affect diabetes treatment. It could be created a non-immunological recellularized scaffold with functional islets, β cells or human-induced pluripotent stem cells (hiPSC) derived to β cells and endothelial cells subsequently orthotopically transplanted in the patient whose cells were harvested. In this scenario, animal source could provide a hypothetical unlimited pool of scaffold to repopulate and stem cells (mesenchymal or iPS) can be used for the repopulation.

A bioengineered approach could so provide an unlimited source of transplantable pancreas, eliminating the organ shortage, and, at the same time, the use of patient-own cells could avoid (or limit) the use of immunosuppressive drug regimes.

The entire decellularization development must be balanced according to the final desired quality of the extracellular matrix. This quality is currently evaluated by histological staining, DNA content, collagens and proteoglycans assessment and tridimensional imaging technologies (scanning electron microscopy). This assessment provides crucial data not only on the effective cellular removal but also about the biological and structural properties of the decellularized matrix intended to seed.

If decellularization technology has provided satisfying results in term of pancreas decellularization and ECM maintenance, recellularization strategies seems to be the most important hurdle, still not yet overcome.

These, together with engineering significant advances for the manufacturing of dedicated bioreactors, are the next steps to go.

4.4. Recellularization technology

Recellularization strategies play a key role for the creation of a functional bioengineered organoid. Working recellularization requires a proper cell source (considering both mature or stem cell origin), an optimal seeding method and a long-term culture system (available with appropriate bioreactors). Scaffold recellularization needs tree different cell group, belonging to parenchymal, vascular and supporting types respectively with different tasks and responsibilities.

Parenchymal cells are responsible for the effective organ function whereas vascular cells must entirely cover the vascular extracellular matrix providing a suitable blood flux (after transplantation) both as inlet (with oxygen and nutrients delivering) as well as in outlet for metabolic wastes spill.

In this regards, vascular coverage during recellularization is essential for the success, and it must be achieved covering as much ECM as possible. Missing ECM parts, which are not protected by vascular cells, will directly expose ECM to the blood flow, resulting in an almost instant activation of the coagulative cascade and in the formation of blood clots. This situation will bring to a subsequently blood stoppage directed to the entire portion downstream the clot, with final seeded cells death [119].

Finally supporting cells must provide an active sustain for parenchymal and vascular cellular families.

Pancreas scaffold recellularization can be performed via vascular perfusion (using the same principle utilized to remove cells). Inlet pancreatic vasculature can be mainly accessed by splenic and pancreaticoduodenal artery.

Pancreatic inlet can also exploit the presence of the pancreatic duct via a retrograde flow. Pancreatic outlet vasculature relies on the portal and the splenic vein.

4.5. Cell-on-scaffold technology toward bioartificial pancreas: state-of-art

Cell-on-scaffold technique has been already examined in order to produce a pancreatic organoid.

One of the first pioneeristic studies that explored the possibility of using ECM matrix as a template for islets seeding has been proposed by De Carlo et al. in 2010. In their report 240-μm slices of rat pancreatic and hepatic acellular were decellularized by a detergent-based protocol (4% sodium deoxicholate was used) and then seeded with islets after static culture conditions (37°C, 95% O_2 and 5% CO_2 in RPMI-1640 medium added with 5.6 mM glucose, 100 IU/ml of penicillin and 100 μg/ml of streptomycin). Furthermore pancreatic islets (N = 50) were seeded on five acellular matrix and cultured under standard conditions. After 7 days, the islet-matrix complexes were inserted into synthetic, tubular PVA/PEG devices and prepared for *in-vivo* implant. Rats were then made diabetic and PVA/PEG devices (containing islets and matrix) were implanted. Results have demonstrated, in an *in-vivo* follow-up up to 6 weeks, how pancreatic devices reacted to glucose acute stimula with insulin delivery and a decreasing dose of daily insulin needed to maintain the euglycaemic condition [120]. To our knowledge, in the same year, Ott's lab proposed for the first time a rat pancreatic whole pancreas scaffold (obtained via detergent perfusion) afterwards seeded with human islets and supporting human MSCs. Results were very fascinating showing effective tridimensional growth of cells on the matrix with good response to glucose stimula [121]. Both these studies have shown important starting lines for the use of this technology to produce a functional bioengineered pancreas. However they have also highlighted significant limitations, some of which have been already overtaken.

Herein we describe the most important findings regarding the use of a whole organ pancreatic scaffold from different models with the aim to manufacture a transplantable bioengineered pancreas.

4.5.1. Small animal models

Small animal model represent the baseline for basic science in terms of costs and prospective translational results. For this reason small animal model has been the first to be investigated.

One of the first complete proof-of-concepts of a murine whole organ pancreas decellularization has been offered by Goh et al. in 2013 [122]. After a midline laparotomy the pancreas was carefully harvested preserving intact vascular inlets that have been then cannulated and used as inlets for retrograde perfusion with a flow of 8 ml/min. A multistep protocol with 0.5% SDS and 1% Triton X-100 has been chosen. Cells for repopulation (AR42J acinar cell line for the exocrine component and MIN-6 β cell for the endocrine one) were cultured in standard static condition and then seeded on the acellular scaffold. AR42J cells were injected through the main pancreatic duct (30 × 10^6) cells whereas MIN-6 β cells (30 × 10^6) via the hepatic vein. The seeded pancreas was cultured under static conditions at 37°C with a 95% air/5% CO_2 atmosphere for 5 days and then implanted by a dorsal subcutaneous pocket of an adult mouse for biocompatible preliminary tests. Results have demonstrated an optimal feasibility about decellularization with preservation of ECM composition. Secondly recellularization was evaluated in terms of cellular engraftment, survival and functionality by Immunohistochemistry (IHC). Reconstructed pancreas showed a homogenous distribution of cell types with a minimal apoptosis rate (detected by TUNEL staining) less than 18% and robust cellular functionality expressed by C-peptide staining positivity and finally confirmed by up-regulation of insulin genes. After 14 days, subcutaneous implantation pancreatic organoid was harvested and established as biocompatible organoid able to positive regulate a neoangiogenetic action. This paper has set the bases for all the subsequent experiments.

Recently a group headed by Struecker has refined the entire process providing, for the first time, a proof-of-concept for the repopulation of the decellularized rat pancreas with functional islets of Langerhans [123]. Briefly rat pancreas was decellularized via vascular perfusion (1% Triton X: 0.5% SDS and 1% Triton X-100) with a flow-rate around 10 ml/min and then repopulated with approximately 2000 islets via the pancreatic duct to test viability and functionality of the islets after the process. *Ex-vivo* TUNEL staining and glucose-stimulated insulin secretion (GSIS) revealed how islets were viable and functional after the injection inside the acellular scaffold.

Both the presented studies have analyzed the opportunity to use mature and already differentiated cells to repopulate pancreas whole organ scaffold.

A further progress toward the creation of an ECM-pancreas can be accomplished exploiting the use of stem cells driven in their growth and specific differentiation by precise ECM biological cues. Thus stem cells can be considered a potent and encouraging cell source for tissue engineering [124]. Accordingly to this concept, Wan et al. has just proposed a study about the possibility to culturing iPSCs derived pancreatic β cells on decellularized ECM [125]. After peristaltic artery perfusion with detergents, decellularized ECM-based pancreas were seeded with already differentiated into β cells like iPSCs (approx. 3 × 10^6) by two different methods (vascular perfusion and multipositional parenchymal injection). In-vitro continuous monitoring exhibits maintained insulin, C-peptide and glucagon expression. Additionally, insulin level expression in the perfusated media was twofold higher than those levels obtained by traditional bidimensional culturing.

4.5.2. Large animal models

Even if encouraging results have been reported for small animal model, it is mandatory to aim to a functional bioengineered pancreas with clinically relevant size.

That way the use of a large animal model has been investigated.

The main difficulty in the switching from small to large animal is mainly due to the higher stiffness of the pancreatic parenchyma and to the larger volume that has to be decellularized. In this regards, decellularized protocols could not be too aggressive in order to preserve ECM native characteristics. In 2013, research group headed by Orlando published the proof-of-concept for the achievement of a porcine decellularized whole organ scaffold [126]. Vascular-based retrograde peristaltic perfusion via superior mesenteric vein and pancreatic duct with nonionic detergent (1% Triton X-100) guarantied pancreatic cellular removal. Pancreatic extra-cellular matrix biocompatibility was tested by the static short-term (7 days) cellular seeding with human amniotic fluid stem cells. This type of cells has been also proposed as a potential source of insulin-secreting cells [127]. To validate the hypothetical capacity of acellular ECM-based pancreas like scaffold to sustain endocrine pancreatic function, porcine pancreas scaffold was repopulated with porcine islets and insulin secretion was *ex-vivo* measured under different glucose stimula and at several time points. Results proposed showed a significant similarity to physiological circumstances with higher insulin release in response to higher glucose concentration. Katsuki et al. have also proposed analogous results in 2016 [128] with the creation of a portioned repopulated porcine pancreas.

4.5.3. Discarded human pancreas as a source of ECM

The use of discarded human organs has been explored as a possible source of organ to decellularize. This opportunity is based on the incredible amount of organ that, annually in the USA, are retrieved for transplantation purposes but finally discarded for various reasons. Orlando et al. described this scenario envisioning [129], in a future not so distant, that a discarded organ will be use to build an acellular human scaffold subsequently repopulated by patients own cells. The final result will be a bioengineered pancreas composed by human-derived extracellular matrix as well as recellularization cell type. The hypothesis to recycle discarded human organs for organ bioengineering projects, hinges on the theory that decellularized human ECM could be the perfect environment for human cells during recellularization.

In 2015, our group has explored the use of discarded pancreas as a source from which obtain acellular whole organ scaffolds [130]. Compared to data achieved on the porcine model, an additional inlet access was demonstrated crucial switching from a two-way perfusion (pancreatic duct and superior mesenteric vein) to a three-way perfusion system (pancreatic duct, superior mesenteric artery and splenic artery). This adjustment brought to a more homogenous distribution of the detergent used (1% Triton X-100) permitting a relative more gentle approach (Triton X-100 is considered less aggressive than SDS). Staining and DNA content confirmed the cellular removal as well as the preservation of the most important pancreatic ECM elements and growth factor that enriched the human native pancreatic parenchyma. Besides important results about static seeding with human islets and dynamic peristaltic seeding with human primary pancreatic endothelial cells, crucial information has been obtained by the analysis of ECM immune properties. In fact, to the best of our knowledge, it is the first time that human pancreas ECM documents an immunosuppressive T-reg promoting properties paving the way for its possible use as immunosuppressant (**Table 1**).

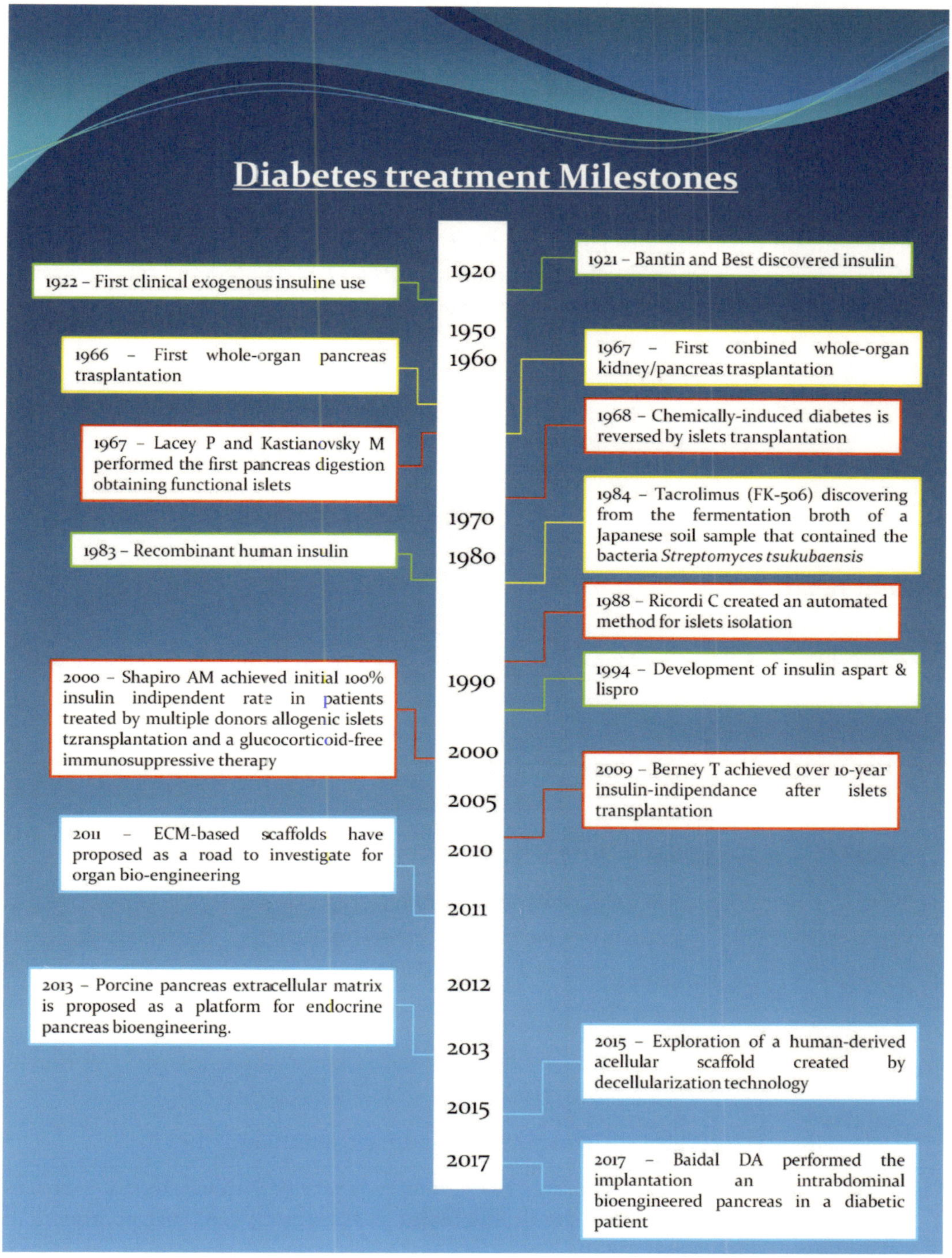

Table 1. Timeline of the major advances in insulin treatment, islets and pancreas transplantation and pancreas bioengineering.

5. Conclusion and future perspectives

In the last decade, regenerative medicine and organ bioengineering have accomplished important progress toward the manufacturing of a functional, bioengineered pancreas, exploring different options both in terms of platform for decellularization as well as for cell type repopulation source. As of now, the emergent use of stem cells appears as an exciting and encouraging field to discover. Several crucial hurdles are not yet overcome and require important advancements, above all *in-vivo* short- and long-term functional testing. Despite these obstacles, cell-on-scaffold technology holds a huge potential in order to solve the problem of pancreas shortage creating a solid transplantable alternative.

Acknowledgements

Andrea Peloso is supported by an Investigator Fellowship from Collegio Ghislieri, Pavia, Italy.

Author details

Andrea Peloso[1,3]*, Antonio Citro[2], Graziano Oldani[3], Szandra Brambilla[4], Lorenzo Piemonti[2] and Lorenzo Cobianchi[1]

*Address all correspondence to: andreapeloso@hotmail.it

1 IRCCS Policlinico San Matteo, Department of General Surgery, University of Pavia, Pavia, Italy

2 Diabetes Research Institute, IRCCS San Raffaele Scientific Institute, Milan, Italy

3 Division of Transplantation, Department of Surgery, Geneva University Hospitals, Geneva, Switzerland

4 University of Pavia, Pavia, Italy

References

[1] Global Report on Diabetes. Geneva: World Health Organization; 2016

[2] Centers for Disease Control and Prevention. National Diabetes Statistics Report: Estimates of Diabetes and Its Burden in the United States, 2014. Atlanta, GA: US Department of Health and Human Services. p. 2014

[3] International Diabetes Federation. IDF Diabetes Atlas. 6th ed. Brussels, Belgium: International Diabetes Federation; 2013

[4] Li G. The long-term effect of lifestyle interventions to prevent diabetes in the China Da Qing Diabetes Prevention Study: A 20-year follow-up study. Lancet. 2008;**371**:1783-1789

[5] DeFronzo RA, Tripathy D, Schwenke DC, et al. Pioglitazone for diabetes prevention in impaired glucose tolerance. The New England Journal of Medicine. 2011;**364**:1104-1115

[6] NCD Risk Factor Collaboration (NCDRisC). Worldwide trends in diabetes since 1980: a pooled analysis of 751 population-based studies with 4.4 million participants. Lancet. 2016 Apr 9;**387**(10027):1513-1530

[7] Aghazadeh Y, Nostro MC. Cell therapy for Type 1 diabetes: Current and future strategies. Current Diabetes Reports. 2017 Jun;**17**(6):37. DOI: 10.1007/s11892-017-0863-6

[8] Nilsson B, Ekdahl KN, Korsgren O. Control of instant blood-mediated inflammatory reaction to improve islets of Langerhans engraftment. Curr Opin Organ Transplant. 2011;**16**:620-626

[9] Chatterjee S, Khunti K, Davies MJ. Type 2 diabetes. Lancet. 2017;**389**:2239-2251

[10] http://www.cdc.gov/diabetes

[11] Dall TM, Roary M, Yang W, Zhang S, Chen YJ, Arday DR, Gantt CJ, Zhang Y. Health care use and costs for participants in a diabetes disease management program, United States, 2007-2008. Preventing Chronic Disease. 2011 May;**8**(3):A53 Epub 2011 Apr 15

[12] www.who.org—Global Report on Diabetes

[13] NCD Risk Factor Collaboration (NCDRisC). Worldwide trends in diabetes since 1980: a pooled analysis of 751 population-based studies with 4.4 million participants. Lancet. 2016 Apr 9;**387**(10027):1513-30

[14] IDF Diabetes Atlas, 6th ed. Brussels: International Diabetes Federation; 2013

[15] Rowley WR, Bezold C, Arikan Y, Byrne E, Krohe S. Diabetes 2030: Insights from yesterday, today, and future trends. Population Health Management. 2017 Feb;**20**(1, 1):6-12

[16] www.diabetes.org/research-and-practice/student-resources/history-of-diabetes.html

[17] Orlando G, Stratta RJ, Light J. Pancreas transplantation for type 2 diabetes mellitus. Current Opinion in Organ Transplantation. 2011;**16**:110-115

[18] Kim KW. Diabetes Research and Clinical Practice. 2004;**66**(Suppl 1):S11

[19] Haas L, Maryniuk M, Beck J, et al. Standards Revision Task Force. National standards for diabetes self-management education and support. Diabetes Care. 2012, 2014;**37**(Suppl. 1): S144-S153

[20] Bell KJ, Toschi E, Steil GM, Wolpert HA. Optimized mealtime insulin dosing for fat and protein in type 1 diabetes: Application of a model-based approach to derive insulin doses for open-loop diabetes management. Diabetes Care. 2016;**39**:1631-1634

[21] van den Berghe G, Wouters P, Weekers F, Verwaest C, Bruyninckx F, Schetz M, Vlasselaers D, Ferdinande P, Lauwers P, Bouillon R. Intensive insulin therapy in critically ill patients. The New England Journal of Medicine. 2001;**345**:1359-1367

[22] Bagshaw SM, Egi M, George C, Bellomo R. Australia New Zealand Intensive Care Society Database Management Committee: Early blood glucose control and mortality in critically ill patients in Australia. Critical Care Medicine. 2009;**37**:463-470

[23] Yeh HC, Brown TT, Maruthur N, et al. Comparative effectiveness and safety of methods of insulin delivery and glucose monitoring for diabetes mellitus: A systematic review and meta-analysis. Annals of Internal Medicine. 2012;**157**:336-347

[24] Kelly WD, Lillehei RC, Merkel FK, et al. Allotransplantation of the pancreas and duodenum along with the kidney in diabetic nephropathy. Surgery. 1967;**61**:827-837

[25] Gruessner A, Sutherland DE. Pancreas transplantation in the United States (US) and non-US as reported to the United Network for Organ Sharing (UNOS) and the International Pancreas Transplant Registry (IPTR). Clinical Transplants. 1996:47-67

[26] Gruessner AC, Gruessner RW. Pancreas transplantation of US and Non-US cases from 2005 to 2014 as reported to the United Network for Organ Sharing (UNOS) and the International Pancreas Transplant Registry (IPTR). The Review of Diabetic Studies. 2016;**13**:35-58

[27] Stegall MD, Simon M, Wachs ME, et al. Mycophenolate mofetil decreases rejection in simultaneous pancreas-kidney transplantation when combined with tacrolimus or cyclosporine. Transplantation. 1997;**64**:1695-1700

[28] Navarro X et al. Influence of pancreas transplant on cardiorespiratory reflexes, nerve conduction and mortality in diabetes mellitus. Diabetes. 1990;**39**(7):802-806

[29] Zehr PS et al. Pancreas transplantation: Assessing secondary complications and life quality. Diabetologia. 1991;**34**(Suppl 1):S138-S140

[30] Zehrer CL, Gross CR. Quality of life of pancreas transplant recipients. Diabetologia. 1991;**34**(Suppl 1):S145-S149

[31] Gruessner AC, Gruessner RW. Pancreas transplant outcomes for United States and non United States cases as reported to the United Network for Organ Sharing and the International Pancreas Transplant Registry as of December 2011. Clinical Transplants. 2012:23-40

[32] http://onlinelibrary.wiley.com/doi/10.1111/ajt.13665/full

[33] Stratta RJ, Fridell JA, Gruessner AC, Odorico JS, Gruessner RW. Pancreas transplantation: A decade of decline. Current Opinion in Organ Transplantation. 2016 Aug;**21**(4):386-392

[34] Srinivasan P, Huang GC, Amiel SA, Heaton ND. Islet cell transplantation. Postgraduate Medical Journal. 2007;**83**:224-229

[35] American Diabates Association. Pancreas transplantation for patients with type 1 diabetes: American Diabetes Association. Diabetes Care. 2003;**23**:117

[36] Ballinger WF, Lacy PE. Transplantation of intact pancreatic islets in rats. Surgery. 1972;**72**:175-186

[37] Ricordi C, Lacy PE, Scharp DW. Automated islet isolation from human pancreas. Diabetes. 1989;**38**(Suppl. 1):140-142

[38] Ricordi C. Islets transplantation: A brave new world. Diabetes. 2003;**52**(7):1595-1603

[39] Pyzdrodrowski KL, Kendall DM, Halter JB, et al. Preserved insulin secretion and insulin independence in recipients of islets autograft. The New England Journal of Medicine. 1992;**327**:220-226

[40] Shapiro A. Islets transplantation in type 1 diabetes: Ongoing challenges, refined procedures and long-term outcome. The Review of Diabetic Studies. 2012;**9**:385-406

[41] Shapiro AM et al. Islet transplantation in seven patients with type 1 diabetes mellitus using a glucocoticoid-free immunosuppression regime. The New England Journal of Medicine. 2000;**343**:230-238

[42] Pileggi A, Cobianchi L, Inverardi L, Ricordi C. Overcoming the challenges now limiting islet transplantation: A sequential, integrated approach. Annals of the New York Academy of Sciences. 2006 Oct;**1079**:383-398

[43] Villinger P et al. Prevention of bleeding after islet transplantation: Lesson learned from a multivariate analysis of 132 cases at a single center institution. American Journal of Transplantation. 2005;**5**:2992-2998

[44] Brennan DC et al. Portal vein thrombosis complicating islet transplantation in a recipient with the factor V Leiden mutation. Transplantation. 2004;**78**:172-173

[45] Ryan EA, Paty BW, Senior PA, Shapiro AM. Risk and side effects of islets transplantation. Current Diabetes Reports. 2004;**4**:304-309

[46] Ratner BD, Hoffman AS, Schoen FJ, Lemon J. Biomaterials Science: A Multidisciplinary Endeavor. Biomaterials Science. 2nd ed. Elsevier; 2004. Academic Press 2004 San Diego, California 92101-4495 eBook ISBN: 9780080470368

[47] Ratner BD. The biocompatibility manifesto: Biocompatibility for the twenty-first century. Journal of Cardiovascular Translational Research. 2011;**4**(5):523-527

[48] Sachs F. Mechanical transduction in biological systems. Critical Reviews in Biomedical Engineering. 1988;**16**(2):141-169

[49] Engler AJ, Sen S, Sweeney HL, Discher DE. Matrix elasticity directs stem cell lineage specification. Cell. 2009;**126**:677-689

[50] Chan BP, Leon KW. Scaffolding in tissue engineering: general approaches and tissue-specific considerations. Eur Spine J. 2008;**17**:467-479

[51] Langer R, Vacanti JP. Tissue engineering. Science. 1993;**260**(5110):920-926

[52] Hollister SJ. Porous design for tissue engineering. Nature Materials. 2005;**4**:518-524

[53] Nof M, Shea LD. Drug-releasing scaffolds fabricated from drug-loaded microspheres. Journal of Biomedical Materials Research. 2002 Feb;**59**(2):349-356

[54] Jaklenec A, Wan E, Murray ME, Mathiowitz E. Novel scaffolds fabricated from protein-loaded microspheres for tissue engineering. Biomaterials. 2008 Jan;**29**(2):185-192 Epub 2007 Oct 24

[55] Yamada N, Okano T, sakai H, Karikusa F, Sawasaki Y, et al. Thermo-responsive polymeric surfaces; control of attachment and detachment of cultured cells. Macromolecular Rapid Communications. 1990;**11**:571-576

[56] Kushida A, Yamato M, Konno C, Kikuchi A, et al. Decrease in culture temperature releases monolayer endothelial cell sheets together with deposited fibronectine matrix from temperature-responsive culture surfaces. Journal of Biomedical Materials Research. 1999;**45**:355-362

[57] Okano T, Yamada N, Okuhara M, Sakai H, et al. Mechanism of cell detachment from temperature-modulated, hydrophilic-hydrophobic polymer surfaces. Biomaterials. 1995;**16**:297-303

[58] Veriter S, Gianello P, Dufrane D. Bioengineered sites for islets cell transplantation. Current Diabetes Reports. 2013;**13**:745-755

[59] Shimizu H, Ohashi K, Utoh R, Ise K, et al. Bioengineering a functional sheet of islet cells for the treatment of diabetes mellitus. Biomaterials. 2009;**30**:5943-5949

[60] Matsuura K, Utoh R, Nagase K, Okano T. Cell sheet approach for tissue engineering and regenerative medicine. Journal of Controlled Release. 2014;**190**:228-239

[61] Soon-Shiong P et al. Insulin indipendence in a type 1 diabetic patient after encapsulated islets transplantation. Lancet. 1994;**343**:950-951

[62] Bisceglie VV. Uber die antineoplastiche Immunitat. Zeitschrift fur Krebsforschung. 1933;**40**:141-158

[63] Scharp DW, Marchetti P. Encapsulated islets for diabetes therapy: History, current progress and critical issues requiring solution. Advanced Drug Delivery Reviews. 2014;**67-68**:35-73

[64] Barkai U, Rotem A, de Vos P. Survival of encapsulated islets: More than a membrane story. World Journal of Transplantation. 2016 Mar 24;**6**(1):69-90

[65] de Vos P, Hamel AF, Tatarkiewicz K. Considerations for successful transplantation of encapsulated pancreatic islets. Diabetologia 2002; **45**:159-173

[66] Chandy T, Mooradian DL, Rao GH. Evaluation of modified alginate-chitosan-poliethylene glycol microcapsules for cell encapsulation. Artificial Organs. 1999;**23**:894-903

[67] Desai NP, Sojomihardjo A, Yao Z, Ron N, et al. Interpenetrating polymer networks of alginate and polyethylene glycol for encapsulation of islets of Langherans. Journal of Microencapsulation. 2000;**17**:677-690

[68] Goosen MF, O'Shea GM, Gharapetian HM, Chou S, et al. Optimization of microencapsulation parameters: Semi-permeable microcapsule sas a bioartificial pancreas. Biotechnology and Bioengineering. 1985;**27**:146-150

[69] O'Shea GM, Goosen MF, Sun AM. Prolonged survival of transplanted islets of Langerhans encapsulated in a biocompatible membrane. Biochimica et Biophysica Acta. 1984;**804**:133-136

[70] Cobianchi L, Fornoni A, Pileggi A, Molano RD, Sanabria NY, Gonzalez-Quintana J, Bocca N, Marzorati S, Zahr E, Hogan AR, Ricordi C, Inverardi L. Riboflavin inhibits IL-6 expression and p38 activation in islet cells. Cell Transplantation. 2008;**17**(5): 559-566

[71] Souza YE et al. Islet transplantation in rodents: Do encapsulated islets really work? Arquivos de Gastroenterologia. 2011;**48**:146-152

[72] Duvivier-Kali VF, Omer A, Lopez-Avalos MD, et al. Survival of microencapsulated adult pig islets in mice in spite of an antibody response. American Journal of Transplantation. 2004;**4**:1991-2000

[73] Algire GH, Legallais FY. Recent developments in the transparent-chamber technique as adapted to the mouse. Journal of the National Cancer Institute. 1949;**10**(2):225-253

[74] Strautz RL. Studies of hereditary-obese mice (obob) after implantation of pancreatic islets in Millipore filter capsules. Diabetologia. 1970;**6**(3):306-312

[75] de Vos P, Spasojevic M, Faas MM. Treatment of diabetes with encapsulated islets. Advances in Experimental Medicine and Biology. 2010;**670**:38-53

[76] Dufrane D, Goebbels RM, Saliez A, et al. Six-month survival of microencapsulated pig islets and alginate biocompatibility in primates: Proof of concept. Transplantation. 2006;**81**:1345-1353

[77] Youhwan K, Hyojin K, Ik KK, Kwanwoo S. Extracellular matrix revisited: Roles in tissue engineering. International Neurourology Journal. 2016 May;**20**:S23-S29

[78] Peloso A, Dhal A, Zambon JP, Li P, Orlando G, Atala A, Soker S. Current achievements and future perspectives in whole-organ bioengineering. Stem Cell Research & Therapy. 2015 Jun;**6**(1):107

[79] Gattazzo F, Urciuolo A, Bonaldo P. Extracellular matrix: A dynamic microenvironment for stem cell niche. Biochimica et Biophysica Acta. 2014 Aug;**1840**(8):2506-2519

[80] Ahmed M, Charles Ffrench-Constant C. Extracellular matrix regulation of stem cell behavior. Current Stem Cell Reports. 2016;**2**:197-206

[81] Linetsky E, Bottino R, Lehmann R, Alejandro R, Inverardi L, Ricordi C. Improved human islet isolation using a new enzyme blend, liberase. Diabetes. 1997;**46**:1120

[82] Ryan EA, Lakey JR, Rajotte RV, Korbutt GS, Kin T, Imes S, et al. Clinical outcones and insulin secretion after islets transplantation in the Edmonton protocol. Diabetes. 2001;**50**:710

[83] Crapo PM, Gilbert TW, Badylak SF. An overview of tissue and whole organ decellularization processes. Biomaterials. 2011 April;**32**(12):3233-3243

[84] Dong X, Wei X, Yi W, Gu C, Kang X, Liu Y, et al. RGD-modified acellular bovine pericardium as a bioprosthetic scaffold for tissue engineering. J Mater Sci Mater Med. 2009; **20**:2327-2336

[85] Reing JE, Brown BN, Daly KA, Freund JM, Gilbert TW, Hsiong SX, et al. The effects of processing methods upon mechanical and biologic properties of porcine dermal extracellular matrix scaffolds. Biomaterials. 2010;**31**(33):8626-8633 [PubMed: 20728934]

[86] Arnold T, Linke D. The use of detergents to purify membrane proteins. Current Protocols in Protein Science. 2008;**4**(4):1-4

[87] Peloso A, Katari R, Zambon JP, Defrancesco A, Moore A, Holton C, Mogul A, Manzia TM, Orlando G. Abdominal organ bioengineering: Current status and future perspectives. Minerva Chirurgica. 2015 Feb;**70**(1):43-55

[88] Rieder E, Kasimir MT, Silberhumer G, et al. Decellularization protocols of porcine heart valves differ importantly in efficiency of cell removal and susceptibility of the matrix to recellularization with human vascular cells. The Journal of Thoracic and Cardiovascular Surgery. 2004;**127**:399-405

[89] Peloso A, Petrosyan A, Da Sacco S, Booth C, Zambon JP, O'Brien T, Aardema C, Robertson J, De Filippo RE, Soker S, Stratta RJ, Perin L, Orlando G. Renal extracellular matrix scaffolds from discarded kidneys maintain glomerular morphometry and vascular resilience and retains critical growth factors. Transplantation. 2015 Sep;**99**(9):1807-1816

[90] He M, Callanan A, Lagaras K, Steele J, Stevens MM. Optimization of SDS exposure on preservation of ECM characteristics in whole organ decellularization of rat kidneys. Journal of Biomedical Materials Research—Part B: Applied Biomaterials. 2016 Apr;**105**(6):1352-1360

[91] https://pubchem.ncbi.nlm.nih.gov/compound/Triton_X

[92] Meyer SR, Chiu B, Churchill TA, Zhu L, Lakey JR, Ross DB. Comparison of aortic valve allograft decellularization techniques in the rat. Journal of Biomedical Materials Research. Part A. 2006;**79**(2):254-262

[93] Yang B, Zhang Y, Zhou L, Sun Z, Zheng J, Chen Y, et al. Development of a porcine bladder acellular matrix with well-preserved extracellular bioactive factors for tissue engineering. Tissue Engineering. Part C, Methods. 2010;**16**(5):1201-1211

[94] Kasimir MT, Rieder E, Seebacher G, Silberhumer G, Wolner E, Weigel G, et al. Comparison of different decellularization procedures of porcine heart valves. The International Journal of Artificial Organs. 2003;**26**(5):421-427

[95] Du L, Wu X, Pang K, Yang Y. Histological evaluation and biomechanical characterisation of an acellular porcine cornea scaffold. Br J Ophthalmol. 2011;**95**:410-414

[96] Funamoto S, Nam K, Kimura T, Murakoshi A, Hashimoto Y, Niwaya K, et al. The use of high-hydrostatic pressure treatment to decellularize blood vessels. Biomaterials. 2010;**31**(13):3590-3595

[97] Tudorache I, Cebotari S, Sturz G, Kirsch L, Hurschler C, Hilfiker A, et al. Tissue engineering of heart valves: Biomechanical and morphological properties of decellularized heart valves. The Journal of Heart Valve Disease. 2007;**16**(5):567-573

[98] Wainwright JM, Czajka CA, Patel UB, Freytes DO, Tobita K, Gilbert TW, et al. Preparation of cardiac extracellular matrix from an intact porcine heart. Tissue Engineering. Part C, Methods. 2010;**16**(3):525-532

[99] Gui L, Chan SA, Breuer CK, Niklason LE. Novel utilization of serum in tissue decellularization. Tissue Engineering. Part C, Methods. 2010;**16**(2):173-184

[100] Keane TJ, Swinehart IT, Badylak SF. Methods of tissue decellularization used for preparation of biologic scaffolds and in vivo relevance. Methods. 2015;**84**:25-34

[101] Roth SP, Glauche SM, Plenge A, Erbe I, Heller S, Burk J. Automated freeze-thaw cycles for decellularization of tendon tissue—A pilot study. BMC Biotechnology. 2017 Feb 14;**17**(1):13

[102] Grauss RW, Hazekamp MG, Oppenhuizen F, van Munsteren CJ, Gittenberger-de Groot AC, DeRuiter MC. Histological evaluation of decellularised porcine aortic valves: Matrix changes due to different decellularisation methods. European Journal of Cardio-Thoracic Surgery. 2005;**27**(4):566-571

[103] Pellegata A, Asnaghi MA, Zonta S, Zerbini G, Mantero S. A novel device for the automatic decellularization of biological tissues. The International Journal of Artificial Organs. 2012 Mar;**35**(3):191-198

[104] Fu Y, Fan X, Tian C, Luo J, Zhang Y, Deng L, Qin T, Lv Q. Decellularization of porcine skeletal muscle extracellular matrix for the formulation of a matrix hydrogel: A preliminary study. Journal of Cellular and Molecular Medicine. 2016 Apr;**20**(4):740-749

[105] Van Deijen JH, Hulstaert CE, Woltters GH, van Schilfgaared R. Significance of the peri-insular extracellular matrix for islet isolation from the pancreas of rat, dog, pig and man. Cell and Tissue Research. 1992;**267**:139

[106] Deijnen JHV, Suylichen PTV, Wolters GH, Schilfgaarde RV. Distribution of collagen type I, type III and type V in the pancreas of rat, dog, pig and man. Cell and Tissue Research. 1994;**277**:115

[107] Cheng J, Raghunath M, Whitelock J, Poole-Warren L. Matrix components and scaffolds for sustained islet function. Tissue Engineering: Part B. 2011;**17**:235-247

[108] Pinkse GGM, Bouwman WP, Jiawan-Lalai R, Terpstra OT, Bruijn JA, de Heer E. Integrin signaling via RGD peptides and anti-beta1 antibodies confers resistance to apoptosis in islets of Langerhans. Diabetes. 2006;**55**:312

[109] Kaido T, Yebra M, Cirulli V, Rhodes C, Diaferia G, Montgomery AM. Impact of defined matrix interactions on insulin production by cultured human beta-cells. Effects on insulin content, secretion and gene transcription. Diabetes. 2006;**55**:2723

[110] McCall-Culbreath KD, Zutter MM. Collagen receptor integrins: Rising to challenge. Current Drug Targets. 2008;**9**:139

[111] Parnaud G, Hammar E, Rouiller DG, Armanet M, Halban PA, Bosco D. Blockade of beta1 integrin-laminin-5 interaction affect spreading and insulin secretion of rat beta-cells attached on extracellular matrix. Diabetes. 2006;**55**:1413

[112] Kantengwa S, Baetens D, sadoul K, Buck CA, Halban PA, Rouiller DG. Identification and characterization of alpha 3 beta 1 integrin on primary and transformed rat islet cells. Experimental Cell Research. 1997;**237**:394

[113] Wang R, Li J, Lyte K, Yashpal NK, Fellows F, Goodyer CG. Role for beta1 integrin and its associated alpha3, alpha5 and alpha6 subunits in development of the human fetal pancreas. Diabetes. 2005;**54**:2080

[114] Kaido T, Yebra M, Cirulli V, Montgomery AM. Regulation of human beta-cell adhesion, motility and insulin secretion by collagen IV and its receptor alpha1beta1. The Journal of Biological Chemistry. 2004;**279**:53762

[115] Iozzo RV. Proteoglycans. Ney York: CRC Press; 2000

[116] Mitsou I, Multhaupt HAB, Couchman JR. Proteoglycans, ion channels and cell-matrix adhesion. The Biochemical Journal. 25 May 2017;**474**(12):19

[117] Guyette JP, Gilpin SE, Charest JM, Tapias LF, Ren X, Ott HC. Perfusion decellularization of whole organs. Nature Protocols. 2014;**9**:1451-1468

[118] JE1 A-H, Ko IK, Atala A, Yoo JJ. Decellularization for whole organ bioengineering. Biomedical Materials. 2013 Feb;**8**(1):014106

[119] Peloso A, Ferrario J, Maiga B, Benzoni I, Bianco C, Citro A, Currao M, Malara A, Gaspari A, Balduini A, Abelli M, Piemonti L, Dionigi P, Orlando G, Maestri M. Creation and implantation of acellular rat renal ECM-based scaffolds. Organogenesis. 2015;**11**(2):58-74

[120] De Carlo E, Baiguera S, Conconi MT, Vigolo S, Grandi C, Lora S, Martini C, Maffei P, Tamagno G, Vettor R, Sicolo N, Parnigotto PP. Pancreatic acellular matrix supports

islet survival and function in a synthetic tubular device: In vitro and in vivo studies. International Journal of Molecular Medicine. 2010 Feb;**25**(2):195-202

[121] Conrad C, Schuetz C, Clippinger B, Vacanti JP, Markmann JM, PhD MD, Ott HC. Bioengineered endocrine pancreas based on decellularized matrix and mesenchymal stem cell/islet cell cocolture. Journal of the American College of Surgeons. 2010;**211**(3):S62

[122] Goh SK, Bertera S, Olsen P, Candiello JE, Halfter W, Uechi G, Balasubramani M, Johnson SA, Sicari BM, Kollar E, Badylak SF, Banerjee I. Perfusion-decellularized pancreas as a natural 3D scaffold for pancreatic tissue and whole organ engineering. Biomaterials. 2013 Sep;**34**(28):6760-6772

[123] Napierala H, Hillebrandt KH, Haep N, Tang P, Tintemann M, Gassner J, Noesser M, Everwien H, Seiffert N, Kluge M, Teegen E, Polenz D, Lippert S, Geisel D, Reutzel Selke A, Raschzok N, Andreou A, Pratschke J, Sauer IM, Struecker B. Engineering an endocrine Neo-Pancreas by repopulation of a decellularized rat pancreas with islets of Langerhans. Scientific Reports. 2017 Feb 2;**7**:41777

[124] Rana D, Zreiqat H, Benkirane-Jessel N, Ramakrishna S, Ramalingam M. Development of decellularized scaffolds for stem cell-driven tissue engineering. Journal of Tissue Engineering and Regenerative Medicine. 2017 Apr;**11**(4):942-965

[125] Wan J, Huang Y, Zhou P, Guo Y, Wu C, Zhu S, Wang Y, Wang L, Lu Y, Wang Z. Culture of iPSCs derived pancreatic β-like cells in vitro using decellularized pancreatic scaffolds: A preliminary trial. BioMed Research International. 2017;**2017**:4276928. DOI: 10.1155/2017/4276928

[126] Mirmalek-Sani SH et al. Porcine pancreas extracellular matrix as a platform for endocrine pancreas bioengineering. Biomaterials. 2013;**34**:5488-5499

[127] Chun SY, Mack DL, Moorefield E, SH O, Kwon TG, Pettenati MJ, et al. Pdx1 and controlled culture conditions induced differentiation of human amniotic fluid-derived stem cells to insulin-producing clusters. Journal of Tissue Engineering and Regenerative Medicine. 2012

[128] Katsuki Y, Yagi H, Okitsu T, Kitago M, Tajima K, Kadota Y, Hibi T, Abe Y, Shinoda M, Itano O, Takeuchi S, Kitagawa Y. Endocrine pancreas engineered using porcine islets and partial pancreatic scaffolds. Pancreatology. 2016 Sep-Oct;**16**(5):922-930

[129] Gifford S, Zambon JP, Orlando G. Recycling organs-growing tailor-made replacement kidneys. Regenerative Medicine. 2015 Nov;**10**(8):913-915

[130] Peloso A, Urbani L, Cravedi P, Katari R, Maghsoudlou P, Fallas ME, Sordi V, Citro A, Purroy C, Niu G, McQuilling JP, Sittadjody S, Farney AC, Iskandar SS, Zambon JP, Rogers J, Stratta RJ, Opara EC, Piemonti L, Furdui CM, Soker S, De Coppi P, Orlando G. The human pancreas as a source of protolerogenic extracellular matrix scaffold for a new-generation bioartificial endocrine pancreas. Annals of Surgery. 2016 Jul;**264**(1):169-179